For a complete listing of titles in the
Artech House Technology Management and Professional Development Library,
turn to the back of this book.

Applied Systems Engineering

Second Edition

Applied Systems Engineering

Second Edition

Ian Faulconbridge
Michael Ryan

ARTECH
HOUSE

BOSTON | LONDON
artechhouse.com

Library of Congress Cataloging-in-Publication Data
A catalog record for this book is available from the US Library of Congress.

British Library Cataloguing in Publication Data
A catalogue record for this book is available from the British Library.

ISBN 13: 978-1-68569-133-2

Cover design by Iain Hill

Artech House
685 Canton Street
Norwood, MA 02062

10 9 8 7 6 5 4 3 2 1

Contents

Preface

The need to manage complexity is now commonplace in almost all fields of endeavor. Complex systems such as cars, airplanes, airports, financial systems, and communications networks commonly involve millions of hours of work by thousands of people from a wide range of disciplines spread across dozens of companies in dozens of countries. At the same time, the need to accommodate rapid changes in the marketplace has created considerable pressure on traditional engineering processes. Therefore, it is little wonder that we have become used to hearing about the difficulties associated with complex projects: cost and schedule overruns, dramatic failures to achieve requirements, and project cancelations.

These issues cannot be addressed by simply ensuring that each of the contributing disciplines pays more attention to their individual professions. Complex technical systems can only be developed effectively by addressing the complete system life cycle.

First, business and stakeholder needs and requirements must be defined formally and then combined into an integrated set of needs and requirements to provide a comprehensive description of the functionality (a functional architecture) of the system to be developed. These functional needs and requirements are then analyzed and elaborated to create a functional description of subsystem requirements, which are then allocated to physical configuration items to provide a physical architecture of the system. The aim of developing the physical subsystems and configuration items is to decompose the complex system to a combination of well-defined elements that can be designed and then built by manageable teams using extant processes and extant procedures.

However, the subsequent development of these separate physical elements must be managed, so that they are tested and integrated into the final system to be verified and validated and delivered. To be successful, the entire process must be planned, recorded, and managed.

This book provides a basic but complete coverage of the discipline known as systems engineering. We offer a framework encapsulating the entire systems engineering discipline, clearly showing where the multitude of associated activities fits within the overall effort, providing an ideal vehicle for understanding the complicated nature of the discipline. We take a top-down approach that introduces the philosophical aspects of the discipline and provides a framework within which the reader can assimilate the associated activities. Without such a reference, the practitioner is left to ponder the plethora of terms, standards, and practices that have been developed independently and often still lack cohesion, particularly in nomenclature and emphasis.

Although the field of systems engineering is often viewed as dry, detailed, complicated, acronym-intensive, and uninteresting, the discipline holds the solution to delivering complex technical projects on time and within budget while avoiding many of the failures of the past. The principal intention of this book is to cover all aspects of the discipline providing a framework for the consideration of the many issues associated with the engineering of complex systems.

Second, we describe a wide and seemingly complicated field in a simple, easily digested manner that is accessible to a wide spectrum of readers, from students to professionals, from novices to experienced practitioners. This book is directed at a wide audience and aims to be a valuable reference for all professions associated with the management of complex technical systems and their associated projects: project managers, business analysts, systems engineers, quality assurance representatives, operators, integrated logistic support practitioners, maintainers, and many specialist disciplines.

Chapter 1 introduces systems engineering and related issues, and Chapter 2 provides the framework for the remainder of the book. Chapters 3 and 4 examine in more detail the issues associated with conceptual design and preliminary design, respectively. Emphasis is given to these early activities, as they have the greatest impact on the development of the system. Chapters 5, 6, and 7 examine, respectively, detailed design and development, construction and/or production, and operational use and support. Chapter 8 examines the broad topic of systems engineering management and details some of the associated activities. Some of the more common and popular systems engineering standards are introduced in Chapter 9, and their application to engineering management and process is explained. Chapter 10 explains the interrelationship between the systems engineering effort and other closely related disciplines such as project management, business analysis, quality management, and integrated logistics support management. Finally, Chapter 11 discusses the relationship between systems engineering methodologies and each of a number of acquisition and development methods.

Systems engineering is a broad discipline, and its application to different projects always requires individual and independent thought. There is never a single solution that will work with all projects, and there is rarely a solution for any given project that is either completely right or wrong. This book aims to introduce the main systems engineering issues to the reader to facilitate some of that individual and independent thought.

1

Introduction to Systems Engineering

While elements of systems engineering [1] are recognizable in all major engineering ventures throughout history, the discipline is relatively young. The term was first used in Bell Telephone Laboratories in the 1940s in the context of taking a systems view when engineering telecommunications networks. Conference and journal papers [2] began using the term in the 1950s, but many were still more related to engineering a system rather than systems engineering, as we would use the term today. Modern use began to emerge with three seminal papers from Cole [3], and then a number of authors [4] began to define the practice of not just engineering a system but also the need for management and integration of specialist disciplines. The first book on systems engineering was published in 1957 [5] and the discipline has been taught at universities since the 1960s.

More formal methodologies and practices began to emerge from experience gained in the US Department of Defense (DoD) acquisition programs in the late 1940s and early 1950s when, for the first time, the scope of system acquisition had begun to outstrip the ability of traditional engineering practices to cope with complex and challenging user requirements that tended to be incomplete and poorly defined. Additionally, most programs entailed high technical risk because they involved a wide variety of technical disciplines and the use of high technology. Following a number of program failures, the discipline of systems engineering emerged to help to avoid, or at least mitigate, some of the technical risks associated with complex system acquisition. Since that time, systems engineering processes and methodologies have continued to develop and are now widely applied throughout the life cycles of modern systems, not just in the traditional defense and aerospace domains but also in nontraditional domains such as transportation and health.

Today, systems engineering provides the framework within which complex systems can be adequately defined, analyzed, specified, designed, manufactured, tested, operated, supported, and ultimately retired. The focus of systems engineering is on a strong interdisciplinary approach to the system as a whole across the whole life cycle. Project management, business analysis, quality assurance, integrated logistics support, and a wide variety of specialized engineering disciplines are but a few of the many disciplines that are part of a coordinated systems engineering effort.

Throughout the following chapters, we use examples wherever possible to illustrate and reinforce the key concepts as we introduce them. To aid an understanding of the systems engineering processes and activities, we also use two simple examples that flow through the whole book. In doing so, we do not intend to capture the entire design process of those systems. Rather, they have been chosen as convenient examples that can be readily recognized by readers from a wide variety of disciplines and specialties. That is, readers are not forced to become domain experts in a particular field just to understand the illustration—the majority of readers can immediately understand the system context, the business needs, stakeholder needs, and system needs; the subsequent requirements; the interface issues; technical performance measures; the logical-to-physical translation; broad trade-off analyses; and the physical configuration items involved in the final design. It should also be noted that we do not at all suggest that the aggregation of examples throughout the text represents an adequate or complete design for either system; the available space prohibits the inclusion of sufficient detail, which would also obscure the general lessons that are to be illustrated by the examples. Further, we often illustrate by leaving an aspect of an artifact unfinished.

Example 1.1: Introduction to Aircraft Example

An aircraft operator (ACME Air) has identified a business need for a medium-sized aircraft to replace the aging platform that it currently operates over domestic routes and some short international routes.

Example 1.2: Introduction to Domestic Security Alarm Example

Another division of ACME Industries, ACME Alarms, has a business need to develop a domestic security alarm. The company proposes to sell the alarm to the domestic market to compete in price and functionality for all forms of domestic dwelling such as houses, flats, and apartments. The alarm is to be capable of being installed by the customer, should they so desire, and must be able to operate in a back-to-base monitored mode as well as a stand-alone mode.

1.1 What Is a System?

Before we begin to address the discipline associated with engineering a system, we need to consider what is meant by a system—particularly since "system" is perhaps one of the most overused words in the English language. There are physical systems such solar systems, river systems, railway systems, satellite systems, communication systems, information systems, pulley systems, and nervous systems, just to name a few. There are philosophical systems, social systems, religious systems, gambling systems, banking systems, systems of government, and many more. The word is even used for more esoteric examples such as the consideration of individual and social behavior as part of a system of purposeful events [6]. Before we continue, therefore, it is important that we consider what we mean by a system in the context of systems engineering.

1.1.1 The Definition of a System

The common aspect of the use of system in these varied contexts stems from its early use (and its Greek root) to refer to the whole (or the set) that results when a number of things have been grouped together in a particular manner, for a particular reason. ISO/IEC/IEEE 15288 [7] defined a system in its purest form as: "an arrangement of parts or elements that together exhibit behaviors or meaning that the individual constituents do not."

We will see shortly that, from the perspective of the design of an engineered system, it is more useful to use a slightly narrower definition that a system is: a combination of interacting elements organized to achieve one or more stated purposes.

This latter definition implies that a system comprises internal system elements with interconnections (interactions) between elements and, by the very act of identifying the system in which we are interested, an external system boundary is implied.

So, as illustrated in Figure 1.1, when we draw the boundary around selected system elements, we define the system of interest (SOI), which consists of those system elements and their interconnections that exist within the defined boundary as well as the external interfaces (inputs/outputs) across

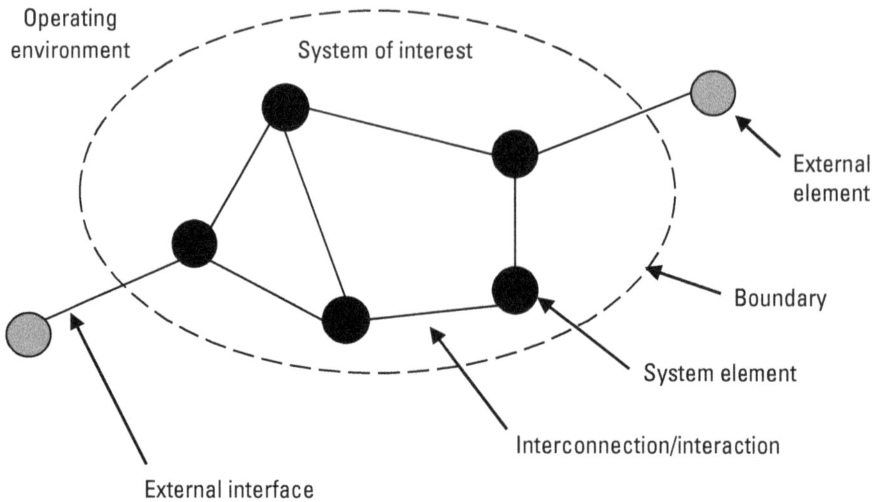

Figure 1.1 An SOI: its elements, interconnections, boundary, and interfaces to external elements in the operating environment.

the system boundary to external elements that exist in an external operating environment (or perhaps in a related system).

The purpose of the system is encapsulated in a single sentence and is called its mission. A single mission sentence, which must be clearly stated by business management and stakeholders, represents the starting point of the design process as well as providing the basis for the ultimate test of the system's fitness for purpose once it has been fielded.

As noted earlier, we narrow the general use of the term "system" to be an "engineered system," which achieves its mission to provide a solution to a business problem or opportunity. This is very important because it has two major implications:

- When we refer to a system as comprising system elements that are interconnected to achieve the system's mission, we imply that all three of those principal aspects result from conscious choice. That is, we are referring to systems that have been deliberately engineered (designed), and hence our interest in systems engineering.
- A system that has been engineered to achieve a specified mission (to solve a particular problem) must be able to perform that mission with relative autonomy—that is, it must be managerially and operationally independent (and may well have been procured independently). We

return to this issue shortly when we discuss the differences between systems and subsystems (and between systems and systems of systems).

1.1.2 Types of Systems

There are numerous ways to classify systems—here we identify the four main classifications so that we can be clear as to which type of system to which we refer when considering systems engineering and therefore in our discussions in the remainder of this book in which our focus is on engineered systems:

- *Closed or open systems.* An open system interacts with its operating environment—it accepts inputs from that environment across its boundary and returns outputs across the same boundary to the external environment. A closed system is isolated from its external environment and is not useful. We are only interested in useful systems, which are therefore open.
- *Natural or human-made/human-modified systems.* Natural systems contain natural elements and are the result of natural processes; human-made systems come into existence through the efforts of humans and may contain human-made elements or natural elements adapted to human-designed purposes (called human-modified systems). The systems engineering for natural systems is certainly not conducted by humans, so we are only interested in those things that humans can do, that is, human-made systems and the human contribution to human-modified systems.
- *Physical or conceptual systems.* Physical systems exist in a physical form; conceptual systems, such as economic, political and religious systems do not have a physical form. We focus here on physical systems comprising combinations of integrated hardware and software elements.
- *Precedented or unprecedented systems.* In a precedented system, similar systems (or, at least, most system elements) have been produced before. An unprecedented system is one that has not been previously produced, nor have any of the system elements. Systems that comprise mostly unprecedented elements are the result of scientific research and development effort, which do not lend themselves to being part of a project that can be managed with a robust scope, schedule and budget. Here we focus on systems that comprise largely precedented elements, that is, those to which engineering, systems engineering and project management are appropriate.

The wide variety of combinations of the above (and other) characteristics can lead to a large number of types of systems, each of which has markedly difficult properties. It is important to recognize that this book and most of the standards discussed refer to engineered systems: open, physical systems that are human-made/human-modified from largely precedented elements.

1.1.3 A System and Its Environment

Although our consideration of a boundary in Figure 1.1 has taken into account the system's relationship with its external elements, sometimes we need to be cognizant of an even wider context so, as illustrated in Figure 1.2, an SOI might be considered as part of a wider SOI (WSOI) within an operating environment, which can be conceived as being part of a wider environment [8].

We will see throughout the remaining chapters that, as illustrated in Figure 1.3, systems engineering is responsible in the first instance for describing the SOI functionally as part of its environment, extracting during conceptual design a stand-alone description of the SOI in a manner that can be contracted with a developer, developing the SOI through the acquisition project, and then introducing the SOI into service back in its environment as a physical solution.

Figure 1.3 demonstrates that our articulation of the SOI with its boundary and interfaces is a critical step in bringing the system into being. Even with a perfect acquisition process, any deficiencies in the initial description will result in delivering a system that is not fit for purpose because it was not sufficiently described in the first place.

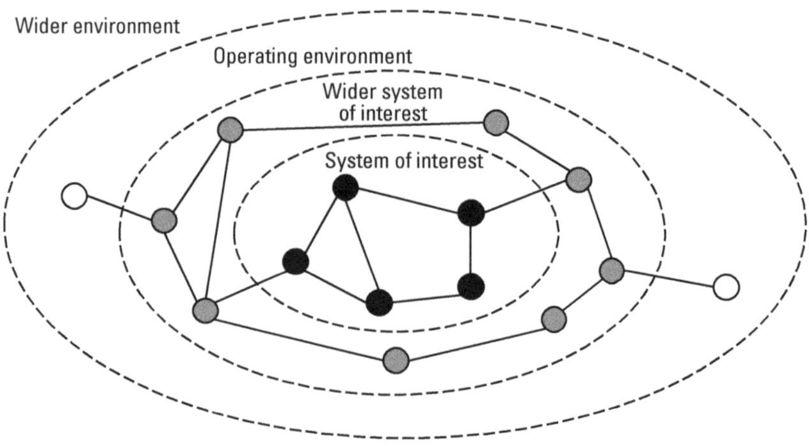

Figure 1.2 An SOI, which might be considered as part of wider SOI, which can be conceived as being part of an operating environment, as part of a wider environment.

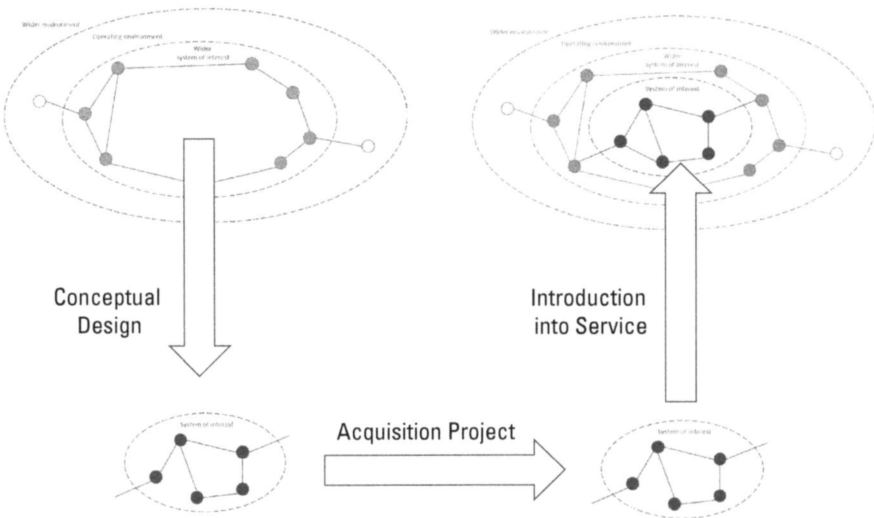

Figure 1.3 An SOI described functionally as part of its environment, extracted as a stand-alone description during conceptual design, developed through the acquisition project, and then introduced into service back in its environment as a physical solution.

1.1.4 A System as a Product

In a physical sense, the term system is sometimes considered to be synonymous with product, that is, we say that the project is delivering a system or is delivering a product. However, a system is normally considered to comprise a number of products. Figure 1.4 shows that EIA-632 [9] defines a system as comprising operational products (end products such as the aircraft in our Aircraft System example) and enabling products (such as test, training, support, maintenance, facilities, and disposal products).

1.1.5 A System as a Capability—A Capability System

The EIA-632 concept in Figure 1.4 illustrates that we must acknowledge that an SOI is much more than an aggregation of hardware or software products and must be described in terms of all of its constituent elements, including: the major hardware and software products, the organization within which the system will be fielded, the personnel who will interact with it in many ways, the collective training systems required, as well as the facilities, data, and support (including supplies), and the operating procedures and organizational policies required to keep the system in service. The system is fully defined by

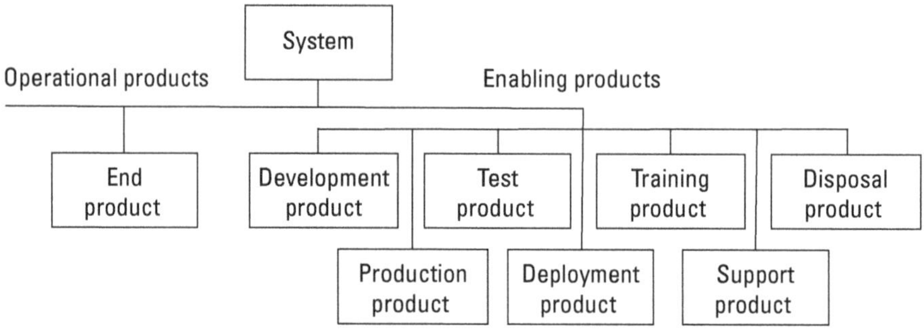

Figure 1.4 EIA-632 building block concept of a system comprising operational products and enabling products.

the combination of these resources operating in an operational environment in order to achieve some purpose. In that sense then, we could define a system as delivering an operational capability.

Therefore, it is common, particularly in defense environments, to refer to the system at this level as a capability system and its management as capability management. In the US DoD, the acronym DOTMLPF-P refers to the capability system elements of doctrine, organization, training, materiel, leadership and education, personnel, and facilities—policy [10]. In NATO, the acronym DOTMLPFI refers to doctrine, organization, training, materiel, leadership, personnel, facilities, and interoperability [11]. In Australia, the capability system is considered to comprise fundamental inputs to capability (FIC): command and management, organization, collective training, major systems, personnel, facilities and training areas, supplies, support, industry, and data [12]. In the United Kingdom, the Defence Lines of Development (DLOD) refer to concepts and doctrine, equipment, information, infrastructure, interoperability, logistics, organization, personnel, and training [13]. In Canada, the acronym PRICIE refers to personnel, research and development, infrastructure, concepts and doctrine, information technology, and equipment (or, more completely, personnel, leadership, and individual training; research and development and operational research; infrastructure, environment, and organization; concepts, doctrine, and collective training; information management and technology; and equipment and support) [14].

In business, it is common to refer to the system at this level as an asset and to refer to its management as asset management [15]. The commercial asset management standard ISO 55000 [16] referred to the elements of an asset as people (competence, roles, leadership); processes (management systems, planning); information (data, documentation); physical assets; financial resources; and support systems (information technology (IT), facilities, infrastructure).

Having acknowledged all of the elements of a capability system (an asset), it must also be recognized that each of those elements will most probably have a different acquisition cycle; for example, people are "acquired" in a different manner to that in which the major equipment will be developed and each element of the capability may even be acquired by a different acquisition element in the organization. In the remainder of this text, for ease of description, we focus on the acquisition of the major equipment element of the capability system. However, we must always keep in mind that this acquisition is being undertaken in parallel with the acquisitions of the other elements of the desired capability and that all the elements must be brought back together prior to introduction into service to field an operational capability/asset. We discuss this in more detail and provide an illustration in Chapter 2.

Example 1.3: Capability System Elements for Our Aircraft Example

Capability elements for our aircraft system example could include, but not be limited to:

- Major equipment. *The most tangible part of the system is the hardware itself. The aircraft will be produced, distributed, and sold to operators who will then use the aircraft in a number of different ways such as domestic and international operations. Software is also now a critical item within most systems. The aircraft is likely to use hardware and software to control a range of functions from engine management, through navigation and environmental control systems, to the communications and flight-control systems.*
- Personnel. *Air crew are required to operate the system. Ground crew are required to maintain and support the fleet of aircraft. All these personnel will require individual training to some degree.*
- Support. *Maintenance facilities and equipment are required for routine maintenance and repairs throughout the aircraft's life. Materials are required to operate the system, including fuel, lubricants, and other consumables such as tires and spare parts.*
- Facilities. *Other facilities such as terminals are also necessary to operate the aircraft and its support systems.*
- Organization, policies, and procedures. *ACME must conform to a significant number of regulations and will need appropriate organizational structures, policies, and procedures to be able to operate the aircraft effectively.*

- Collective training. *Although air crew and ground crew will be trained for their individual roles, they will require collective training as an integrated team throughout the system life cycle.*
- Data. *Data is required to maintain and operate the aircraft. Data could include maintenance information such as specifications and drawings and operational information such as user manuals and instructions.*

1.1.6 Logical and Physical Descriptions of a System

A system can be described in two broad ways—in logical terms and in physical terms. A logical description (historically often referred to as a functional description) of a system articulates what the system will do, how well it will do it, how it will be verified, under what conditions it will perform, and what other systems will be involved with its operation. A physical description relates to the system elements and explains what the elements are, how they look, and how they are to be manufactured, integrated, and verified. The logical description contains the "whats" of the system, and the physical description contains the "hows."

It is worth noting here that in the context of systems engineering, the term "physical" is used to differentiate this hierarchy from the functional hierarchy described earlier, and that physical elements within a system may include elements of software that have been designed to operate on hardware elements to deliver prescribed functionality. The term "functional" is related to what the system is to do (and how well it is to do it, and how it is proposed to be verified) and the term "physical" is related to how the system operates and is supported.

The two descriptions are valid independent descriptions of a system, and it is very important that a system is described both logically and physically, focusing first on the logical description:

- It is axiomatic that we develop the logical description first. In order to determine whether any particular physical implementation (that is, how we are going to implement the elements of the system) is appropriate, we first must understand (from the logical architecture) what it is that we want the system to do (that is, what purpose it serves). We therefore need to focus on the logical description (what) first, from which a series of candidate physical descriptions (how) can be developed, one of which can be selected as the preferred physical solution.
- We also must not allow the way in which we implement current physical systems to color unnecessarily the way in which we might describe

future systems. An initial focus on the logical description therefore allows us to provide novel solutions to new (or even old) problems; if we focus on the physical description initially, we would always tend to solve new problems with old physical implementations.

- Upper-level trade-offs and feasibility analyses must be conducted at the logical level before deciding on the physical implementation; if not, significant waste may result from the selection of physical solutions that either perform unnecessary functions or do not possess critical functionality.
- A logical description is ideally suited to the interface between the solution and the business case that drove its implementation. While it is often possible for the business case to be met directly by an obvious physical solution, it is better for business management to transition from the business case into a more detailed logical description of what is required before considering how to achieve it in a physical sense. The definition of the logical description before the development of an appropriate physical description therefore moves from the business case to the final physical solution in controlled verifiable steps.
- The logical description changes slowly; the physical description changes much faster, particularly as the pace of technological change quickens. Arguably, for example, the need for, and upper-level logical description of, an internal combustion engine has changed little (other than performance requirements) over the last two centuries, while the physical implementations have changed dramatically. That is, while the purpose of the engine subsystem, as part of the car system, has not changed over the years, the physical implementation is obviously very different.

In the development of a system, therefore, there are at least two architectural views: a system logical architecture, and a system physical architecture. These two descriptions are of the same system, so they must be related. We will see later how the logical architecture, as outlined in the requirements breakdown structure (RBS), is mapped onto the physical architecture as represented by the hardware and software configuration items contained in the system breakdown structure (SBS) in the project's work breakdown structure (WBS) during acquisition.

1.1.7 A Systems Approach

When considering the system, either from a logical or a physical perspective, a systems approach focuses on the system as a whole, linking the value judgments of what is desired to solve the problem to the design decisions which are based on what is feasible to implement the solution. Systems approaches draw on the wide range of knowledge appropriate to the SOI and its wider context, to produce desired outcomes without adverse, unintended consequences.

Systems approaches are either soft or hard:

- Soft systems approaches are suited to addressing problems involving incomplete data and unclear goals drawing on subjective and humanist philosophies with constructivist and interpretivist foundations. Soft systems approaches comprise:
  - Creative thinking tools including brainstorming, morphological analysis, de Bono's six thinking hats, and Van Grundy's creative problem-solving techniques;
  - Systems thinking methodologies and techniques including cognitive mapping, influence diagramming, soft systems methodology, strategic choice approach, iterative planning, strategic assumptions surfacing and testing, viable systems model (method), system dynamics modeling, rich pictures, systems thinking, causal loop diagramming, Actifeld, and field anomaly relaxation.
- Hard systems approaches are suited to solving well-defined problems with reliable data and clear goals, using analytical methods and quantitative techniques. Hard systems approaches comprise systems engineering and engineering.

Soft systems approaches and hard systems approaches are complementary. However, as Figure 1.5 implies, soft systems approaches are useful in assisting to gain an understanding of the problem space early on; they focus on creative thinking, human perspectives, and systemic interrelationships. These methodologies focus on the understanding of problems in ill-structured, ambiguous, or socially complex situations. They involve stakeholder engagement, dialogue, and iterative problem structuring, making them particularly useful in organizational change, policy development, and social systems. Techniques such as rich pictures, mind mapping, and causal loop diagrams help to visualize multiple perspectives and emergent properties of systems, leading to adaptive and context-sensitive solutions.

In contrast, hard systems approaches rely more on structured, quantitative, and analytical techniques rooted in the systems engineering and engineering disciplines. These approaches focus on designing, optimizing, and managing well-defined systems with measurable parameters. They employ

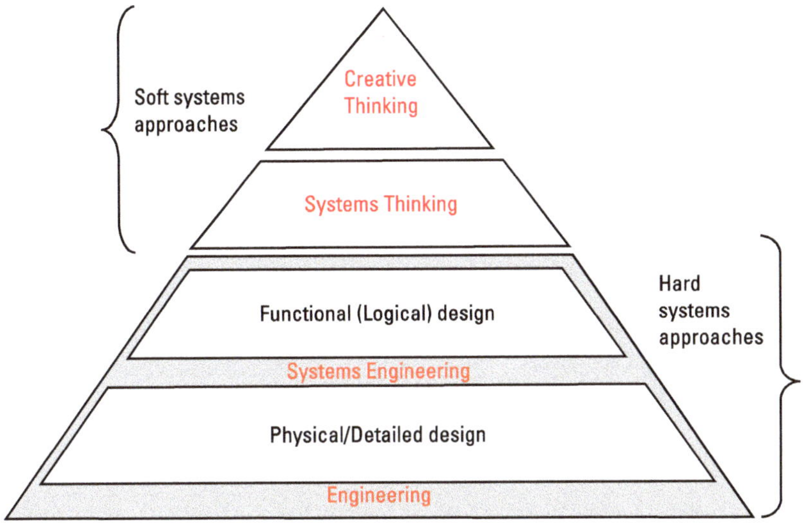

Figure 1.5 The hierarchy of soft and hard systems approaches: creative thinking, systems thinking, systems engineering, and engineering.

mathematical modeling and analysis, simulations, rigorous methodologies, decision analysis, and modeling.

Together, these systems approaches enhance problem-solving capabilities. Soft systems thinking fosters creativity and adaptability, while hard systems thinking provides rigor and precision. Integrating both approaches enables a holistic understanding of complex challenges, balancing technical feasibility with social and organizational dynamics. A systems approach therefore means that the design process remains focused on the problem/opportunity as well as the solution.

1.1.8 Hierarchical Descriptions of a System

We saw earlier that ISO/IEC/IEEE 15288 [17] defined a system as a combination of system elements that interact to achieve a defined mission. Since each of these system elements will need to perform functions allocated to it so that it can contribute to the systems mission, we can consider the system to be a hierarchical composition of system elements, as illustrated in Figure 1.6.

The system elements in Figure 1.6 can be logical elements or physical elements, which support the concepts of a logical architecture and a physical architecture as we discussed in the preceding section.

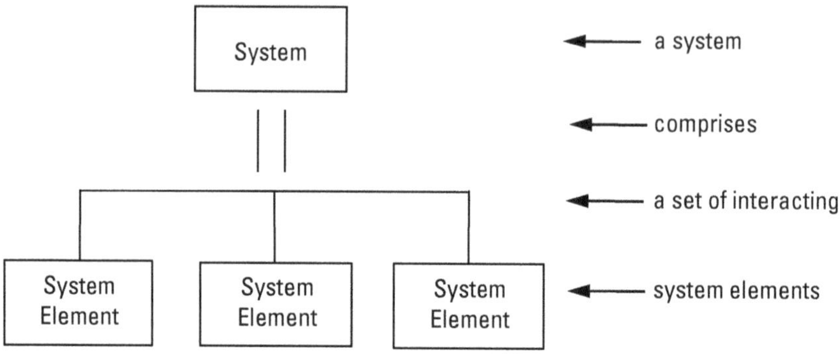

Figure 1.6 A system comprises a hierarchical set of interacting system elements.

1.1.8.1 Logical Hierarchy

In a logical description of a system, as illustrated in Figure 1.7, the system's mission can be broken down into a hierarchical structure of its major functions so that the logical description or architecture can therefore be depicted as a functional hierarchy.

1.1.8.2 Physical Hierarchy

In a physical sense, system elements are also normally described in a hierarchy using at least a four-layer hierarchy, which we will use to illustrate the concept throughout the remining chapters. The top-level entity is the system that comprises a number of subsystems that comprise a number of assemblies

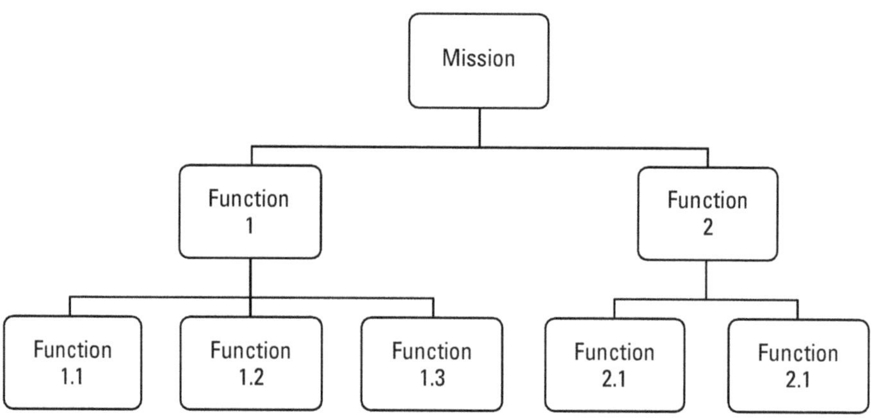

Figure 1.7 A logical (or functional) hierarchy of system functions.

that comprise a number of components. Although these terms are perhaps the most common, there are others in use. For example, Figure 1.8 illustrates what is perhaps the most complete physical hierarchy of system elements as defined by IEEE-STD-1220 [18], which also adds entities such as products, subassemblies, subcomponents, and parts to our simple four-layer taxonomy.

If we consider that there is a level of hierarchy above a system, ISO/IEC/IEEE 15288 [19] considers an SOI to comprise a combination of interacting system elements, some of which may be systems in their own right, as illustrated in Figure 1.9.

When the SOI consists only of system elements that are systems in their own right, the SOI is commonly called a *system-of-systems (SoS)* [20]. Figure 1.10 illustrates how the hierarchy of Figure 1.9 can be delineated into groupings of systems and SoS. However, to have a useful view, we need a better understanding of the ways in which system elements can be grouped.

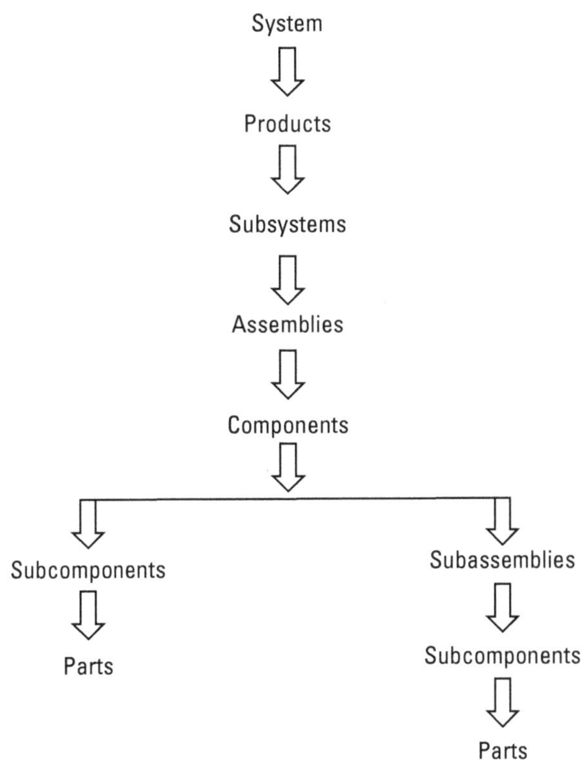

Figure 1.8 Hierarchy of the physical elements of a system from IEEE-STD-1220.

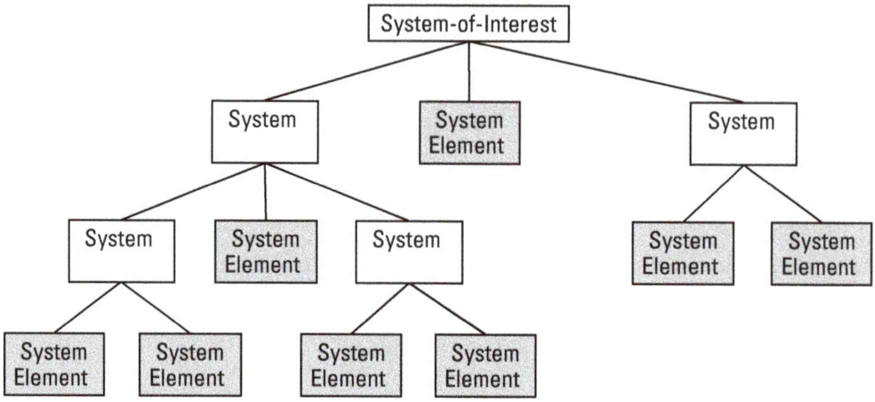

Figure 1.9 Hierarchy of the elements of an SOI. (After: ISO/IEC/IEEE 15288.)

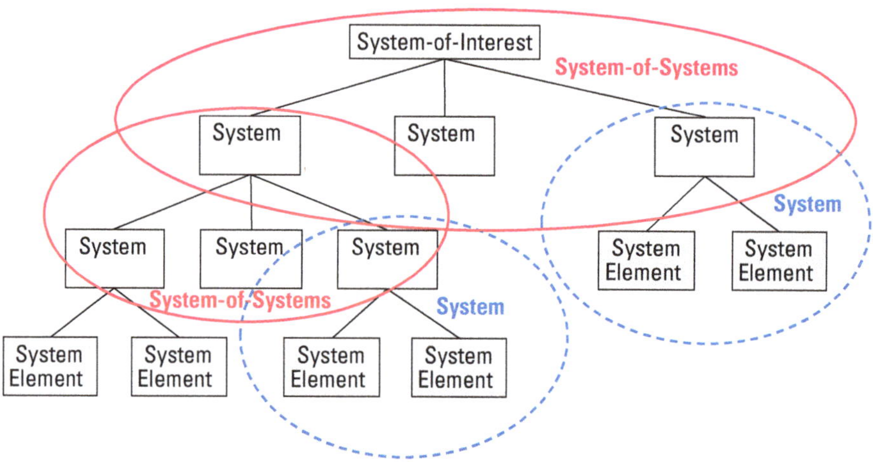

Figure 1.10 Hierarchy of system elements collected as systems-of-systems or as systems. (After: ISO/IEC/IEEE 15288.)

1.1.9 Collections of Systems

If a system comprises a set of interacting elements, there are a number of ways in which those elements can be collected. Unfortunately, we are not always very careful about the language we use to describe those connections. First, we are not always careful to make the distinction between a system and a subsystem. The difference matters because we simply cannot refer to any grouping of elements as a system. For example, an aircraft system may contain, among other elements, the engine, which may consist of assemblies such as fuel tanks, pumps and lines, turbines, compressors, gear boxes, and hydraulic pumps. Commonly, the engine manufacturer will refer to the engine as the engine system. If we continue such a use of the term to the lowest level of components, a nut and bolt can be considered a fastening system because those two elements interact for that purpose. Consequently, if we use the term in that manner, everything is a "system," which is not very useful since almost everything would therefore be a system and all systems comprise systems which themselves contain systems and the term becomes almost meaningless and, worse perhaps, so does the term system-of-systems, which could even then able to be applied to two bolts and two nuts (that is to two "systems") used to join two plates, the end result of which would be another system-of-systems [21].

So, although we should always take a systems approach to design at all levels, not all combinations of elements are usefully considered to be systems. To use the term more effectively in our consideration of systems engineering, an implicit part of the definition of a system is that it must be able to stand alone in its own right, that is, it is managerially and operationally independent. With that qualification, an engine is not a system; it is only ever useful as an element of a system (that is, as a subsystem) and has no purpose other than to be a good servant of the parent system and so is optimized for that effect. We can therefore be more careful about the use of the terms system and subsystem by noting:

- A system is a set of tightly coupled, codependent elements (subsystems) optimized at the system level for the permanent purpose of the system. Systems are managerially and operationally independent and will no doubt have independent life cycles (they will almost certainly have been procured separately).
- A subsystem is a set of tightly coupled, codependent elements (components) optimized at the system level for the permanent purpose of the system; they remain permanently part of the system. Subsystems are not independent and only exist to serve the system; they will have been designed by the system designer to be part of the system. Further, since the system is to be optimized, the subsystems are invariably suboptimal.

Note again that, although we state that a subsystem is strictly speaking not a system, we should always take a systems approach to its design. IEEE-STD-1220 calls the systems approach the systems paradigm.

Since a system can be viewed as a system element, there are then three types of ways in which systems can be grouped [22]:

- *Portfolio-of-systems (PoS).* A PoS is a collection of systems managed jointly with a unified budget to fulfill one or more missions (that may not be related). Member systems interact to share (compete for) resources. For example, ACME Air may group all facilities, buildings, and infrastructure into a PoS for ease of management; further, if they operate in three cities, they may have three facilities' PoS, one in each city.
- *System-of-systems (SoS).* As we saw at the end of the previous section, an SoS is a collection of independent systems that interact for a common purpose, normally to produce an operational effect. The constituent systems of an SoS are loosely coupled and, since they are independent and can join or leave the SoS over time, have been optimized for their own purpose before joining the SoS. An individual system may be a member of more than one SoS, even at the same time.
- *Family-of-systems (FoS).* A collection of systems (also commonly called a product line) that are grouped because they are jointly designed, developed, and manufactured even though they may not interact in operations.

Having introduced the various manners of grouping, in this book we are particularly interested in three levels of combination: subsystem, system, and SoS. As illustrated simplistically in Figure 1.11, the three groupings have similar architectures, in that each comprises elements that are interconnected.

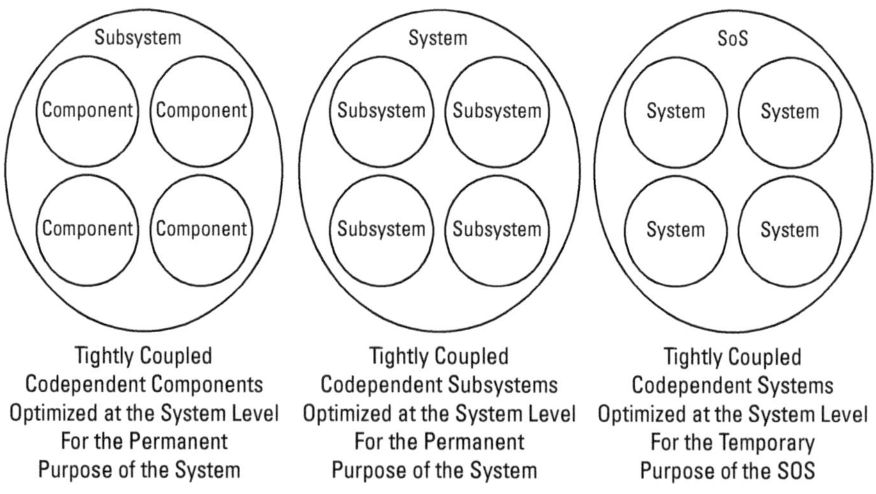

Figure 1.11 Elements of a subsystem, a system, and a system-of-systems.

This similarity in upper-level architecture can lead to the misapprehension that the three groupings are the same. However, as noted in the descriptions above, there are differences, despite a similar upper-level architecture:

- *SoS.* SoS elements are systems in their own right, so that they are managerially independent and operationally independent and have been optimized for their own purpose before contributing to the purpose of the SoS. Further, since the constituent systems must be allowed to be optimized for their own purpose, the resultant SoS will invariably be suboptimal. Individual systems in an SoS can come and go and be replaced readily as they are loosely coupled, and can belong to one or more higher-level SoS, even at the same time.
- *System.* A system comprises subsystems that are not independent and only exist to serve the parent system; subsystems are therefore invariably suboptimal (from their perspective) since it is the system that is to be optimized, not the constituent subsystems. An individual system can be a member of one or more SoS, even at the same time.
- *Subsystem.* Subsystems comprise components that are not independent and only exist to serve the system as good servants of the subsystem; again, it is the system that is to be optimized, not the constituent subsystems. An individual subsystem is only ever a member of the one system and cannot readily be removed from the operational system.

The attributes of subsystems, systems, and SoS are summarized in Table 1.1.

An important observation from Table 1.1 (the last row in particular) is the central position of system design; note that it is the system that is optimized

Table 1.1
Attributes of a Subsystem, a System, and an SoS

Attribute	**Subsystem**	**System**	**SoS**
Purpose	System	System	SoS
Element Type	Component	Subsystem	System
Integration	Tightly Coupled	Tightly Coupled	Loosely Coupled
Element Interaction	Codependent	Codependent	Independent
Period	Permanent	Permanent	Temporary
Optimization Level	System	System	System

at each level of design: subsystem, system, and SoS. Here, therefore, we focus on the system level.

Again, we should emphasize that, just because we have been pedantic about the definition of a system does not mean that we recognize that design at all levels should make use of a systems approach, even for those combinations of elements that are not systems per se.

1.1.10 Problem Domain and Solution Domain

Since a system can be considered as the solution to a problem, it is common to consider the activities being undertaken throughout the life of the system to be in either the problem domain (problem space) where we use predominantly logical descriptions, or the solution domain (solution space) where we use predominantly physical descriptions.

Activities in the problem domain (including production of the logical architecture) are generally considered to be the responsibility of the customer (the acquirer or tasking activity); activities in the solution domain (including the physical architecture) are generally considered to be the responsibility of the organization implementing the system (the developer or the performing activity).

1.2 System Life Cycle

Now that we understand systems a little better, we need to consider that a system has a life. At some point in time it does not exist, it is brought into being, it is used and supported, and then it is disposed of once it can no longer serve the purpose (the mission) for which it was created. Throughout the life of a system, there are a number of phases and activities, each of which builds on the results of the preceding phase or activity. The sum of all these activities is called a system life cycle, which can be described using a model that represents the conceptualization of the original business needs for the system, and then its subsequent realization, utilization, evolution, and ultimate disposal [23].

1.2.1 A Generic System Life-Cycle Model

There is no single accepted agreement on how many phases exist in a system life cycle or what those phases are called. There are a considerable number of generic life-cycle stages in various life-cycle viewpoints; for example, the INCOSE Systems Engineering Handbook [24] provides six example life cycles, each of which has utility in a particular organizational context. For the discussion in this book, we use the system phases illustrated in Figure 1.12, and for further detail during the acquisition phase, we use the life-cycle stages defined by Blanchard and Fabrycky [25] and MIL-STD-499B [26] and illustrated in Figure 1.13.

As shown in Figure 1.12, a generic system life cycle can be divided into four very broad phases:

- *Preacquisition phase.* The life cycle begins in the preacquisition phase with an idea for a system being generated, commonly through business planning. Early consideration of the possible solution options results in the confirmation of the early business needs for the system, which are elaborated by a business case that justifies expenditure of organizational resources on acquisition and utilization of the system. In some instances, the preacquisition phase may determine that it is not feasible or cost-effective to proceed to acquisition (due to technology limitations or funding shortfalls, for example). In that sense then, the preacquisition phase is where organizations expend research and development effort to ensure that only feasible, cost-effective, justifiable projects are taken forward to acquisition.
- *Acquisition phase.* The business needs for the system provide the input for the acquisition phase which is focused on bringing the system into being and into utilization. This would normally involve defining the system in terms of business needs and requirements, stakeholder needs

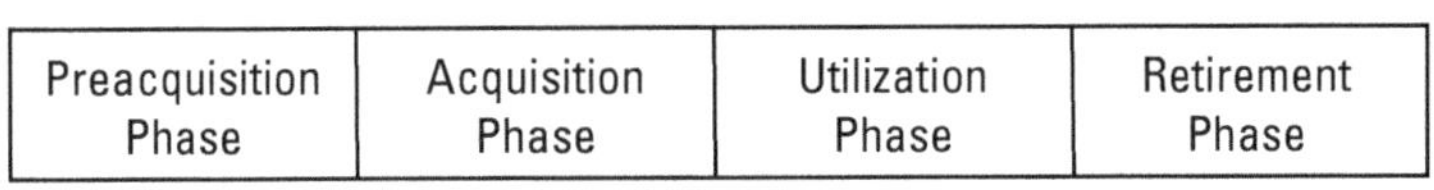

Figure 1.12 The four phases of a generic system life cycle.

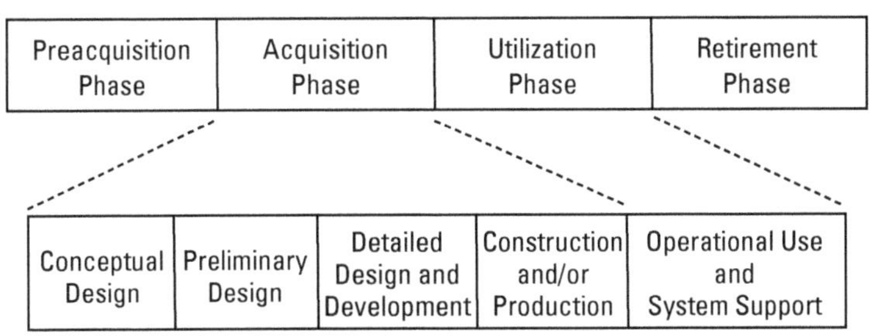

Figure 1.13 The five major activities in the acquisition phase and the utilization phase of the system life cycle. (After: Blanchard and Fabrycky.)

and requirements, and system requirements and then engaging an implementation partner to develop and deliver the system.

- *Utilization phase.* The system remains in service during the utilization phase until the business has no further need for it, or it no longer can meet the functions required of it by the organization, or it is no longer cost-effective to keep it in service. During utilization, the system may undergo a number of modifications and upgrades to rectify performance shortfalls, to meet changing operational requirements or external environments, to enable ongoing support for the system to be maintained, or to enhance current performance or reliability.
- *Retirement phase.* Following operational use and system support, the system is eventually phased out and retired from service. The system life cycle concludes with the retirement phase. If the business need for the capability still exists in the organization, the conclusion of one system's life cycle marks the start of another and the process begins again. If the system has reached the end of its economic life, the life cycle ends in disposal.

1.2.2 Parties Involved

There are a number of parties involved throughout the system life cycle and there is not any standard taxonomy for their names. To be clear, we introduce in this section the nomenclature used throughout the remaining chapters.

The customer organization is managed by enterprise management who set the direction for the organization and for business management who are responsible for managing the activities conducted by the operations element of the organization (undertaken by the operators, sometimes called users).

The systems employed within the organization are acquired by the acquisition element (also called the acquirer [27], or tasking activity [28]) of the organization under the auspices of a project (managed by a project manager). Project managers are supported by a number of related disciplines including business analysis, systems engineering, requirements engineering, specialist engineering disciplines, quality assurance, and integrated logistic support. Operators are supported in their operation of the system by the support element of the organization, which supports, sustains, and maintains the system throughout its life. In addition to the operational, acquisition, and support staff, there are many others within the customer organization who have a stake in the successful implementation of the project. These stakeholders can include representatives from the management, financial, operations, supply, maintenance, and facilities areas of the organization.

The system is obtained from a supplier [29] (also called the performing activity [30]) who may deliver the system off-the-shelf or may develop it, in which case they are often called the developer [31]. The supplier (developer or performing activity) may be internal to the customer (the acquirer) organization. If the development of the system is undertaken in-house, the acquisition element of the organization (the acquirer) will engage with the development element (the developer) to develop the system. It is increasingly common these days for the supply or development to be undertaken by an outside organization called the contractor, which is the entity responsible for supplying the system (perhaps by designing and developing it) to meet the customer requirements. The relationship between the customer and the contractor varies with each project, but, for each project, the relationship is formally defined by the terms and conditions of the contract between the two parties. In many cases, the contractor is not able to perform all the work required and devolves packages of work to a number of subcontractors. The terms and conditions relating to this work are described in the relevant subcontract.

The responsibility for the various phases of the system life cycle is spread across the enterprise (or organization) within which the eventual system will operate. The initial preacquisition phase is the responsibility of enterprise management, who conduct business planning and establish the business case for the projects required to support an organization. A project is then established with a project charter providing authority to a project manager to expend organizational resources on the acquisition of the system. Systems engineering is an important discipline that is responsible for performing the technical management of the project throughout acquisition and then to the capability (product) manager during utilization. Once acquisition is complete, and the system is introduced into service, it is operated by the users and supported by the support element. Note that, rather than considering that each party has a role to play in a particular stage, it is best to consider that all parties are involved at all stages in the life cycle, with the roles and responsibilities of each party shifting in emphasis between stages.

1.3 Acquisition and Utilization Phases

For these two major phases, the acquisition phase and the utilization phase, of the system life cycle, we use the five life-cycle activities in Figure 1.13, which are based on those defined by Blanchard and Fabrycky [32]. In the acquisition phase, the four activities are conceptual design, preliminary design, detailed design and development, and construction and/or production. In the utilization phase, the activities are operational use and system support, which are undertaken in parallel. While there is no standard taxonomy, we choose these activities as a framework here because they are generally accepted in the systems engineering community, and for the following reasons:

- These activities emphasize that a system begins with perceived business needs and finishes with retirement and, ultimately, disposal—the cradle-to-grave approach.
- There is a clear delineation between the acquisition phase and the utilization phase of a system, which recognizes that there is a period of transition required as the system is transferred from the acquirers to the operators/users.
- The activities show the necessary detail in the early stages of the acquisition phase (particularly in conceptual design and preliminary design) where the application of systems engineering methodologies and practices have the potential to make the most significant contribution.
- Importantly, within the acquisition phase, the activities also differentiate clearly between the problem domain, which contains the logical description of the system (the product of conceptual design), and the solution domain, which contains the physical description of the system (the products of preliminary design and detailed design and development) and then the realized system itself (the product of construction and/or production).
- Additionally, the separation of the early system design into conceptual design (what and why) and preliminary design (how) is very important since the responsibility in most programs transitions from the customer/acquirer for conceptual design to the contractor/developer for preliminary design.

1.3.1 Acquisition Phase

Each of the activities shown in Figure 1.13 for the acquisition phase is described in more detail in the following sections and chapters, which outline the major tasks undertaken and the main artifacts produced in each, an overview of which is shown in Figure 1.14.

1.3.1.1 Conceptual Design

Conceptual design focuses on the development of a set of clearly defined requirements, at the system level, and in logical (functional) terms. Although

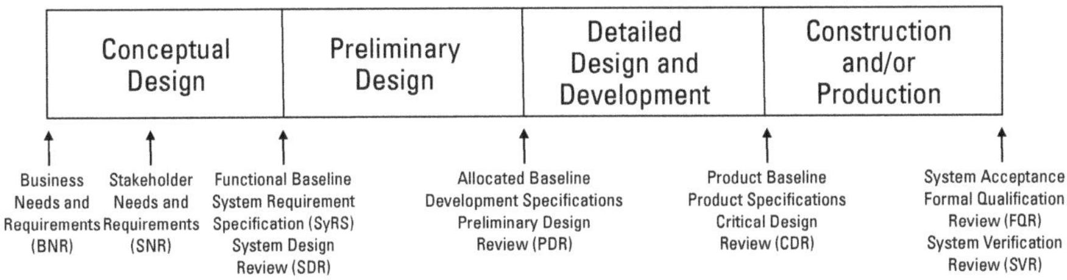

Figure 1.14 Acquisition phase activities and the major artifacts and reviews associated with each.

clearly defining the requirements of the system would seem a logical (and essential) first step, it is often poorly done and is commonly the direct cause of problems later in the development process. Business managers and stakeholders sometimes, understandably, prefer to describe their requirements in loose and ambiguous terms to protect themselves from subsequent changes in their needs and their business environment. The conceptual design process aims to avoid this ambiguity by providing a formal process by which the business needs and requirements (BNR) [33] are articulated and confirmed by business management, and then elaborated by stakeholders at the business operations level into a set of stakeholder needs and requirements (SNR) [34], which are further elaborated by business analysts and requirements engineers into a set of system requirements in the system requirement specification (SyRS) [35]. There may be one SyRS for the entire capability system, but it is more likely that there is one SyRS for each of the constituent elements of capability, that is, one for each of the major equipment, personnel, support, training, facilities, and so on. As noted earlier, each of these constituent capability elements (elements of an asset) may be developed independently, perhaps through separate contracts. Again, we discuss this aspect in more detail in Chapter 2 and illustrate the concept in Figure 2.2.

The BNR, SNR, and the SyRS are the key elements of what is called the functional baseline (FBL), which describes the whats and whys of the system. The FBL therefore contains a system-level logical architecture that meets the business and stakeholder needs and requirements.

Conceptual design ends with the system design review (SDR), which finalizes the initial FBL. The SDR provides a formalized check of the logical design; communicates that design to the major stakeholders; confirms external interface and interoperability issues; confirms the BNR, SNR, and the SyRS; and provides a formal record of design decisions and design acceptance.

1.3.1.2 Preliminary Design

The aim of preliminary design is to convert the FBL into an upper-level physical definition of the system configuration or architecture (the hows of the system). Preliminary design is therefore the stage where logical design is translated into physical design, where focus shifts from the problem domain into the solution domain. The result of preliminary design is a subsystem-level design known as the allocated baseline (ABL) in which the logical groupings defined in the FBL have been defined in more detail and then regrouped and allocated to subsystem-level physical groupings—elements of which are called configuration items (CIs)—which combine to form the upper-level physical design of the system. At the center of the ABL are a series of development specifications, which contain the subsystem-level requirements grouped by subsystem/CI as well as the interface requirements among those subsystems/CIs.

The ABL is called as such because the system-level requirements that are logically grouped in the FBL are allocated at this next baseline into physical groupings (subsystems/CIs). The ABL therefore represents a subsystem-level architecture (couched in physical terms) that meets the requirements of the system-level architecture (couched in logical terms) contained in the FBL.

The initial ABL is formalized at the preliminary design review (PDR). The PDR ensures the adequacy of the preliminary design effort prior to focusing on detailed design. PDR assesses the technical adequacy of the proposed solution in terms of technical risk and the likely satisfaction of the FBL. PDR also investigates the identification of subsystem/CI interfaces and the compatibility of each of the subsystems/CIs.

1.3.1.3 Detailed Design and Development

The ABL developed during preliminary design is used as the basis to complete development of the individual subsystems, assemblies, and components in the system. Prototyping may occur and the system design is verified by test and evaluation. The result of the detailed design and development process is the initial establishment of the product baseline (PBL) as the system is now defined by the numerous products (subsystems, assemblies, and components) making up the total system (as well as the requisite materials and processes for manufacturing and construction). The definition of the system at this stage should be sufficiently detailed to support the commencement of construction and/or production activities, which can be confirmed by a production readiness review.

The PBL is established at the critical design review (CDR), which is the final design review resulting in the official acceptance of the design and the subsequent commencement of construction and/or production activities; CDR evaluates the detailed design; determines readiness for production/

construction; and ensures design compatibility, including a detailed understanding of all external and internal interfaces.

1.3.1.4 Construction and/or Production

The final stage within the acquisition phase is construction and/or production. System components are produced in accordance with detailed design specifications in the PBL and the system is ultimately constructed in its final form. Formal test and evaluation activities (acceptance tests) will be conducted to verify that the system is meeting its requirements as it is being developed (developmental test and evaluation [DT&E]) and that the final system meets the requirements in the SyRS (acceptance test and evaluation [AT&E]).

Construction and/or production, and the acquisition phase, ends with the formal qualification review (FQR), system verification review (SVR), or equivalent upon which the acquirer accepts the system from the developer. The FQR/SVR is informed by the results of AT&E.

1.3.2 Utilization Phase

On acceptance from the supplier, the system moves into the utilization phase. Further testing in conjunction with all elements of the capability (equipment, people, processes, training, facilities, support)—operational test and evaluation (OT&E)—validates that the system meets its original business needs and the system is suitable to enter operational service. The major activities during this phase are operational use and system support as well as continuous OT&E activities to validate that the capability system still meets the business needs. Systems engineering activities may continue during the utilization phase to support any modification that may be required to rectify performance shortfalls, to meet changing operational requirements or external environments, to enable ongoing support for the system to be maintained, or to enhance current performance or reliability.

1.3.3 Retirement Phase

The system life cycle ends with retirement of the system in the retirement phase, which may well overlap with the introduction into service of the replacement system should the organization still require the capability. Functions associated with retirement include transportation and handling, decomposition, and processing of the retiring system. If the system is at the end of its useful life at the end of the life cycle, retirement will mean disposal because the system cannot be reused in any way. A retirement concept should be developed during the early stages of the acquisition phase. If considered early, disposal and phase-out issues will form some of the criteria against which the system is designed (often called "design for disposability"). However, it is important that system designers focus on retirement, as well as the more limiting issues of disposal; planning for disposal is important, but a system may retire from several life cycles before it is ultimately disposed of at the end of life.

Example 1.4: The RAAF F-111 Aircraft

> *An excellent example of an extended life cycle is the F-111 aircraft operated by the Royal Australian Air Force (RAAF). Following a mid-life upgrade in the 1990s involving both avionics and engine modifications, the RAAF had originally expected the F-111 to remain in service until 2020. However, the aircraft was retired earlier than planned as part of the transition to the F/A-18F Super Hornets, with its final withdrawal occurring in 2010. Given that the origins of the F-111 can be traced back to the 1950s, the life cycle of the RAAF F-111s still exceeded 50 years.*

1.4 Systems Engineering and Development Approaches

It should be noted that the generic system life cycle illustrated in Figure 1.12 shows the phases and activities in sequence and is not intended to represent any particular development model or acquisition model. Throughout the early chapters of this book, we describe systems engineering without discussing in great detail the development and acquisition context within which it might be undertaken. We present the life-cycle activities sequentially because it is the best way to explain the activities and artifacts of systems engineering. In doing so, we assume what is generally referred to as a linear sequential or waterfall approach to system development. However, there are a number of other development approaches to implementing the activities of the system life cycle in Figure 1.12, such as the incremental, iterative, spiral, evolutionary, or Agile acquisition models [36], each of which has strengths and weaknesses depending on the nature of the system under development. The selection of a suitable development approach is a critical planning activity early in a system life cycle.

However, for simplicity in the early chapters, the linear sequential (waterfall) approach to system development is assumed in order to provide a logical, sequential flow of activities and deliverables that support the provision of a readily understandable explanation of systems engineering. Additionally, the waterfall approach can be considered to be the basic building block upon which alternative approaches such as incremental, iterative, evolutionary, spiral, and Agile development are built [37, 38]. A solid understanding of waterfall development is therefore useful, regardless of which other development methodology is employed.

We discuss these issues in much more detail in Chapter 11 in which we consider systems engineering as part of the major acquisition and development approaches.

1.5 Verification and Validation and Test and Evaluation (T&E)

The entire systems engineering process aims to produce a system that is both verified against the descriptions produced during the systems engineering process, and validated against the original needs, goals, and objectives that initiated the system development in the first case. Often these two associated aims are combined into the term *verification and validation* (V&V). Verification ensures that we have "built the system right" and validation ensures that we have also "built the right system" [39]. A well-managed approach to test and evaluation (T&E) aims to support the delivery of a system that is both verified and validated. Other systems engineering management functions including technical reviews and configuration audits also support the overall objectives of verification and validation.

There are three major categories of T&E that are applied to coincide broadly with verification and validation required in the activities of the acquisition phase, the transition between the acquisition and utilization phases, and the retirement phase.

- *Developmental test and evaluation (DT&E).* DT&E refers to the T&E activities undertaken during the acquisition phase to support the design and development effort. DT&E activities may also occur during the utilization phase to support activities such as the development of modifications. DT&E mostly supports verification activities to ensure that the system element being developed is meeting, or has met, the design requirements of that system element.
- *Acceptance test and evaluation (AT&E).* AT&E represents the formal verification conducted to enable the acquirer to verify that the system meets the SyRS and to accept it from the developer. AT&E effectively forms the boundary or transition between the acquisition phase and the utilization phase and, as such, is an important contribution to FQR/SVR, which verifies that the delivered system meets the SyRS.

- *Operational test and evaluation (OT&E).* Once the system is accepted from the developer, OT&E may be conducted under realistic operational conditions by operational personnel in order to validate the capability system, that is, the major equipment combined with operational users, collective training, facilities, support, and so on. OT&E is normally conducted for a period following acceptance of the system by the acquirer from the developer (that is, after AT&E) and to validate that the system meets its BNR/SNR before introduction into service. OT&E is also conducted continuously throughout the system life cycle in order to determine continued fitness for purpose and to identify the need for any modifications and upgrades.

1.6 What Is Systems Engineering?

There are many definitions of systems engineering, each of which is subtly different because it tends to reflect the particular focus of its source and its time. The following are some of the definitions from relevant standards and documents.

> "Systems engineering is the management function which controls the total system development effort for the purpose of achieving an optimum balance of all system elements. It is a process which transforms an operational need into a description of system parameters and integrates those parameters to optimize the overall system effectiveness" [40].

> "An interdisciplinary collaborative approach to derive, evolve, and verify a life cycle balanced system solution which satisfies customer expectations and meets public acceptability" [41].

> "An interdisciplinary approach encompassing the entire technical effort to evolve and verify an integrated and life cycle balanced set of system, people, product, and process solutions that satisfy customer needs. Systems engineering encompasses: the technical efforts related to the development, manufacturing, verification, deployment, operations, support, disposal of, and user training for, system products and processes; the definition and management of the system configuration; the translation of the system definition

into work breakdown structures; and development of information for management decision making" [42].

> "Systems engineering is the selective application of scientific and engineering efforts to: transform an operational need into a description of the system configuration which best satisfies the operational need according to the measures of effectiveness; integrate related technical parameters and ensure compatibility of all physical, functional, and technical program interfaces in a manner which optimizes the total system definition and design; and integrate the efforts of all engineering disciplines and specialties into the total engineering effort" [43].

> "Systems engineering is an interdisciplinary, comprehensive approach to solving complex system problems and satisfying stakeholder requirements" [44].

The authoritative definition should come from the International Council on Systems Engineering (INCOSE): ". . .a transdisciplinary and integrative approach to enable the successful realization, use and retirement of engineered systems, using systems principles and concepts, and scientific, technological and management methods" [45, 46].

Although each of these definitions has a slightly different focus, a number of common themes are evident, so we focus on those in the following sections.

1.6.1 The Top-Down Approach

Traditional engineering methods are based on a bottom-up approach in which known components are combined into assemblies and then into the subsystems from which the system is then constructed. The system is then evaluated for the desired properties and the design is modified in an iterative manner until the system meets the desired needs. This approach is valid and extremely useful for relatively straightforward problems that are well defined. Unfortunately, complicated problems cannot be solved effectively with the bottom-up approach.

Systems engineering takes a systems approach by addressing the system as a whole, which facilitates an understanding of the system, its mission, goals, and objectives, its environment, and its interfaces. Once system-level requirements are understood, architectural options are developed and a preferred architecture is chosen. The system is then broken down into subsystems and the subsystems further broken down into assemblies and then into components until a complete understanding is achieved of the system from top to bottom. This top-down approach is a very important element of managing the development of complicated systems. By initially viewing the system as a whole and then progressively breaking the system into smaller elements, the interaction between the components can be understood more thoroughly, which assists in identifying and designing the necessary interfaces between components (internal interfaces) and between this and other systems (external interfaces). For example, Figure 1.15 illustrates the EIA-632 [47] approach to top-down development from a product perspective.

However, it must be recognized that, although design is conducted top-down, the system is implemented using a bottom-up approach, as illustrated in Figure 1.16 for a simple four-tier system. That is, one of the aims of system engineering is to provide a rigorous, reproducible process by which the complex system can be broken down into a series of simpler components that can then be designed and built using the traditional bottom-up engineering approach. Importantly, the second principal facet of systems engineering is therefore to provide a process by which the components, assemblies, and subsystems can be integrated and verified as achieving the desired system purpose.

Note that, as discussed in Section 1.1.8, the terms system, subsystem, assembly, and component are relative. Each system comprises subsystems that consist of assemblies that consist of components. However, each subsystem can be treated as if it were a system in its own right, which has subsystems, assemblies, and components. IEEE-STD-1220 calls this the system paradigm, which is synonymous with the systems approach as we discussed it earlier. Consequently, while the building-block concept for top-down development is very useful, it is often a source of confusion among novices if the associated terms are allowed to be used in a relative manner. Here, as outlined earlier, we avoid that confusion by maintaining distinctions among the terms SoS, system, and subsystem while always keeping the systems approach in mind.

1.6.2 Requirements Engineering

The development of a complete and accurate set of system requirements is fundamental to project success and is a primary focus of the early systems engineering effort (recognizing that a complete description may not be possible at the start of a system development). The life cycle of a system begins with business needs and stakeholder needs, which are ultimately transformed into a large number of statements of requirement that form the basis for the logical design and subsequently elaborated further to form the physical architecture.

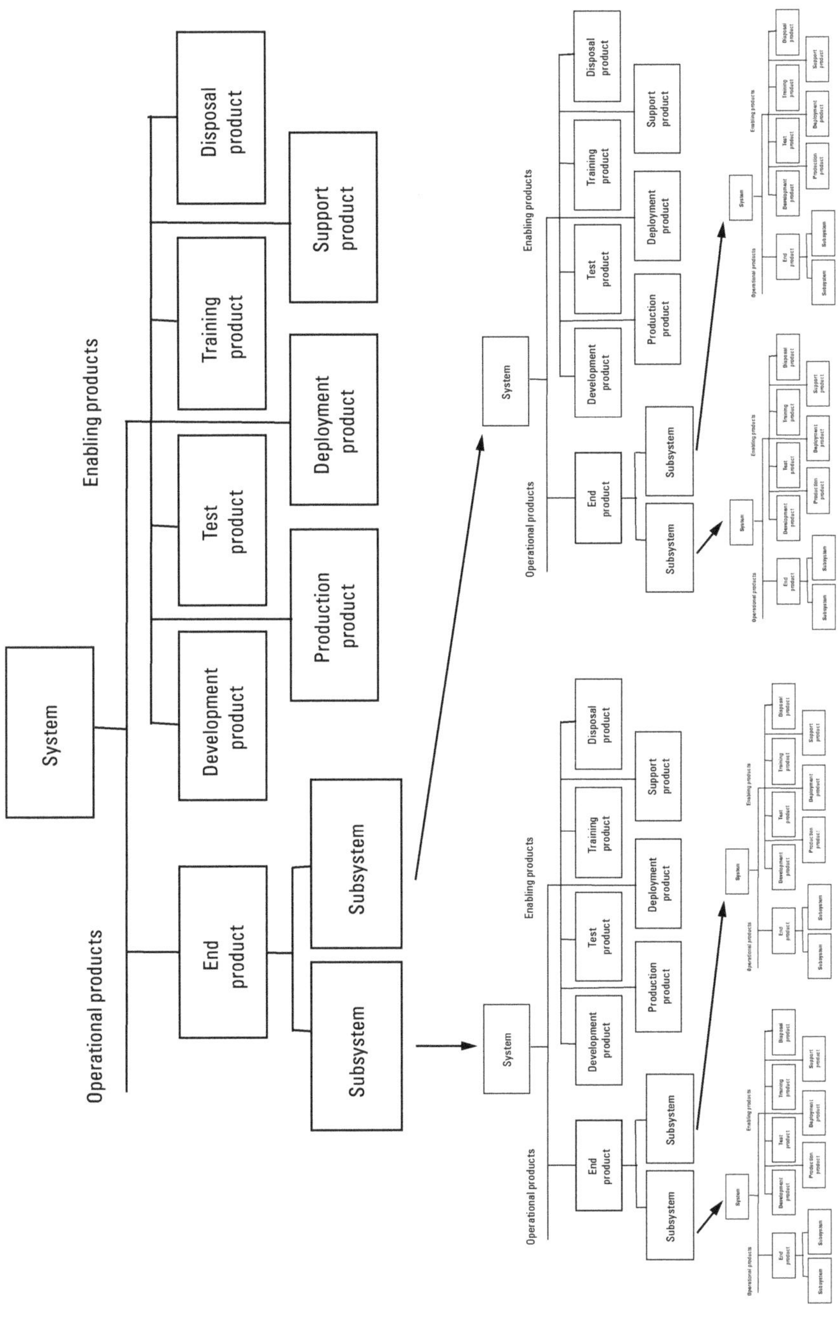

Figure 1.15 EIA-632 building block concept for top-down development.

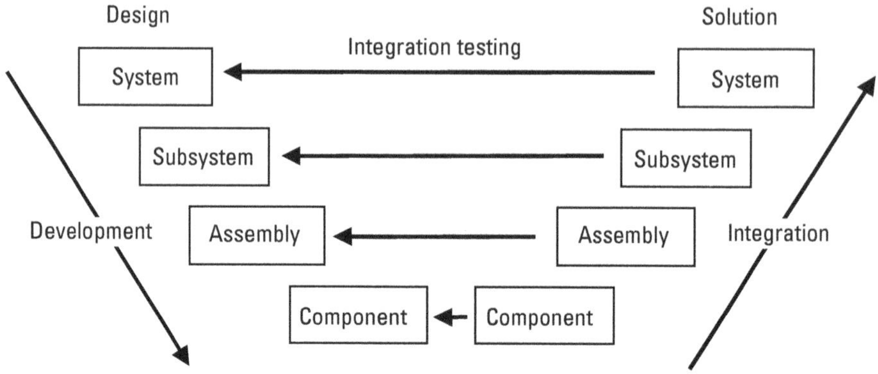

Figure 1.16 Bottom-up integration of a system—the system Vee diagram.

These transitions must be managed by a rigorous process, called business analysis or requirements engineering, which is aimed at ensuring that all relevant requirements are included (and all irrelevant requirements excluded). The establishment of correct requirements is fundamental to the success of the subsequent design activities. Poor requirements cannot be rectified by good design, so it invariably follows that rigorous development of requirements is essential for the acquisition to be successful. Note that we are not referring here to any time period in particular; the same principle applies to a full development period in waterfall development, to a build in evolutionary development, or to a sprint in Agile (see Section 1.4 and Chapter 11).

Once a set of requirements has been established, the systems engineering process then focuses on the derivation and decomposition of these requirements from the system level down to the lowest constituent component (sometimes referred to as requirements flowdown). This process involves elicitation, analysis, definition, validation, and management of requirements.

Requirements traceability is also an essential element of effective management of complex projects. Through traceability, design decisions can be traced from any given system-level requirement down to a detailed design decision (called forward traceability). Similarly, any individual design decision must be able to be justified by being associated with at least one higher-level requirement (called backward traceability). This traceability is important since the customer must be assured that all requirements can be traced forward and can be accounted for in the design at any stage. Further, any aspect of the design that cannot be traced back to a higher-level requirement is likely to represent unnecessary work for which the customer is most probably paying an unnecessary premium. Traceability also supports the configuration management (change control) process, especially the investigation of the impact of the change (see Chapter 8). Support for requirements traceability is a feature of the top-down approach that provides a mechanism by which it can be guaranteed that requirements can be satisfied at any level. A bottom-up approach cannot provide the same guarantee.

Chapter 2 provides more detail on the body of knowledge of requirements practice.

1.6.3 Focus on Life Cycle

Systems engineering is focused on the entire system life cycle and takes this life cycle into consideration during decision-making at any point. Unfortunately, it is still too common to consider solutions only in the light of the issues associated with the acquisition phase and to pay little attention to through-life support aspects. It is proper for project managers and their teams to focus on the acquisition phase of the project and on the development of a quality system that meets its scope while minimizing cost and schedule. However, a lack of attention to whole-of-life considerations can often lead to larger-than-expected costs in the utilization phase that must be met from budgets that are insufficient to keep systems in service. A life-cycle focus is on the capability system throughout its life cycle, not just on the product throughout its acquisition.

The significance of focusing on the system life cycle is that decisions made early in conceptual design are informed by the activities proposed to be conducted later in the acquisition phase and particularly in the utilization phase. For example, the design of an aircraft airframe must take into account the maintenance and operation of that airframe during the utilization phase; it would be pointless to design the best airframe in the world if it did not have the necessary access points to allow maintenance personnel to service it or operators to operate it in the intended environment.

Other examples of utilization phase requirements that impact equipment design or selection during the acquisition phase include reliability and availability. Reliability normally refers to the ability of the equipment to operate without failure for a given period. Availability is a measure of the degree to which a system is in an operable condition when required at some random point in time. Again, a superior design is pointless unless the system can meet specified minimum levels of reliability and availability and, for that matter, all the other "-ilities."

Economic factors provide arguably the most compelling evidence to support the focus on life cycle as opposed to an emphasis on the end product itself. In short, a life-cycle focus can save money in the long term. History has shown that a large proportion of total life-cycle cost (LCC) for a given system stems from decisions made early in the acquisition phase of the project. To that end, the acquisition phase presents the maximum opportunity to reduce the total LCC, the total cost of ownership (TCO), of a system. Poor requirements cannot be rectified by good design, so it invariably follows that rigorous development of requirements is essential for the acquisition to be successful. Put slightly differently, it is impossible for a project manager to manage scope, time, and quality to achieve project success if the scope of the project is fundamentally flawed; success or failure is therefore defined early in a project's life, knowingly or not.

Example 1.5: Life-Cycle Focus

> *As a simple analogy to demonstrate the concept (and problems) of a product focus as opposed to a life-cycle focus, consider a motor vehicle manufacturer that has introduced a new model. The new family sedan is a roomy, good-looking, high-performance vehicle that costs half the price of similar models. Strong interest seems ensured with buyers attracted by the vehicle's features as well as the cost-effectiveness of the purchase.*
>
> *However, on further investigation, it transpires that the low purchase price has been achieved by a design that incorporates poor-quality components that require iterative, regular maintenance resulting in running costs that are five times higher than competing models.*
>
> *This example continues to be relevant today where many people are choosing to switch from internal combustion engine vehicles (ICEVs) to battery electric vehicles (BEVs). At the time of this writing, some of the older BEVs around the world have already reached the end of their original battery's useful life. The owners of these BEVs are discovering that the cost of replacing batteries in these BEVs is significant compared to the new car purchase price. This information is no doubt impacting on true life-cycle cost of BEV ownership. Clearly, a decision to purchase the new vehicle (whether it is an ICEV or a BEV) must focus on more than just the purchase price of the vehicle and must take into account the system life-cycle aspects of operation, maintenance and support, and residual resale value.*

Example 1.6: Life-Cycle Focus

> *For our aircraft example, ACME Air must consider much more than the purchase price of the new aircraft. If a low purchase price has been achieved by a design that incorporated poor-quality components that require regular maintenance, the running costs may well be much higher than competing design alternatives. Clearly, the choice of an aircraft must focus on total cost*

of ownership of the aircraft, which, in addition to the purchase price, must consider the system life-cycle aspects of operation, maintenance, and support. The aircraft is acquired once but is operated and supported for thousands of hours. It will not take long for the operating, support, and maintenance costs to exceed the cost of acquisition.

1.6.4 System Optimization and Balance

A system architecture must represent a balance among the large number of requirements and constraints that cover a wide range of factors such as environmental, economic human factors, moral, ethical, social, cultural, and psychological. A simple, but essential, example is the balance between just two design factors: cost and performance; if either is allowed to dominate, it will generally be at the expense of the other.

A balance in system design must also be struck across the life cycle. Metrics such as cost-effectiveness must be measured across all life-cycle phases, not just acquisition. It is often the case that savings made in acquiring the system are then countered by significant maintenance and repair costs in service. Although, as a corollary, it does not follow necessarily that paying more for a system during acquisition will guarantee that the support costs are lower.

Systems engineering performs a very important role in system design in ensuring that there is a balance among the various system components because it is essential that optimization and balance are managed at the system level. As we discuss in Chapter 4, it does not necessarily follow that the combination of optimized subsystems leads to an optimized system. It is not normally useful, therefore, to allow the designers of subsystems to optimize their part of the system in isolation of system-level considerations; put another way, it is generally necessary for system-level considerations to constrain subsystem-level design in order for the system to be optimized.

It follows that a number of subsystems, if not all subsystems, may need to be suboptimal (or at least constrained in some manner) to allow their combination (the system) to be optimal. Therefore, a further advantage of the top-down approach in systems engineering is that system optimization and balance can be achieved as a by-product of the design process.

Example 1.7: System and Subsystem Optimization

Consider the impact of incorporating an F1 engine in a small family saloon car; the engine may well be optimized for performance, but it will destroy the remainder of the drive train that has been designed for a much less powerful engine. Additionally, the F1 engine is capable of propelling the family car far faster than is safe given its suspension and brakes and much faster than is allowed by legal speed limits. In general, if a subsystem (in this case the engine) is to be a useful servant of a system (in this case, the car), its performance has to be defined (constrained) by the system designer, not optimized by the subsystem designer.

1.6.5 Integration of Disciplines and Specializations

Systems engineering aims to manage and integrate the efforts of a multitude of technical disciplines and specialties to ensure that all needs and requirements are adequately addressed. Rarely is it possible for a complicated system to be designed by a single discipline.

Systems engineering focuses on the definition of the tasks to be completed by these disparate disciplines and specialties and then on the provision of the management required to integrate their efforts to produce a system that meets the business needs. In modern system developments, this function is all the more important because of the complexity of large projects and their contracting mechanisms, and the geographic dispersion of contractor and subcontractor personnel across the country and around the world.

Example 1.8: Integration of Disciplines

Consider our aircraft example. While aeronautical engineers may have a major role, the design, development, and production of a modern aircraft system require a wide variety of engineering disciplines including electrical/electronics, safety and assurance, EMI/EMC, production, metallurgical, and corrosion engineers. Other engineering disciplines are required for testing and for logistics and maintenance support as well as the design and building of facilities such as runways, hangars, refueling facilities, and embarkation and disembarkation facilities. Other nonengineering disciplines are involved in such aspects as marketing, finance, accounting, legal, and environmental.

In short, there could be hundreds, even thousands, of engineers and members of other disciplines involved in the delivery of a single aircraft system. Commonly, complex systems such as modern aircraft can involve millions of hours of work by thousands of people from a wide range of disciplines and backgrounds spread across dozens of countries.

1.6.6 Management

While systems engineering clearly has a technical role and provides essential methodologies for systems development, it is not limited to technical issues and is not simply another engineering process to be adopted. Systems engineering has both a management and a technical role. During acquisition, project management is responsible for ensuring that the system is delivered on time and within budget and meets the expectations of customers. The trade-offs and compromises implicit in those functions are informed by the products of systems engineering, which must be managed to be efficient. Additionally, the scope of the project is defined by the work breakdown structure, which is the result of the systems engineering effort. During acquisition, systems engineering and project management are therefore inextricably linked. These issues are discussed further at the end of this chapter and in more detail in Chapter 10.

1.7 Systems Engineering Relevance

Systems engineering principles and processes are applicable (albeit to varying degrees) to a wide range of projects. For example, EIA-632 [48] states that the standard, as a systems engineering standard, is intended to be applicable to:

> "the engineering or the reengineering of:
>
> a) commercial or non-commercial systems, or part thereof;
> b) any system, small or large, simple or complex, software-intensive or not, precedented or unprecedented;
> c) systems containing products made up of hardware, software, firmware, personnel, facilities, data, materials, services, techniques, or processes (or combinations thereof);
> d) a new system or a legacy system, or portions thereof."

It is difficult to imagine a system or its attendant project that does not fit into the above description, which can therefore effectively be summarized to say that systems engineering is applicable to all systems and therefore to all projects. However, a project is defined as producing a unique product, which means that the application of systems engineering to each project will also be unique to that project. Consequently, it is critical to understand the merits of systems engineering and apply them in a tailored manner, cognizant of the relative size, complexity, and risks associated with each undertaking. At one end of the spectrum are large complex projects making use of leading-edge developmental technology. These projects typically involve large sums of money, long time frames, and significant risks. At the other end of the spectrum are small projects making use of extant techniques and existing technology. These projects typically involve short periods, low costs, and a minimum of risk. Clearly different levels of systems engineering are applicable to each of these types of projects and the aspects thar we consider here must be tailored for each individual project.

The most obvious application of systems engineering principles and methodologies is in projects that are large and complicated. However, smaller and less complex projects can also benefit from the tailored application of systems engineering. In short, systems engineering practices, such as project management practices, are always relevant but must always be tailored to be useful.

Example 1.9: Systems Engineering Relevance to Large Projects

> *A classic example cited by the Software Engineering Institute in their Systems Engineering Capability Maturity Model (SE-CMM) [49] demonstrates the need for good systems engineering. The Tacoma Narrows bridge was a very long suspension bridge with a flexible roadway that collapsed in 1940 due to strong winds that had set up an aerodynamic oscillation.*
>
> *It is believed that the problems associated with the bridge could have been avoided or managed more effectively had systems engineering principles and processes been applied rigorously during the project. While adhering to good systems engineering practices would not necessarily ensure that designers are more likely to know the parameters of a particular problem, rigorous developmental approaches are more likely to draw out the relevant risk areas of a system (such as environment-related risks in this case).*

The failures of system developments are often attributed to misunderstandings, ambiguities, misinterpretations, errors and omissions in what the supplier is attempting to deliver. The result is a system that fails to address the acquirer's problem leading to a breakdown in the relationship between the acquirer and the supplier. Systems engineering specifically targets these problems and therefore is relevant to both parties in any system development.

Within an acquirer's project office, systems engineering is particularly relevant to the technical personnel who are responsible for the application of systems engineering principles as part of the overall project management effort. In most acquirers, systems engineering personnel report to the project manager to apply the tools of systems engineering to assist the project manager in each of the 10 project management knowledge areas. Systems engineering is therefore an essential element of the project manager's ability to acquire a quality system within budget, time, and scope constraints.

Suppliers should also be focused on developing processes and best practices to support the delivery of superior products and services to their customers. Systems engineering offers philosophies, methodologies, and tools that support the consistent development of quality products and services.

1.8 Systems Engineering Benefits

The principal causes of cost and schedule overrun on large-scale projects can be traced to such factors as overzealous advocacy, immature technology, lack of corporate roadmaps, requirements instability, ineffective acquisition strategy, unrealistic project baselines, project office personnel tenure and experience, and inadequate systems engineering [50]. In this book, we focus on the latter (which, incidentally, is the only one of those causes that is directly within the project manager's control during acquisition) because there are a number of potential benefits from the successful implementation of systems engineering processes and methodologies.

The first and most visible benefit is the scope for saving money during all phases of the system life cycle: LCC/TCO savings. While some may argue that the additional requirements imposed by systems engineering can increase costs, these increases are comparatively small and are generally felt in the very early design phases. If applied appropriately, systems engineering can ensure that the savings achieved far outweigh the cost of implementing appropriate procedures and methodologies. Experience indicates that an early emphasis on systems engineering can result in significant cost savings later in the construction and/or production, operational use and system support, and disposal phases of the life cycle [51].

Systems engineering should also assist in reducing the overall schedule associated with bringing the system into service. Systems engineering ensures that the user requirements are accurately reflected in the design of the system helping to minimize costly and time-consuming changes to requirements later in the life cycle. If changes are required, they can be incorporated early in the design and in a controlled manner. The rigorous consideration and evaluation of feasible design alternatives during the design phases of the project promote greater design maturity, which can be achieved earlier.

System failures, cost overruns, and schedule problems are often the direct result of the poor application of business analysis and requirements engineering; poor requirements cannot be rectified by good design. The systems engineering discipline aims to put in place a rigorous process to produce well-defined needs and requirements, adequate levels of traceability between the different levels of technical design back to the original needs and requirements, and system requirements, which are both verifiable and consistent. This requirements-management process must achieve these results without pre-supposing a particular technical solution or placing unnecessary technical constraints on the solution.

Figure 1.17 provides a simplistic illustration of the impact of systems engineering on the system life cycle. Note that the greatest impact is through the rigorous application of processes and methodologies during the early stages of the project where the ease of change is greatest, and cost of modification is lowest. The curve in Figure 1.17 could be relabeled as the ease with which changes can be made throughout the system life cycle. Consequently, systems engineering provides the ideal opportunity to have the greatest impact on a project at a time when adjustments are easiest to make. Additionally, as illustrated in Figure 1.18, the greatest impact of requirements engineering comes at a time when the cost of implementing changes is the lowest.

Systems engineering leads to a reduction in the technical risks associated with the system development. Risks are identified early and monitored throughout the process using a system of technical performance measures and design reviews and audits. Design decisions can be traced back to the original needs and requirements and conflicting needs and requirements can be identified and clarified early, significantly reducing the risk of failure later in the project.

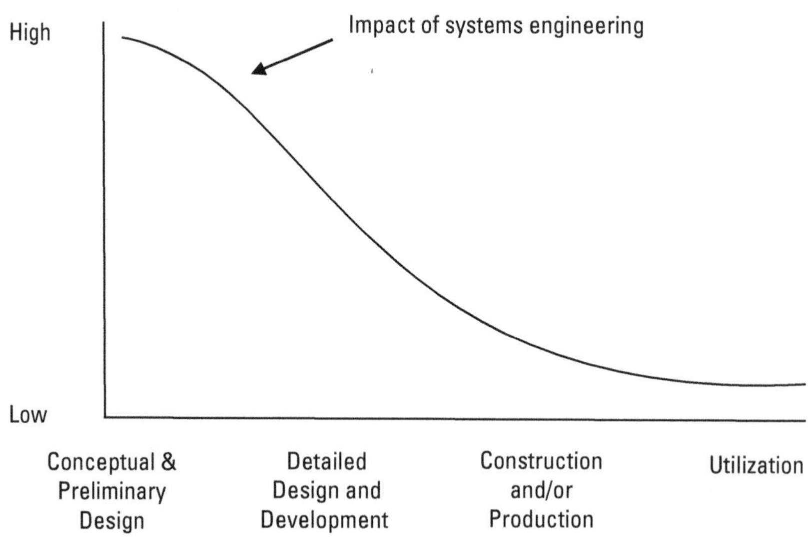

Figure 1.17 Impact of systems engineering on the system life cycle.

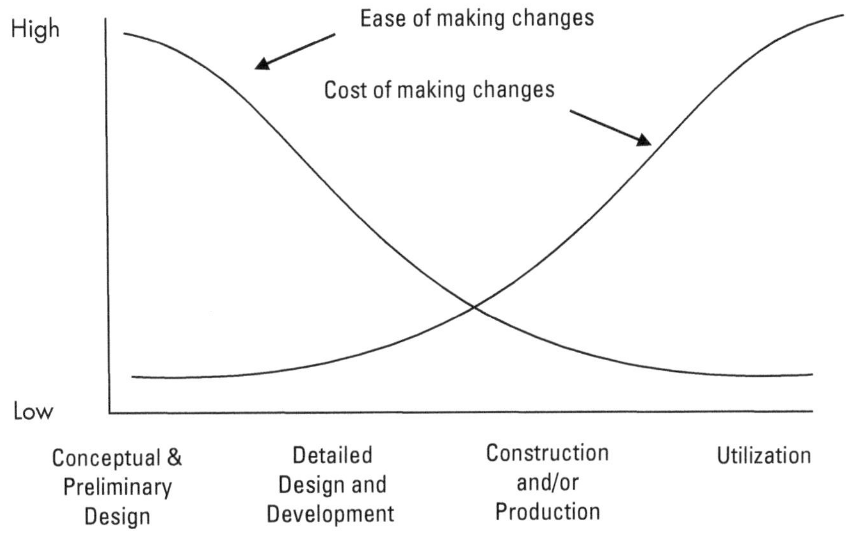

Figure 1.18 Ease and cost of making changes throughout the system life cycle.

Finally, and probably most importantly from a project management perspective during acquisition, the disciplined approach of systems engineering leads to a delivered solution that meets the original intended purpose more completely; systems engineering makes an essential contribution to all the knowledge areas in the Project Management Body of Knowledge (we discuss this in more detail in Chapter 10).

1.9 Analysis, Synthesis, and Evaluation

All extant systems engineering standards and practices extol processes that are built around an iterative application of analysis, synthesis, and evaluation [52]. The iterative nature of the application is critical to systems engineering processes. Initially, the process is applied at the systems level; it is then reapplied at the subsystem level and then the assembly level and so on until the entire development process is complete. During the earlier stages, the acquirer is heavily involved; in the latter stages, the developer is mainly responsible for the continuing effort, which is monitored by the acquirer.

Prior to detailing the individual activities within the systems engineering processes, it is worth considering the basic foundations of the analysis-synthesis-evaluation loop illustrated in Figure 1.19. This concept is neither new

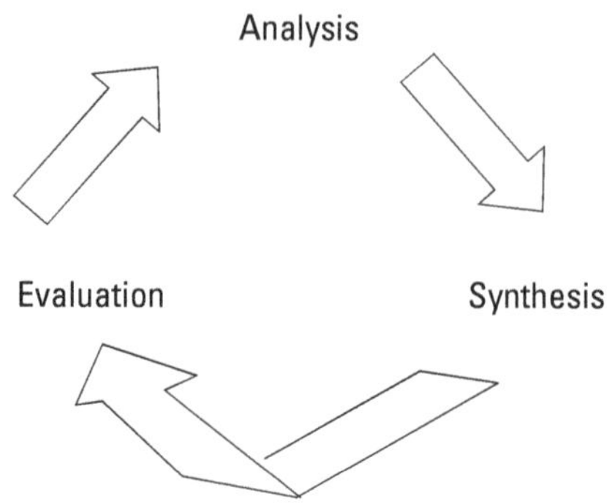

Figure 1.19 The analysis-synthesis-evaluation iteration.

nor terribly sophisticated; it is simply a good, sound approach to problem solving. While applicable in any domain, the loop is particularly fundamental to systems engineering.

1.9.1 Analysis

Analysis commences with a statement of perceived need or a set of customer requirements. During conceptual design, analysis investigates these needs and requirements and identifies the essential functions that the system must perform in order to meet the needs. Requirements analysis at the system level aims to answer the what, how well, and the why questions relative to the system design. Analysis activities continue throughout the subsequent stages of the life cycle to help in defining lower-level requirements, often called decomposed or derived requirements associated with physical aspects of system design.

A basic tool that can be used during analysis is the functional flow block diagram (FFBD), which shows the logical sequences and relationships of the functions of the system at the system level. The system-level functions detailed in the FFBD are further developed at subsystem level and below, using more detailed FFBDs. Examples of FFBDs are presented in Chapters 2, 3, and 4. As the design becomes progressively more detailed, the system-level functions will be individually investigated and have an FFBD produced and so on until a very detailed set of functions and their interrelationships have been identified.

Once all the functions associated with the system have been identified, the requirements associated with each function can be defined. These requirements could include performance parameters such as speed, altitude, and accuracy; interoperability requirements detailing other systems with which the system under development must operate; and interface requirements to describe the necessary outputs expected from the system and the inputs to the system. Depending on the particular design phase, these requirements and functions may also be grouped in accordance with some sort of logical criteria and then allocated to a particular physical component of the system. That is, the component becomes responsible for the satisfaction of those requirements by performing the functions assigned to it. The allocation of requirements forms a description of the system elements and architecture and therefore assists in the process of synthesis or design.

1.9.2 Synthesis

Analysis revolves around what is required as well as how well and why. Synthesis, or design, then determines how. Synthesis is possibly the most widely recognized role of a professional engineer. Synthesis is the process where creativity and technology are combined to produce a design that best meets the stated system requirements. The term synthesis is more appropriate than design in the systems engineering context as it hints at the evolutionary nature of design and development.

In the early stages of the systems engineering processes, synthesis is limited to defining completely the functional design of the system and then considering all possible technical approaches using the results from the requirements analysis effort. From this consideration, the best approach is selected, and the process moves to the next level of detail. Later in the systems engineering processes, the selected design concept is synthesized further until ultimately, the complete system design is finalized.

Although synthesis is the creative part of the systems engineering effort, the following chapters describe a number of tools that are available to the engineer to ensure that all alternatives are considered and the most suitable alternative is ultimately selected.

1.9.3 Evaluation

System cost and risk are directly associated with requirements and design. Evaluation is the process of investigating the trade-offs between requirements and design, considering the design alternatives, and making the necessary decisions. The process of evaluation continues throughout all stages of the systems engineering effort, ultimately determining the system's satisfaction of original requirements. Trade-off analysis is one of the tools available to the system designer in performing evaluation of competing requirements or designs; we discuss trade-off analysis in Chapter 4.

The outcome of the evaluation is selection or confirmation of the desired approach to design. Discrepancies are also identified if applicable and may result in further analysis and synthesis. The basic analysis-synthesis-evaluation loop described in Figure 1.19 is applied iteratively throughout the system life cycle, as illustrated in Figure 1.20.

1.10 A Systems Engineering Framework

Discussions on systems engineering can quickly become complicated due to the broad mandate of the discipline, the wide range of types of systems, the complexity and interrelationship of the many systems engineering activities, and the relationships with other disciplines throughout the system life cycle.

The ability to understand a complex subject such as systems engineering is greatly enhanced by a solid framework within which concepts can be considered. Examples of such a structure include the Project Management Body of Knowledge [53], which provides a clear framework within which to consider the many facets of project management, and the Software Engineering Body of Knowledge [54] that performs a similar role for the software engineering discipline. Without an equivalent framework, the broad scope of systems engineering soon becomes confusing given the complexity of its components and their many interrelationships. There are a number of excellent systems engineering standards available today that contribute to the elements of a suitable framework, but each standard contains complexity, (differing) terminology, and detail that requires substantial interpretation. The entry level of many students, junior engineers, and project managers therefore does not readily allow the use of such standards as effective frameworks within which to examine systems engineering.

A systems engineering framework [55] (illustrated in Figure 1.21) has been synthesized by the authors through a thorough survey of existing systems engineering publications and standards, and through the authors' experience in teaching systems engineering at a range of levels. The main aim of the structure is to provide a simple construct within which the systems engineering discipline can be understood and implemented.

The framework illustrates the relationship of the three main elements of systems engineering processes, systems engineering management, and systems engineering tools and places them in context with related disciplines such as traditional engineering disciplines, project management, integrated logistics support, and quality assurance. The terms systems engineering management

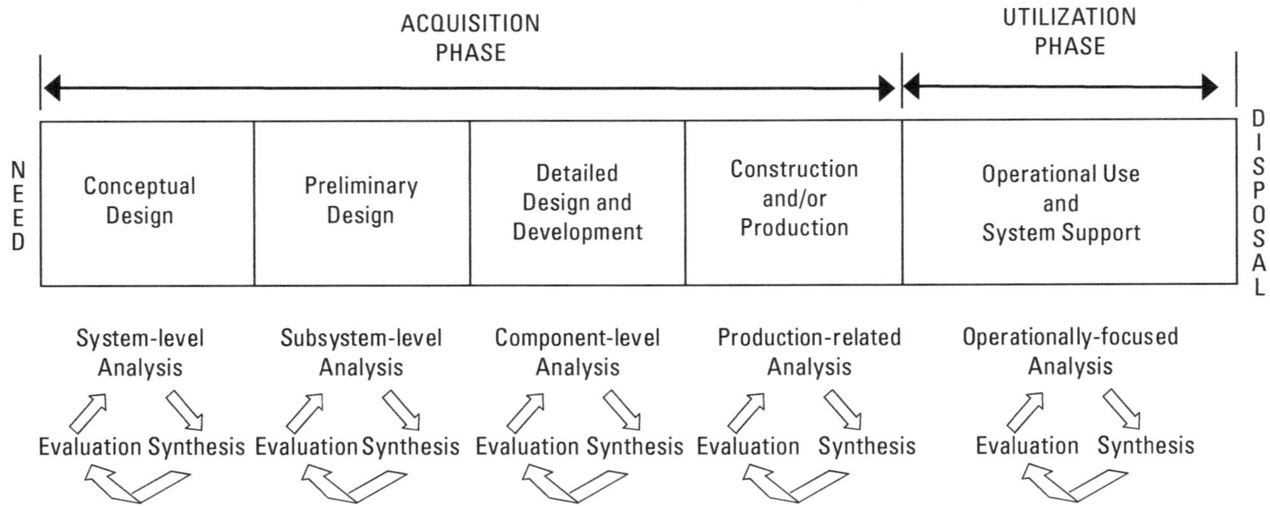

Figure 1.20 Iterative application of the analysis-synthesis-evaluation loop throughout acquisition and utilization.

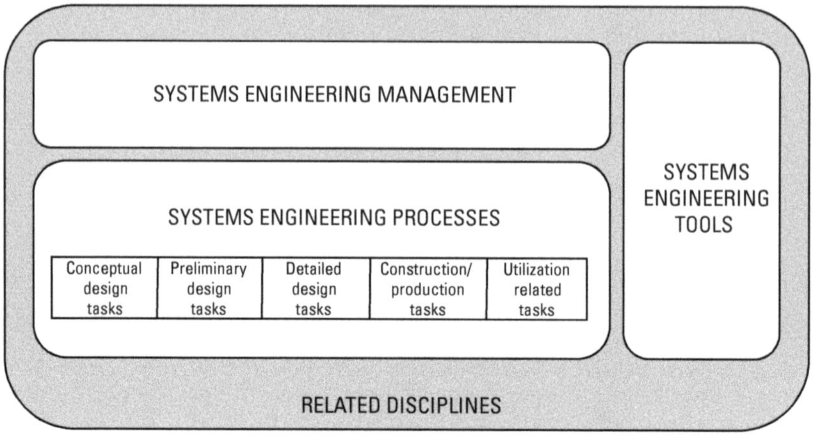

Figure 1.21 A framework for the consideration of systems engineering.

and systems engineering processes are sometimes used interchangeably. Here we make a distinction between the two.

We present the systems engineering processes as the hows of systems engineering, the application of which forms the foundation of the systems engineering effort. Over the top of these processes sits the systems engineering management function, which is responsible for directing the systems engineering effort, monitoring and reporting that effort to the appropriate areas, and reviewing and auditing the effort at critical stages in the entire process. These two elements are supported by a range of tools and all elements exist in the context of related disciplines.

The systems engineering framework provides an excellent structure within which to examine and explain the complex discipline of systems engineering. Experience in undergraduate and graduate courses as well as professional education short courses [56] has shown that the framework provides an excellent means of communicating the complexities and interrelationships of the systems engineering discipline, particularly to those who do not have a significant amount of project or systems experience. Those new to the discipline can successfully grasp the fundamental concepts of systems engineering within a relatively short timeframe.

1.10.1 Systems Engineering Processes

Systems engineering processes, activities, and tasks are divided into the life-cycle stages within which they typically occur. In this book, we do not attempt to detail exhaustively all systems engineering processes. Instead, we concentrate on the intent and main aim of each phase of the system life cycle and examine some of the likely techniques that may be used to achieve that aim. We place particular emphasis on the acquisition phase of the life cycle, as it is the phase during which systems engineering can have the most impact on a system.

1.10.2 Systems Engineering Management

Systems engineering management is an overarching activity responsible for directing the systems engineering and project management efforts, monitoring and reporting that effort to the appropriate areas, and reviewing and auditing the effort at critical stages in the entire process; such management is the key to success of the entire systems engineering effort. In Chapter 8, we address the major systems engineering management elements of technical reviews and audits, test and evaluation, technical risk management, configuration management, the use of specifications and standards, and systems engineering management planning.

1.10.3 Systems Engineering Tools

Many tools exist to assist systems engineering processes and management, including a range of techniques and methods and a number of standards. Here we describe the most popular tools without repeating information contained elsewhere in standards and other documents. However, it is not the intended purpose of this book to repeat information contained elsewhere in the many standards and artifacts; we simply demonstrate their utility in systems engineering and project management.

Throughout the book we present generic process tools such as requirements breakdown structures (RBS), functional flow block diagrams (FFBD), work breakdown structures (WBS), trade-off analyses, prototyping and simulation, and model-based systems engineering (MBSE) as examples of tools that may be applied to the systems engineering process effort. We also describe the use of standards as systems engineering management tools. In Chapter 9, we also describe the use of standards as systems engineering management tools, reviewing and summarizing MIL-STD-499B [57], EIA/IS-632 [58], IEEE-STD-1220 [59], EIA-632 [60], SAE1001 [61], and ISO/IEC/IEEE 15288 [62].

1.10.4 Related Disciplines

There are many disciplines (both technical and nontechnical) related to systems engineering. Examples include business analysis, project management, logistics management, quality assurance, requirements engineering, software engineering, hardware engineering, and interface engineering (or integration engineering).

The relationship between the related disciplines and the various facets of systems engineering within a project depends very much upon the discipline in question. Some (such as project management) oversee the whole systems engineering discipline during acquisition, while others (such as hardware and software engineering) sit between systems engineering management and the processes, and others (such as quality assurance) sit alongside the systems engineering effort. Chapter 10 discusses these disciplines and their relationship with systems engineering.

1.11 Summary

In this chapter, we introduced the nature of systems, the systems approach, the logical and physical descriptions of a system in the problem domain and the solution domain, the system life cycle and its constituent activities, and the parties involved in system acquisition. We also examined some definitions of systems engineering and extracted the common themes of top-down design, requirements engineering, focus on life cycle, system optimization and balance, integration of related disciplines and specialties, and management. We then considered the benefits of systems engineering and examined the analysis-synthesis-evaluation loop, which is applicable from the basic engineering activities to the systems engineering activities right across the system life cycle.

In the next chapter, we consider the body of knowledge of requirements engineering, which is an essential aspect of system development. Subsequent chapters then consider each of the major systems engineering processes in some detail.

1.12 Some Qualifications Before We Continue

Before we continue, having provided useful background in this introduction and explained what we are going to focus on in the remainder of the book, it behooves us to also be clear as to what we are *not* going to do (and why).

A project, by definition, produces a unique output. So, while there are general principles involved in the conduct of projects, the constituent activities of an individual project must be tailored into a relatively unique combination for that project to produce the unique output. There are therefore arguably as many ways to implement a particular project, as there are unique systems to be developed. We cannot possibly address here the many combinations of factors associated with complex technical projects, and it would not be useful to attempt to do so because any benefit of an introductory overview would be lost. Although we are acutely aware of these complications, we just do not have the room to address them directly here. So, to navigate the multidimensionality of the project space in a useful manner, we take a direct path and make a number of simplifications, which should be kept in mind when reading the following chapters:

- Most organizations involved in systems development will incorporate a number of development methodologies (perhaps at least a dozen) as part of their systems engineering and project management policies and procedures. Then, as part of the early project management activities, organizational practice requires a project manager to justify the selection of an appropriate methodology as part of their development of a project management plan. Here we use the linear sequential development methodology (commonly known as waterfall development) as a framework as we progress through the chapters. We then provide a description of other major development methodologies (incremental, iterative, evolutionary, spiral, and Agile) in Chapter 9. There are two main reasons for taking this approach in an introductory text.
  - The various elements of systems engineering are much easier to understand on first contact if they are explained sequentially; the following chapters are therefore structured that way. However, just because we have explained the processes sequentially does not mean at all that we suggest that the steps must be implemented that way. For each unique project output, an appropriate project development methodology must be chosen, tailored, and applied.
  - Even when a more elegant development methodology is applied, an understanding of the logical steps is essential, that is, it does not matter whether the project team is currently in a phase in waterfall, an increment in incremental, a build in evolutionary, a spiral in spiral, or a sprint in Agile, the steps we take the reader through here still have to be applied to the fundamental time period (phase, increment, build, spiral, or sprint) of the selected approach.
- We refer to the word complexity throughout the chapters. As we move through the processes, we are focused on the methodologies, tools, and techniques that can be applied to project activities to deal with

complexity in the project output. However, we are not focused on complexity per se and do not address it directly.

- In any system development, there will be a number of what most modern standards call acquirer-supplier arrangements. These are also called acquirer-developer, customer-contractor, and tasking-performing activity arrangements, among a number of other terms. It does not matter whether the development is conducted, as are most government acquisitions, through a major formal tendering/contracting process or, as many in-house developments are, undertaken between two or more groups internal to the organization, or in any other arrangement. In all cases (albeit slightly less formally for in-house development), it is essential that the acquirer-supplier arrangement is clear and unambiguous and that there are agreements for milestones, deliverables, budget, schedule, review and audit, verification and validation, and so on. This remains true (albeit less formally), even for short development periods such as sprints in Agile development methodologies. Consequently, since there can be as many unique contracting arrangements as there are projects bringing unique systems into being, we focus our explanations on the most formal of those arrangements as may be implemented between a large government acquirer and a major supplier. Once the reader is familiar with the issues that drive formal developments, it is relatively straightforward to apply the general principles to less formal arrangements, such as guiding in-house developments.
- One of the most dynamic areas in modern systems engineering is in digital engineering (DE) and the associated model-based systems engineering (MBSE). Here we focus on project artifacts that could be documents, spreadsheets, or databases and on tools that could be requirements management tools (RMT) or models in MBSE. For example, we say that system requirements are recorded in the system requirement specification (SyRS)—while we use the word specification that might be construed to imply document, we do not use the term in that way and expect that the SyRS will be recorded in any form suitable to the organization: from a document to a spreadsheet, a database, a model, or any other digital form, or some combination. So, when we use the term document as a verb we mean to record as either a paper-based document, a database, or a model. We provide a brief description of MBSE in Chapter 4, but we avoid direct reference to DE and MBSE in the remainder of the book for three reasons:
  - Many projects still use paper-based documents to record project activities and will do for years, or even decades to come; we have

    worked with a number of major international companies that still use word processors to capture requirements, contracts, and design outputs. While DE should be, and is being, embraced enthusiastically wherever possible, there is significant associated overhead that requires significant organizational investment. Additionally, with major government contracts that can span a decade or more, many projects will remain in the nondigital or partly digital world for some considerable time.
    - Although the form of the physical artifact changes between the paper-based and the digital environments, the logic of activities and relationships does not change; the same principles apply in both environments. Here, therefore, we focus on these principles and logical steps. For convenience, we describe the resultant artifacts as a logical result of each stage, phase, and activity. We do not at all propose that those artifacts take any particular physical or digital form; that is a matter for the project and the organization it supports.
    - There are already a number of excellent texts and manuals describing the application of MBSE and modeling using SysML; see the major endnotes at the end of Chapter 4.
- Huge advances have been made in the 2023 revisions of ISO/IEC/IEEE 15288, the INCOSE Systems Engineering Handbook, and its attendant manuals and guides to provide a common view of systems engineering in terms of life-cycle processes, methodologies, tools, techniques, and terms. However, there are still many subtle (and sometimes not-so-subtle) differences in the way in which systems engineering is applied in organizations around the world. We are therefore careful to provide a balance to reflect the older terms and processes as well as the new; our interest is for the content of the book to be as relevant to system developments that are just beginning as it is to as those that have been in progress for a decade.
- Since we are concentrating on the relationship between project management and systems engineering, we are necessarily focusing on the acquisition phase of the system life cycle; we address all phases and stages for completeness, but our emphasis is on the role of systems engineering in the projects that bring a system into being.

We invite the reader to seek more detail from the wide range of endnotes that we provide at the end of each chapter or to follow with the two sibling books in this Artech House set: *Managing Complex Technical Projects* [63] and *Requirements Practice in Conceptual Design: A Practical Guide* [64]. In

Managing Complex Technical Projects, we provide an overview recognizing that the reader may come from any project-related discipline and with one of a wide variety of experience levels. Here we provide a much deeper understanding of the body of knowledge of systems engineering across the life cycle. Then, in *Requirements Practice in Conceptual Design: A Practical Guide*, we focus on greater detail again in the acquisition phase and particularly conceptual design since the development of an appropriate set of system requirements is the most important task in system development.

Review Questions

1. Briefly define a system.
2. What is meant by a system of interest (SOI)?
3. What is meant by the terms of operating environment and wider SOI (WSOI)?
4. Briefly describe the system classifications: open or closed; natural or human-made or human-modified, physical or conceptual, and precedented or unprecedented.
5. List the major resources that make up a capability system in the context of systems engineering.
6. A system can be described in both logical and physical terms. Briefly explain why the two descriptions must coexist, what each description provides, and the relationship between the two descriptions.
7. Briefly outline how and why a system is described hierarchically (both logically and physically).
8. Briefly describe the three types of collection of system: portfolio-of-systems (PoS), system-of-systems (SoS), and family-of-systems (FoS).
9. Briefly describe the differences between a subsystem, a system, and a system-of-systems (SoS).
10. Briefly describe the difference between the problem domain and the solution domain.
11. Briefly describe what is meant by a systems approach and what is meant by a soft systems approach and a hard systems approach.
12. Draw a diagram and briefly describe the four broad phases of a generic system life cycle (preacquisition phase, acquisition phase, utilization phase, and retirement phase).
13. Identify the major responsible parties (enterprise management, business management, project management, systems engineering, the

developer (contractor), and users/support) for the major phases of the life cycle.

14. Briefly describe the acquisition phase and the utilization phase of the system life cycle (as proposed by Blanchard and Fabrycky and used in this book) and explain briefly the activities that occur within each, and the major artifacts and reviews associated with each.
15. Briefly describe the three main categories of test and evaluation (T&E): developmental test and evaluation (DT&E), acceptance test and evaluation (AT&E), and operational test and evaluation (OT&E).
16. List and briefly describe these key themes of systems engineering (top-down approach, requirements engineering, focus on life cycle, system optimization and balance, integration of disciplines and specializations, management).
17. Describe the benefits of using systems engineering discipline during the development of a system.
18. Describe diagrammatically the analysis-synthesis-evaluation process that is applied iteratively throughout the system life cycle and explain each of the components.

Endnotes

[1] Further reading on systems engineering can be obtained from the following major sources:

Aslaksen, E. W., *Designing Complex Systems: Foundations of Design in the Functional Domain*, Boca Raton, FL: CRC Press, 2008.

Aslaksen, E. W., *The Systems Concept and its Application to Engineering*, New York: Springer, 2013.

Aslaksen, E., and R. Belcher, *Systems Engineering*, Upper Saddle River, NJ: Prentice Hall, 1992.

EIA-632, *Processes for Engineering a System*, Washington, DC: Electronic Industries Association (EIA), 2003.

Beam, W., *Systems Engineering: Architecture and Design*, New York: McGraw-Hill, 1990.

Blanchard, B. S., and J. E. Blyler, *Systems Engineering Management*, New York: John Wiley & Sons, 2016.

Blanchard, B., and W. Fabrycky, *Systems Engineering and Analysis*, Upper Saddle River, NJ: Prentice Hall, 2011.

Boardman, J., *Systems Engineering: An Introduction*, Upper Saddle River, NJ: Prentice Hall, 1990.

Buede, D. M., and W. D. Miller, *The Engineering Design of Systems: Models and Methods*, New York: John Wiley & Sons, 2016.

Chestnut, H., *Systems Engineering Tools*, New York: John Wiley & Sons, 1965.

Chestnut, H., *Systems Engineering Methods*, New York: John Wiley & Sons, 1967.

Defense Systems Management College (DSMC), *Systems Engineering Management Guide*, Washington, DC: US Government Printing Office, 1990.

Dickerson, C. E., and D. N. Marvis, *Architecture and Principles of Systems Engineering*, Boca Raton, FL: CRC Press, 2010.

EIA/IS-632, *EIA Interim Standard—Systems Engineering*, Washington, DC: Electronic Industries Association, 1994.

De Weck, O. L., D. Roos, and C. L., Magee, *Engineering Systems*, Cambridge, MA: MIT Press, 2011.

Dori, D., *Model-Based Systems Engineering with OPM and SysML*, New York: Springer, 2016.

Douglass, B. P., *Agile Systems Engineering*, San Francisco, CA: Morgan Kaufmann, 2015.

Eisner, H., *Essentials of Project and Systems Engineering Management*, New York: John Wiley & Sons, 2008.

Faulconbridge, R. I., and M. J. Ryan, *Managing Complex Technical Problems*, Norwood, MA: Artech House, 2026.

Forrester, J., *Principles of Systems*, Cambridge, MA: The MIT Press, 1968.

Gangl, M., et al., *Modern Methods of Systems Engineering: With an Introduction to Pattern and Model Based Methods*, Createspace Independent Publishing Platform, 2011.

Grady, J. O., *Systems Integration*, Boca Raton, FL: CRC Press, 1994.

Grady, J. O., *System Engineering Planning and Enterprise Identity*, Boca Raton, FL: CRC Press, 2010.

Haberfellner, R., et al., *Systems Engineering: Fundamentals and Applications*, Cham, Switzerland: Birkhauser, 2019.

Hall, A., *A Methodology for Systems Engineering*, Princeton, NJ: D. Van Nostrand, 1962.

Heap, G., J. Stanway, and A. Windsor, *A Structured Approach to Systems Development*, New York: McGraw-Hill, 1992.

Hitchins, D., *Advanced Systems Thinking, Engineering, and Management*, Norwood, MA: Artech House, 2003.

Hitchins, D., *Putting Systems to Work*, New York: John Wiley & Sons, 1992.

Hitchins, D., *Systems Engineering: A 21st Century Systems Methodology*, New York: John Wiley & Sons, 2007.

Holt, J., *Systems Engineering Demystified: A Practitioner's Handbook for Developing Complex Systems Using a Model-Based Approach*, Birmingham, UK: Packt Publishing, 2021.

Holt, J., and S. Perry, *SysML for Systems Engineering: A Model-Based Approach*, Herts, UK: Institution of Engineering and Technology, 2013.

Hubka, V., and W. Eder, *Theory of Technical Systems*, Berlin: Springer-Verlag, 1988.

Hunger, J., *Engineering the System Solution: A Practical Guide to Developing Systems*, Upper Saddle River, NJ: Prentice Hall, 1995.

IEEE-STD-1220, *IEEE Standard for Application and Management of the Systems Engineering Process*, New York: IEEE Computer Society, 2005.

INCOSE, *Systems Engineering Handbook: A Guide for System Lifecycle Processes and Activities*, Walden, D. D., et al., (eds.), INCOSE-TP-2003-002-05 2023, Hoboken, NJ: Wiley, 2023.

INCOSE, *Journal of the International Council on Systems Engineering*, Seattle, WA: International Council on Systems Engineering.

INCOSE, *Proceedings of Annual Conference*, Seattle, WA: International Council on Systems Engineering,

ISO/IEC/IEEE 15288, *Systems and Software Engineering—System Lifecycle Processes*, 2023.

Jackson, S., *Systems Engineering for Commercial Aircraft*, Surrey, England: Ashgate, 2015.

Jamshidi, M., *Systems of Systems Engineering: Principles and Practice*, Boca Raton, FL: CRC Press, 2009.

Jenney, J., et al., *Modern Methods of Systems Engineering*, CreateSpace Independent Publishing Platform, 2011.

Kamrani, A. K., and M. Azimi, *Systems Engineering Tools and Methods*, Boca Raton, FL: CRC Press, 2011.

Khisty, C. J., and J. Mohammadi, *Fundamentals of Systems Engineering with Economics, Probability, and Statistics*, Upper Saddle River, NJ: Prentice Hall, 2012.

Kossiakoff, A., et al., *Systems Engineering Principles and Practice*, New York: John Wiley & Sons, 2020.

Lacy, J., *Systems Engineering Management: Achieving Total Quality*, New York: McGraw-Hill, 1992.

Larson, W., et al., *Applied Space Systems Engineering*, New York: McGraw-Hill, 2009.

Machol, R., *System Engineering Handbook*, New York: McGraw-Hill, 1965.

Maier, M., and E. Rechtin, *The Art of Systems Architecting*, Boca Raton, FL: CRC Press, 2009.

Martin, J., *Systems Engineering Guidebook: A Process for Developing Systems and Products*, Boca Raton, FL: CRC Press, 1997.

MIL-STD-499B, *Military Standard—Systems Engineering—Draft*, Washington, DC: U.S. Department of Defense, 1994.

Micouin, P., *Model-Based Systems Engineering: Fundamentals and Methods*, New York: Wiley, 2014.

Parnell, G. S., P. J. Driscoll, and D. L. Henderson, (eds.), *Decision Making in Systems Engineering and Management*, New York: John Wiley & Sons, 2008.

Rechtin, E., *Systems Architecting: Creating and Building Complex Systems*, Upper Saddle River, NJ: Prentice Hall, 1991.

Reilly, N., *Successful Systems Engineering for Engineers and Managers*, New York: Van Nostrand Reinhold, 1993.

Sage, A., *Decision Support Systems Engineering*, New York: John Wiley & Sons, 1991.

Sage, A., *Systems Engineering*, New York: John Wiley & Sons, 1992.

Sage, A., *Systems Management for Information Technology and Software Engineering*, New York: John Wiley & Sons, 1995.

Sage, A., and J. Armstrong, *Introduction to Systems Engineering*, New York: John Wiley & Sons, 2000.

Sage, A. and W. Rouse, (eds.), *Handbook of Systems Engineering and Management*, New York: John Wiley & Sons, 2009.

Shishko, R., *ASA Systems Engineering Handbook*, Washington, DC: NASA, 1995.

Stevens, R., et al., *Systems Engineering: Coping with Complexity*, Hertfordshire, England: Prentice Hall Europe, 1998.

Sydenham, P. H., *Systems Approach to Engineering Design*, Norwood, MA: Artech House, 2004.

Thome, B., (ed.), *Systems Engineering: Principles and Practice of Computer-Based Systems Engineering*, New York: John Wiley & Sons, 1993.

Truxal, J., *Introductory Systems Engineering Management*, New York: McGraw-Hill, 1972.

Wasson, C., *System Engineering Analysis, Design, and Development: Concepts, Principles, and Practices*, New York: John Wiley & Sons, 2015.

Westerman, H., *Systems Engineering Principles and Practice*, Norwood, MA: Artech House, 2001.

Wheatcraft, L. S., M. J. Ryan, and T. E. Katz, *INCOSE Needs and Requirements Manual: Needs, Requirements, Verification, Validation Across the Lifecycle*, New York: Wiley, 2024.

Wymore, A., *Model-Based Systems Engineering*, Boca Raton, FL: CRC Press, 1993.

[2] See, for example:

McDonald, G. R., and J. C. Price, "Economics of Trolley Coach Power Supply," *Transactions of the American Institute of Electrical Engineers*, Vol. 70, No. 1, 1951, pp. 687–689.

Rothstein, J., "An Informational Approach to Organization and System Engineering Design," *Transactions of the IRE Professional Group on Engineering Management*, PGEM-1, 1954, pp. 25–29.

Vonpechmann, W., "Method and Systems Engineering," *Industrial and Engineering Chemistry*, Vol. 42, No. 4, 1950, pp. A73–A74.

[3] See:

Cole, R. I., "Management's Role in Research and Development of Electronic Systems," *Proceedings of the I.R.E.*, November 1950, pp. 1252–1253.

Cole, R. I., "Management Aspects of Electronic Systems Engineering," *Proceedings of the Institute of Radio Engineers*, Vol. 39, No. 12, 1951, pp. 1492–1493.

Cole, R. I., "The Electronic 'Systems' Engineer (Systems Project Director)," *Proceedings of the I.R.E.*, November 1952, pp. 260–261.

[4] See, for example:

Quarles, G. G., "The Modern Torpedo: A Case Study in Systems Engineering," *Journal of the American Society for Naval Engineers*, Vol. 68, No. 4, 1956, pp. 801–804.

Ryan, F. M., "Systems Engineering," *IRE Transactions on Communications Systems*, Vol. 4, No. 3, 1956, p. 1.

Schlager, K. J., "Systems Engineering—Key to Modern Development," *IRE Transactions on Engineering Management*, Vol. EM-3, No. 3, 1956, pp. 64–66.

Montgomery, B. E., "Air-Ground Communications: The ANDB System Development Plan," *IRE Transactions on Communications Systems*, Vol. 4, No. 2, 1956, pp. 48–53.

Monsanto, "Training for Automation," *Chemical and Engineering News*, Vol. 35, No. 28, 1957, p. 38.

Noll, A. H., "Airborne Weapons System Development through Simulation," *SAE National Aeronautic Meeting*, 1957.

[5] Goode, H., and R. Machol, *Systems Engineering: An Introduction to the Design of Large Scale Systems*, New York: McGraw-Hill, 1957.

[6] Ackoff, R. L., and F. E. Emery, *On Purposeful Systems: An Interdisciplinary Analysis of Individual and Social Behavior as a System of Purposeful Events*, London: Tavistock Publications, 1972.

[7] ISO/IEC/IEEE 15288, *Systems and Software Engineering—System Lifecycle Processes*, 2023.

[8] Flood, R. L., and E. R. Carson, *Dealing with Complexity: An Introduction to the Theory and Application of Systems Science*, New York: Plenum Press, 1993.

[9] EIA-632, *Processes for Engineering a System*, Washington, DC: Electronic Industries Association (EIA), 2003.

[10] US Department of Defense, *Manual for the Operation of the Joint Capabilities Integration and Development System*, CJCSI 5123.01H, October 2021.

[11] NATO Standard ADatP-34, *NATO Interoperability Standards and Profile (NISP)*, August 18, 2023.

[12] Australian Government, Defence, *The One Defence Capability System (ODCS) Manual*, 2025.

[13] UK Ministry of Defence, *Acquisition Operating Framework*, version 1.1.13, May 2011.

[14] Canadian Department of National Defence, *Designing Canada's Army of Tomorrow: A Land Operation's 2021 Publication*, 2011.

[15] ISO 55000, *Asset Management—Vocabulary, Overview and Principles*, 2024.

[16] ISO 55000, *Asset Management—Vocabulary, Overview and Principles*, 2024.

[17] ISO/IEC/IEEE 15288, *Systems and Software Engineering—System Lifecycle Processes*, 2023.

[18] IEEE-STD-1220, *IEEE Standard for Application and Management of the Systems Engineering Process*, New York: IEEE Computer Society, 2005.

[19] ISO/IEC/IEEE 15288, *Systems and Software Engineering—System Lifecycle Processes*, 2023.

[20] ISO/IEC/IEEE 21839, *Systems and Software Engineering—System of Systems*, 2019.

[21] This observation perhaps needs a little elaboration. When we have a nut and a bolt and need to join them in a useful way, we do need to take a systems approach to the nature of the two elements and their interface requirements (just as we would in considering the interconnection of a number of subsystems in a system, or a number of systems in an SoS). However, taking a systems approach at any level of engineering design does not mean that any combination at any level is usefully called a "system." That is not universally agreed in the systems engineering community; hence, the term system of interest is used to apply to any elements interconnected together. That is a defendable perspective; however, it does rather fall down when defining a system of systems, which is therefore any combination of elements at the next level up. Nonetheless, the nuance of the debate is not important, and the remainder of this text is applicable regardless of the reader's predilection.

[22] Maier, M. W., "Architecting a Portfolio of Systems," *Systems Engineering*, Vol. 1, No. 4, 2019, pp. 1–13.

[23] ISO/IEC/IEEE 15288, *Systems and Software Engineering—System Lifecycle Processes*, 2023.

[24] INCOSE, *Systems Engineering Handbook: A Guide for System Lifecycle Processes and Activities*, Walden, D. D., et al., (eds.), INCOSE-TP-2003-002-05, Hoboken, NJ: Wiley, 2023.

[25] Blanchard, B., and W. Fabrycky, *Systems Engineering and Analysis*, Upper Saddle River, NJ: Prentice Hall, 2011.

[26] MIL-STD-499B, Military Standard—*Systems Engineering—Draft*, Washington, DC: US Department of Defense, 1994.

[27] EIA-632, *Processes for Engineering a System*, Washington, DC: Electronic Industries Association (EIA), 2003.

[28] MIL-STD-499B, *Military Standard—Systems Engineering—Draft*, Washington, DC: US Department of Defense, 1994.

[29] EIA-632, *Processes for Engineering a System*, Washington, DC: Electronic Industries Association (EIA), 2003.

[30] MIL-STD-499B, *Military Standard—Systems Engineering—Draft*, Washington, DC: US Department of Defense, 1994.

[31] EIA-632, *Processes for Engineering a System*, Washington, DC: Electronic Industries Association (EIA), 2003.

[32] Blanchard, B., and W. Fabrycky, *Systems Engineering and Analysis*, Upper Saddle River, NJ: Prentice Hall, 1998.

[33] ISO/IEC/IEEE 15288, *Systems and Software Engineering—System Lifecycle Processes*, 2023.

[34] ISO/IEC/IEEE 15288, *Systems and Software Engineering—System Lifecycle Processes*, 2023.

[35] ISO/IEC/IEEE 29148 *Systems and Software Engineering—Lifecycle Processes—Requirements Engineering*, 2018.

[36] Further detail on these models can be found in, among others:

Blanchard, B., and W. Fabrycky, *Systems Engineering and Analysis*, Upper Saddle River, NJ: Prentice Hall, 2011.

Martin, J., *Systems Engineering Guidebook: A Process for Developing Systems and Products*, Boca Raton, FL: CRC Press, 1997.

Young, R., *The Requirements Engineering Handbook*, Norwood, MA: Artech House, 2004, pp. 73–78.

[37] Forsberg, K., and H. Mooz, *Systems Engineering Overview*, Cupertino, CA: Center for Systems Management, 1996.

[38] Forsberg, K., and H. Mooz, "Application of the 'VEE' to Incremental and Evolutionary Development," *Proceedings of the NCOSE Conference*, St. Louis, MO, July 1995.

[39] Boehm, B. W., "Verifying and Validating Software Requirements and Design Specifications," *IEEE Software*, 1984.

[40] Defense Systems Management College (DSMC), *Systems Engineering Management Guide*, Washington, DC: US Government Printing Office, 1990.

[41] IEEE-STD-1220, *IEEE Trial-Use Standard for Application and Management of the Systems Engineering Process*, New York: IEEE Computer Society, 1994.

[42] EIA-632, *Processes for Engineering a System*, Washington, DC: Electronic Industries Association (EIA), 2003.

[43] SECMM-95-01, *Systems Engineering Capability Maturity Model*, Version 1.1, Carnegie Mellon University, Pittsburgh, PA: Software Engineering Institute, 1995.

[44] Lake, J. G., "Unraveling the Systems Engineering Lexicon," *Proceedings of the INCOSE Symposium*, Vol. 6, No. 1, 1996, pp. 506–515.

[45] INCOSE Fellows, *Systems Engineering and Systems Definitions*, 2018.

[46] INCOSE, *Systems Engineering Handbook: A Guide for System Lifecycle Processes and Activities*, Walden, D. D., et al., (eds.), INCOSE-TP-2003-002-05, Hoboken, NJ: Wiley, 2023.

[47] EIA-632, *Processes for Engineering a System*, Washington, DC: Electronic Industries Association (EIA), 2003.

[48] EIA-632, *Processes for Engineering a System*, Washington, DC: Electronic Industries Association (EIA), 2003.

[49] CMU/SEI-94-HB-04, *A Systems Capability Maturity Model*, Version 1.0, Pittsburgh, PA: Software Engineering Institute, Carnegie Mellon University, 1994.

[50] Meier, S., "Best Project Management and Systems Engineering Practices in the Preacquisition Phase for Federal Intelligence and Defence Agencies," *Project Management Journal*, March 2008, pp. 59–71.

[51] Blanchard, B., and W. Fabrycky, *Systems Engineering and Analysis*, Upper Saddle River, NJ: Prentice Hall, 1998.

[52] Asimov, M., *Introduction to Design*, Englewood Cliffs, NJ: Prentice Hall, 1962.

[53] Project Management Institute (PMI), *PMBOK, A Guide to the Project Management Body of Knowledge*, Upper Darby, PA: Project Management Institute, Version 6, 2017.

[54] Washizaki, H., (eds.), *Guide to the Software Engineering Body of Knowledge (SWEBOK Guide)*, Version 4.0, IEEE Computer Society, 2024; www.swebok.org

[55] Faulconbridge, R. I., *Systems Engineering Body of Knowledge*, Canberra: Magpie Applied Technology, 2000.

[56] Faulconbridge, R. I., "A Systems Engineering Framework," *Journal of Battlefield Technology*, Vol. 3, No. 2, July 2000, pp. 25–28.

[57] MIL-STD-499B, *Military Standard—Systems Engineering—Draft*, Washington, DC: U.S. Department of Defense, 1994.

[58] EIA/IS-632-1994, *Systems Engineering*, Washington, DC: Electronic Industries Association (EIA), 1994.

[59] IEEE-STD-1220, *IEEE Standard for Application and Management of the Systems Engineering Process*, New York: IEEE Computer Society, 2005.

[60] EIA-632, *Processes for Engineering a System*, Washington, DC: Electronic Industries Association (EIA), 2003.

[61] SAE1001 *Integrated Project Processes for Engineering a System*, SAE International, 2018.

[62] ISO/IEC/IEEE 15288, *Systems and Software Engineering—System Lifecycle Processes*, 2023.

[63] Faulconbridge, R. I, and M. J. Ryan, *Managing Complex Technical Problems*, Norwood, MA: Artech House, 2026.

[64] Ryan, M. J., *Requirements Practice in Conceptual Design: A Practical Guide*, Norwood, MA: Artech House, 2026.

2

Requirements Engineering Framework

2.1 Introduction, Definitions, and Frameworks

We saw in Chapter 1 that systems engineering (particularly during the early processes of conceptual design and preliminary design) has a focus on gathering requirements in a formal systematic way, which is referred to generically as requirements engineering or business analysis. Before considering each of the major systems engineering processes in some detail in subsequent chapters, this chapter provides an overview of the extant requirements engineering body of knowledge because it is an essential aspect of system development; for more details, we leave the reader to consult the many excellent references in [1].

In this section, we consider the terms associated with concepts, needs, and requirements as well as their characteristics and attributes.

2.1.1 Concepts, Needs, and Requirements

When describing a system, we make the distinction between concepts, needs, and requirements. Concepts are typically narrative descriptions of ways in which the organization and its entities expect to manage, acquire, develop, deploy, operate, support, and retire the capability offered by the system. Needs are formal representations of expectations for an entity stated in the language and perspective of stakeholders. Needs are transformed into requirements, which are formal representations (model elements or textual statements) that communicate what an entity must do, and how well it must do it, to realize the intent of the needs. There may be more than one need for any given concept and more than one requirement defined for any need.

As illustrated in Figure 2.1, concepts, needs and requirements exist at a number of levels corresponding to four major views: an enterprise view, in which enterprise leadership set the enterprise strategies and concepts of operations; a strategic view, in which business management derives business concepts and needs and formalizes their requirements; an operational view, in which stakeholders in the business units define their concepts, needs, and requirements; and a systems view, in which system designers define the system, first in functional and then physical terms.

Because terms such as system and subsystem (system element) are level-specific (remember we are following a systems approach using a systems paradigm as discussed in Chapter 1), we use the generic term entity, which is ". . . a single thing to which a need or requirement refers: an enterprise, business unit, project, system, or system element (which could be a product, process, human, or organization)" [2].

As shown in Figure 2.1 (based on [3]), the enterprise, strategic, and operational views are in the problem domain; the system view is in the solution domain. As discussed earlier, the problem domain is the responsibility of those who have ownership of the problem to be solved, so the descriptions of the system are predominantly in the language of the customer, focusing on what they need the system to be able to do, how well it should be done, and why. We saw earlier that these descriptions are called functional (or often logical) descriptions. However, the solution domain is the responsibility of the developer (the performing activity) implementing the solution, most commonly a contractor, so the physical descriptions of the system in that domain are predominantly in engineering and physical terms, focusing on how the problem is to be solved; that is, what it will look like and how it will operate once it has been implemented. In many large organizations, there is a soft boundary between the problem domain and the solution domain in that the customer (the acquirer or tasking activity) prepares a draft system requirement specification (SyRS) that is taken to tender, responded to be tenderers, negotiated with the successful tenderer (the contractor) and then revised to be the SyRS that represents the selected solution of that contractor.

Although they coexist at each level of consideration, concepts, needs, and requirements are different [4, 5]:

- A concept is a written or graphic representation that concisely expresses how an entity will satisfy the problem it was defined to address. A

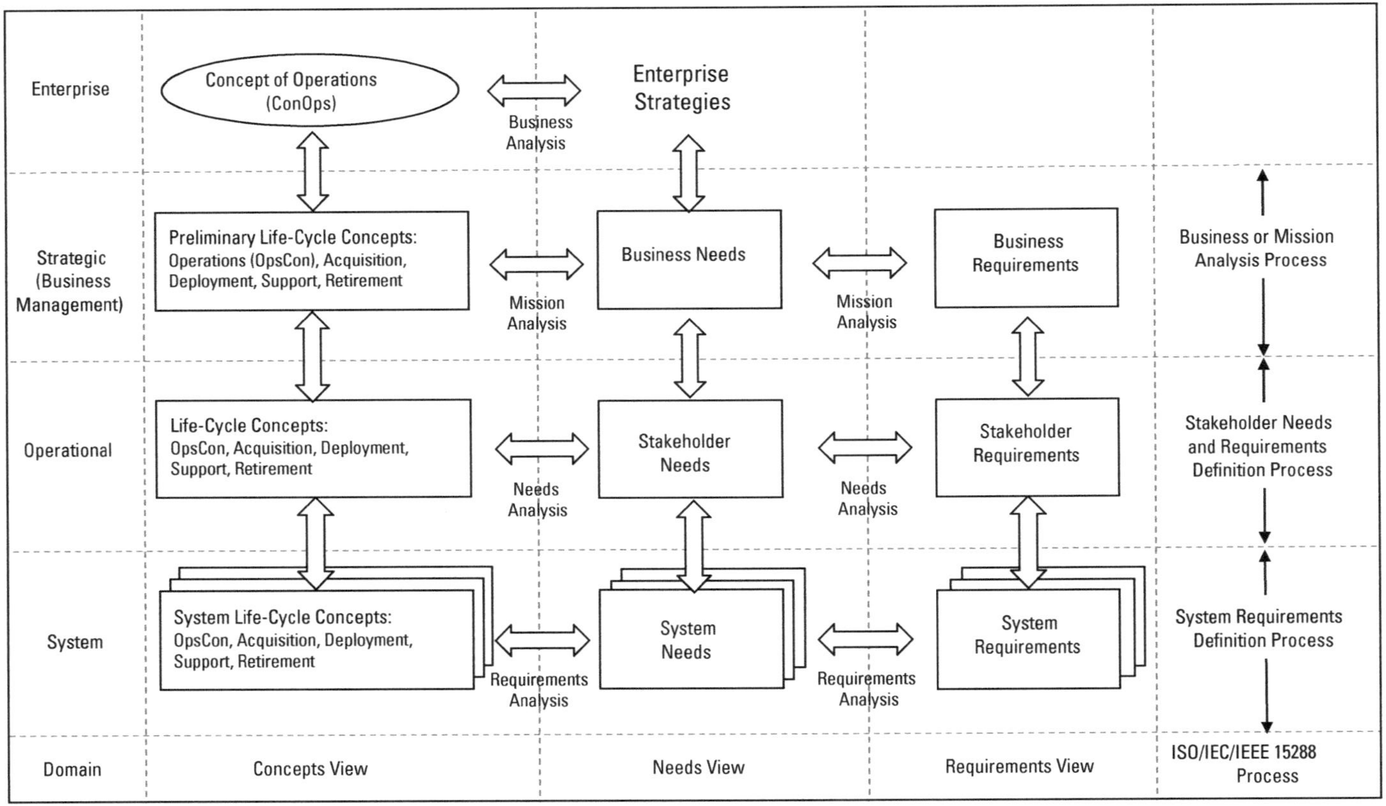

Figure 2.1 Views, domains, and ISO/IEC/IEEE 15288 processes and artifacts associated with the definition of concepts, needs, and requirements.

number of concepts can apply to each entity, so it is useful to develop a minimal set of concepts that define the expectations of all entities. It is also useful to provide some structure to the set of concepts. As illustrated in Figure 2.1, life-cycle concepts include operations concepts, acquisition concepts, deployment concepts, support concepts, and retirement concepts.

- A need is the result of a formal transformation of one or more concepts into an agreed-to expectation for an entity to perform some function or possess some quality (within specified constraints with acceptable risk). Needs are generally capabilities stated as business needs in the language of business management, as stakeholder needs in the language of stakeholders at the business operations level, or as needs of the system. There can be more than one need defined for any one concept.
- A requirement is the result of a formal transformation of one or more needs or parent requirements into an agreed-to obligation for an entity to perform some function or possess some quality (within specified constraints with acceptable risk). Requirements are formal statements that are structured and can be validated. Requirements are elaborated from needs through a process of requirements analysis (also called business analysis or mission analysis [6] at the highest level of application). There can be more than one requirement defined for any one need.

At the highest level, there is an enterprise view in which enterprise leadership sets business vision, and strategic goals and objectives in the form of a concept of operations (ConOps), which is also called a strategic business plan (SBP). At the next two levels, there are the dual views: at the business management level, there are business needs [7] and business requirements [8]; and at the business operations level, there are stakeholder needs [9] and stakeholder requirements [10]. The transformation of needs into requirements occurs at each level. Here we are interested in the three major upper-level processes:

- *Business needs and requirements (BNR) definition process.* At the strategic level, business management begins the process with the guidance contained in the ConOps, from which they define business needs, largely in the form of preliminary life-cycle concepts (PLC), which capture the preliminary acquisition concept, preliminary operational concept (OpsCon) [11], preliminary deployment concept, preliminary support concept, and preliminary retirement concept. Business needs are elaborated and formalized into business requirements, which are recorded in the business requirements specification (BRS) [12], which is also called the business requirement document [13].

- *Stakeholder needs and requirements (SNR) definition process.* Using the ConOps, the preliminary OpsCon, and the other PLC as guidance, stakeholders from the business operations level develop stakeholder needs—in the form of a refined OpsCon and other life-cycle concepts (LC)—and transform them into a formal set of stakeholder requirements, which are recorded in the stakeholder requirement specification (StRS) [14].
- *System requirements definition process.* The requirements in the StRS are then transformed by requirements engineers into system requirements, which are contained in the SyRS [15]. This artifact is also referred to as the solution requirement specification [16], or simply the system specification [17], or design input requirements in the NRM. A number of SyRS may be derived from the StRS; that is, perhaps, an SyRS is produced for each element of the capability described by the StRS (see Figure 2.2). Some organizations may also prepare an individual LC for each of a number of elements of capability that are developed to meet the business needs. The process continues for the system elements.

It should be noted that, in the past, almost every standard, textbook, and academic paper would have a slightly different spin on systems engineering terms, language, and processes. However, with the revision of ISO/IEC/IEEE 15288 in 2023 and the subsequent revision of the INCOSE *Systems Engineering Handbook* and the associated INCOSE manuals and guides (see

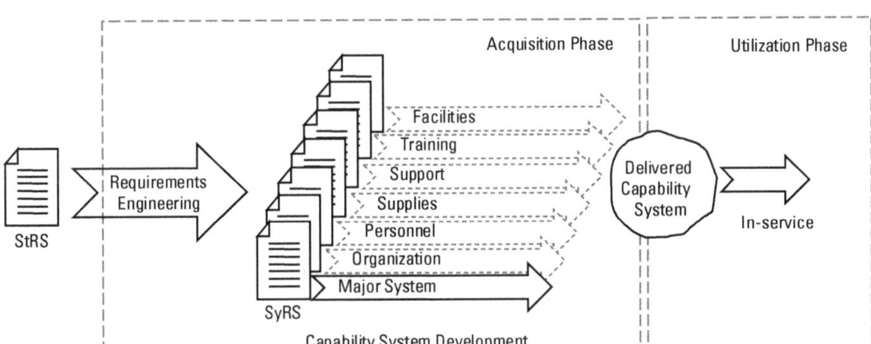

Figure 2.2 An example in which one SyRS is developed for each of the elements of the capability system described by the StRS.

Chapter 9 for a summary), the language and approaches are gradually beginning to converge. The description above is consistent with ISO/IEC/IEEE 15288, and the INCOSE *Systems Engineering Handbook*, and ISO/IEC/IEEE 29148 [18] with the exception that the Business or Mission Analysis Process in ISO/IEC/IEEE 15288 is called the BNR definition process in this text (the activities in the process are consistent).

Additionally, for practical implementation, the INCOSE *Needs and Requirements Manual* [19] combines what ISO/IEC/IEEE 15288 calls the BNR and SNR into an integrated set of needs, which is then transformed into design-input requirements (which ISO/IEC/IEEE 29148 calls the SyRS). Despite a preference for the structure of the *Needs and Requirements Manual*, we conform in this text to ISO/IEC/IEEE 15288 and ISO/IEC/IEEE 29148 and the INCOSE *Systems Engineering Handbook* because systems engineers are certified against those documents and many organizations use them to provide the frameworks for their projects (especially those projects that are already in progress).

In any case, it is of little consequence because the reader can readily make the translation; ISO/IEC/IEEE 15288's BNR and SNR are combined in the NRM's integrated set of needs; and ISO/IEC/IEEE 29148's SyRS is called the design-input specification in the NRM. To assist throughout the remaining chapters, we occasionally use both sets of terms to reinforce the duality.

To help visualize the relationship, Figure 2.3 illustrates the relationship between the artifacts called for by ISO/IEC/IEEE 15288 and ISO/IEC/IEEE 29148, the artifacts from INCOSE's *Needs and Requirements Manual*, and the relevant processes from ISO/IEC/IEEE 15288.

ANSI/AIAA G-043A-2012 [20] identifies that the terms "concept of operations" and "operational concept" are often used interchangeably but notes that an important distinction exists because each has a separate purpose and is used to meet different ends. It is essential that these terms are used so that they are consistent with ANSI/AIAA G-043A-2012 and ISO/IEC/IEEE 29148 [21], as well as the way in which the terms are used in the US DoD and many other defense departments.

ISO/IEC/IEEE 29148 describes the ConOps as: "The ConOps, at the organization level, addresses the leadership's intended way of operating the organization. It may refer to the use of one or more systems, as black boxes, to forward the organization's goals and objectives. The ConOps document describes the organization's assumptions or intent in regard to an overall operation or series of operations of the business with using the system to be developed, existing systems, and possible future systems. This document is frequently embodied in long-range strategic plans and annual operational

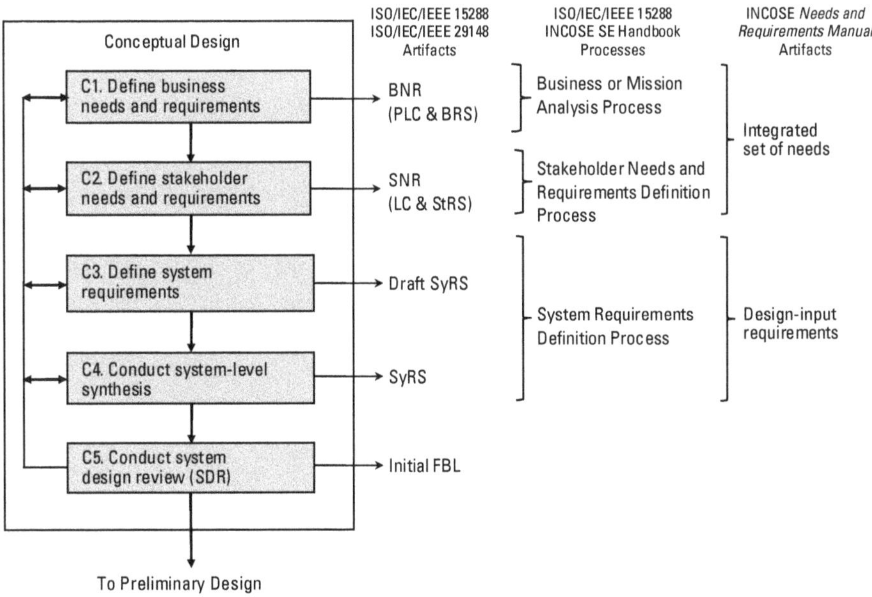

Figure 2.3 The conceptual design processes and needs and requirements artifacts mapped to those called for by ISO/IEC/IEEE 15288 and ISO/IEC/IEEE 29148, and the artifacts from INCOSE's *Needs and Requirements Manual.*

plans. The ConOps document serves as a basis for the organization to direct the overall characteristics of the future business and systems, for the project to understand its background, and for the users of this International Standard to implement the stakeholder requirements elicitation."

ISO/IEC/IEEE 29148 describes the OpsCon as: "A system OpsCon document describes what the system will do (not how it will do it) and why (rationale). An OpsCon is a user-oriented document that describes system characteristics of the to-be-delivered system from the user's viewpoint. The OpsCon document is used to communicate overall quantitative and qualitative system characteristics to the acquirer, user, supplier and other organizational elements."

The ConOps is developed at the enterprise level and provides the context for the OpsCon, which is prepared by business management at the strategic level as part of the C1. Business Needs and Requirements Definition Process. Business management begins with the preparation of the preliminary OpsCon, which summarizes the needs of the system within its business context, the business needs. The preliminary OpsCon is then elaborated and refined into the OpsCon by engagement with stakeholders at the business operations level during the C2. Stakeholder Needs and Requirements Definition Process; the OpsCon therefore contains stakeholder needs.

However, the operational concept in the OpsCon is just one of the concepts required to address the stakeholder needs across the system life cycle. Four more are required:

- The acquisition concept describes the way that the system will be acquired, including aspects such as stakeholder engagement, requirements definition, solicitation and contracting issues, design, production, and verification.
- The deployment concept describes the way the system will be validated, delivered, and introduced into operations.
- The support concept describes the desired support infrastructure and manpower considerations for supporting the system after it is deployed. A support concept addresses operating support, engineering support, maintenance support, supply support, and training support.
- The retirement concept describes the way the system will be removed from operation and retired, including the disposal of any hazardous materials used in or resulting from the process.

PLC are prepared by business management and contain the business needs defined as part of the C1. Business Needs and Requirements Definition Process. PLC are elaborated and refined in much more detail in the LC by stakeholders at the business operations level in the C2. Stakeholder Needs and Requirements Definition Process.

2.1.2 What Is a Requirement?

For our purposes here, a requirement can be defined as a statement of a system service, an attribute or a quality of the system, or a constraint placed on the system. A requirement is therefore:

- Something that the system must do;
- A quality or attribute that it must possess;
- A constraint under which it must operate or be developed.

As described earlier, there are three levels of the requirement statement generated in the BRS, StRS, and SyRS as part of the development of requirements for a system:

- Requirement statements, which are statements of:
  - The services and functions that the system should provide, the things it should do, or some action it should take (often called functional requirements);
  - The qualities, properties, or attributes that the system must possess (often called nonfunctional requirements);
  - Any restrictions or bounds under which the system should operate, or on the way in which the system is to be developed (also called constraints and sometimes also included in the category of non-functional requirements).
- Performance and verification statements;
- Attributes supporting each requirement to facilitate its management;
- Definitions of other systems with which the system must integrate and to which it must interface;
- Information about the application domain within which the system must operate.

The requirements in the BRS and the StRS (the integrated set of needs) are not as formal as those in the SyRS (the design input requirements) but they have the same structure and follow the same rules as presented in the remainder of this chapter.

2.1.3 Requirement Characteristics and Attributes

However it is gathered or elaborated, each requirement is to be examined as it is collected or derived to ensure that it has a number of important characteristics. The single statement of each requirement is not sufficient on its own, however. To support analysis and management, requirements should be recorded with associated attributes. This section examines first the characteristics of a good requirement and then the attributes that should be associated with each requirement.

2.1.3.1 Requirements Characteristics

Each requirement should exhibit the following characteristics (based on [22, 23]):

- *Necessary.* The requirement must be necessary in that the system could not function in the desired way without this requirement. It also follows that a requirement cannot be necessary if it cannot be traced back to a higher-level requirement (and the associated need).

- *Singular.* A requirement cannot be a combination of two or more requirements; that is, the requirements set must be normalized so that requirements do not overlap. This also implies that the requirement is concise and clear.
- *Correct.* That the requirement must be correct is axiomatic; an incorrect requirement will result in a system that does not meet the real needs.
- *Unambiguous.* All readers of the requirement should come to the same understanding of what is meant by the requirement; there can be only one interpretation. This is difficult when the requirement is written in natural language, particularly English, where a single statement can have more than one meaning.
- *Feasible.* A requirement must be achievable using extant design and engineering processes, extant technologies, and extant manufacturing procedures. A requirement should also be achievable within organizational constraints such as budget and schedule unless we take care to be unambiguous.
- *Appropriate to level.* Each requirement must be appropriate to the level at which it is stated. For example, a system-level requirement must be independent of the method of implementation (process) or physical solution; that is, it should say what is required, rather than how it should be implemented.
- *Complete.* The requirement must stand alone and not need any further explanation.
- *Conforming.* Each requirement must conform to a standard formal structure as defined by the organization, or perhaps in accordance with an external standard.
- *Verifiable.* Each requirement must be verifiable, in that it can be proved that the system meets or possesses the requirement. Verification may be by one or more means.

ISO/IEC/IEEE 29148 [24] and the INCOSE *Guide to Writing Requirements* [25] provide more detail on the characteristics of individual requirements.

2.1.3.2 Requirement Attributes

So far, we have been discussing requirement statements. However, a requirement is more than just a requirement statement. The full expression of a requirement includes associated attributes that aid in the management of the requirement (the attributes may not be visible with the requirement statement but may be appended to it in the requirement document or in an accompanying or hosting database). So, we can define a requirement expression as "a requirement expression comprises a requirement statement and a set of associated attributes" [26].

An attribute is ". . . additional information included with a requirement statement, which is used to aid in the management of that requirement" [27]. A number of sources define attributes [28, 29]. The most recent and most authoritative is the INCOSE *Needs and Requirements Manual,* which defines some 46 attributes, organized within four broad categories [30]:

- *Attributes to help to define the requirement and its intent.* Examples include rationale, trace to parent, trace to source, states and modes, and allocation/budgeting.
- *Attributes associated with the system verification or system validation.* Examples include system verification or system validation success criteria, system verification or system validation strategy, system verification or system validation method, system verification or system validation responsible organization, system verification or system validation level, system or system validation verification phase, condition of use, system verification or system validation results, and system verification or system validation status.
- *Attributes to help to manage the needs and requirements.* Examples include unique identifier, unique name, originator/author, date requirement entered, owner, stakeholders, configuration management authority, change proposed, version number, approval date, date of last change, stability/volatility, responsible person, need or requirement verification status, need or requirement validation status, status, status of implementation, trace to interface definition, trace to dependent peer requirements, priority, criticality or essentiality, risk of implementation, risk mitigation, key driving need or key driving requirement, additional comments, and type/category.
- *Attributes to show applicability and allow reuse.* Examples include applicability, region, country, state/province, application, market segment, business unit, and business line.

There is not sufficient space to discuss fully here a taxonomy for attributes or for how such attributes could be defined and used. Here, we simply note the 18 mandatory attributes from the INCOSE *Needs and Requirements Manual* [31]:

- *A1 Rationale.* As each requirement is elicited/elaborated, a rationale must be attached to record the reason for the requirement's existence. Rationale is essential to guide subsequent design.

- *A2 Trace to Parent.* Requirements at one level are implemented at the next level of the system architecture via allocation. A child requirement is one that is an elaboration, derivation, or decomposition of an allocated parent. The achievement of each of the child requirements is necessary to the achievement of the parent requirement.
- *A3 Trace to Source.* There are likely to be a number of sources of requirements. The originator (person, organization, document, or process) must be recorded to ensure that all requirements can be justified. Additionally, sources may need to be consulted when negotiation and compromises are required during requirements analysis.
- *A5 Allocation/Budgeting.* Requirements at one level are allocated or budgeted to parts of the architecture for implementation. There are two types of allocation:
  - Allocation (budgeting) of performance, resources, quantity, or quality from one level to the next level (such as performance, power, mass, reliability, accuracy, precision, and quality).
  - Allocation of responsibility, where requirements at one level are assigned to subsystems and system elements at the next level for implementation.
- *A6 System Verification or System Validation Success Criteria.* The criteria that must be met to provide objective evidence that the system has successfully met the need or requirement.
- *A7 System Verification or System Validation Strategy.* A high-level description of the approach to be used to verify the system meets a design input requirement or to validate the system meets a need.
- *A8 System Verification or System Validation Method.* The method to be used (test, demonstration, inspection, or analysis) to provide objective evidence.
- *A9 System Verification or System Validation Responsible Organization.* The organization that is responsible for system verification and system validation for a need or requirement or sets of needs and requirements for the system.
- *A15 Unique identifier.* Each requirement must be able to be identified as a unique statement. Consequently, when a requirement is recorded, it must be allocated a unique identifier that remains with the requirement throughout the design process. This is essential for requirements management and for requirements traceability, in particular. Although managing the allocation of unique numbers is difficult in a paper-based system, the process is straightforward when conducted within the database underlying an automated requirements management tool.

Each requirement also has performance and verification statements associated with it. While these require separate identifiers, it is useful to use a system whereby the association between the three types of requirements is immediately obvious. One recommended system is to use the same stem number for all three types of requirements and the letters P and V are appended to denote the performance and verification statements associated with the requirement statement. See Chapters 8 and 9 for more details.

- *A18 Originator/Author.* The person defining and entering a need or requirement expressions into the needs/requirements database.
- *A19 Owner.* The person or organizational element that maintains the need or requirement, who has the right to say something about this need or requirement, approves changes to the need or requirement and reports the status of the need or requirement.
- *A28 Need or Requirement Verification Status.* An indication that a need or requirement has been verified.
- *A29 Need or Requirement Validation Status.* An indication that the need or requirement has been validated.
- *A32 Trace to Interface Definition.* This attribute provides a link tracing the interface requirements to where the interaction referenced in the requirements are defined.
- *A33 Trace to Dependent Peer Requirements.* This attribute links requirements that are related to, or dependent on, each other (other than parent-child) at the same level. The relationship with other requirements must be identified to assist in forward and backward traceability.
- *A34 Priority.* Requirements are often stated in terms of their importance to individual stakeholders. The priority of each requirement must therefore be captured with the requirement so that requirements engineers (and designers subsequently) have some design space within which to work during requirements analysis and negotiation. This priority value is allocated from a business management or stakeholder perspective and indicates how important the achievement of each requirement is in the context of the overall system. Stakeholders are tempted to state all requirements as having the highest priority (mostly for the fear that the lower-priority requirements will be subsequently discarded), but they should be encouraged to allow some flexibility within which design can be conducted. The most important (mandatory) requirements are often stated as *shall*; the next most important are stated as *shall, where practicable*; the next most important can be stated as *preferred* or *should*; and the next most important can be

stated as *may*. These levels are often simplified to a three-level (mandatory, highly desirable, desirable) or a two-level (mandatory, optional) scheme. When initially collecting requirements, it may be sufficient to ask stakeholders to rate the requirement's priority as high, medium, or low. A more detailed system could ask them to rate priority on a scale of 1 to 10, which would allow some numerical analysis of priority during subsequent design, if that is desirable. In baseline specifications, priority can therefore be allocated to the requirement in two ways—either in the text of the statement using *shall*, *should*, and *may*; or all statements could be stated using a term such as "shall" and an indication of priority could be appended as an attribute in a numerical form or in the form of high, medium, or low. The latter method is preferred as it avoids errors that may creep in during retyping the requirement when the priority is changed.

- *A35 Criticality or Essentiality.* It is also useful to ask the business management and stakeholders to identify those requirements that are critical in terms of their contribution to system performance. While seemingly related to priority, this category identifies show-stoppers, those requirements that are essential to the success of the overall system. In other words, a requirement that is deemed critical must be met or possessed by the system or else the system cannot be introduced into service. In some cases, it may be useful to have a rating system, but it is most often sufficient to identify each requirement in a binary fashion as critical or noncritical.
- *A36 Risk of Implementation.* Each requirement contributes to overall system risk in its own way and the management of risk is a critical component of systems engineering and project management. It is useful to consider the risk associated with each requirement as it is collected, either by allocating a risk rating (1–10, for example) or by writing an associated risk statement.
- *A37 Risk Mitigation.* This attribute indicates whether a need or requirement is part of a risk mitigation action.

2.1.4 What Not How

In general, business, stakeholder, and system requirements should be statements of what a system should do, rather than how it should do it. However, for a number of reasons, this is often too simplistic in practice. First, it may be essential that the system should perform a function in a particular way (such as for interoperability, or to meet extant standards). We could also specify a particular physical solution to meet broader concepts such as support and commonality with existing solutions. Additionally, a specific statement is often less confusing than an abstract statement of the problem; a specific statement is therefore the best guidance that a conceptual designer can give a lower-level designer. Further, for good business reasons, business management may well dictate not only what is required but also how it is to be implemented. Finally, the people who specify the system are often the domain experts; they are therefore best placed to state how the system should be developed and how it should operate. Still, unless one of those reasons prevails, we should ensure that the BRS, StRS, and SyRS provide logical specifications (because they are in the problem domain), not physical specifications (which belong in the solution domain).

2.1.5 Emergent Properties

Not all properties of a system are exhibited separately by its constituent elements. That is, the system may have properties that only exist when the subsystems are integrated; these are called the emergent properties of a system, which:

- Are those that are possessed by the system as a whole;
- Only emerge after all the individual subsystems have been integrated;
- Cannot be exhibited by parts of the system in isolation;
- Depend on interactions between components, including their environment.

Examples of emergent properties are the "-ilities" such as maintainability, reliability, availability, and usability. Other examples include safety, security, comfort, and ease of use. Emergent properties must therefore be defined from the top down; they cannot be obtained by specifying requirements for individual subsystems and then hoping that the desired properties will exist on integration.

Perhaps the bicycle is one of the best examples of a system that possesses emergent properties, since it can only perform its principal function when all subsystems, including the rider, have been integrated. If any one of the major subsystems is removed, the system cannot function correctly.

We also must consider the possibility of the completed system exhibiting emergent properties that were not considered as part of its design and, in some cases, may be undesirable in the end.

2.2 What Is Requirements Engineering?

Requirements engineering can be described most simply as the process by which we identify a problem's context, locate the business and stakeholder needs and requirements within that context, and deliver specifications that meets those needs [32]. As illustrated in Figure 2.4, that process requires a considerable reduction of the complexity of the real world, and that subset within which the customer organization operates, into a set of requirements that define the scope of the project. There is therefore considerable judgment required in identifying BNR and SNR and then converting them into a SyRS. Further, as Andriole pointed out, most requirements engineering methodologies have been created by people who like precision, diagnosticity, and rigor; yet requirements management is fundamentally a political process, not a technical one [33].

Dorfman and Thayer described requirements engineering as the "science and discipline concerned with analyzing and recording requirements" [34]. Kotonya and Sommerville defined requirements engineering as "the systematic process of eliciting, understanding, analyzing, recording, and managing requirements" [35]. ISO/IEC/IEEE 29148 defined requirements engineering as the "process through which the acquirer and the suppliers of a system discover, review, articulate, understand, and record the requirements on the system and the lifecycle processes . . . [and] . . . is concerned with discovering, eliciting, developing, analyzing, determining verification methods, validating, communicating, recording, and managing requirements" [36]. We elaborate on

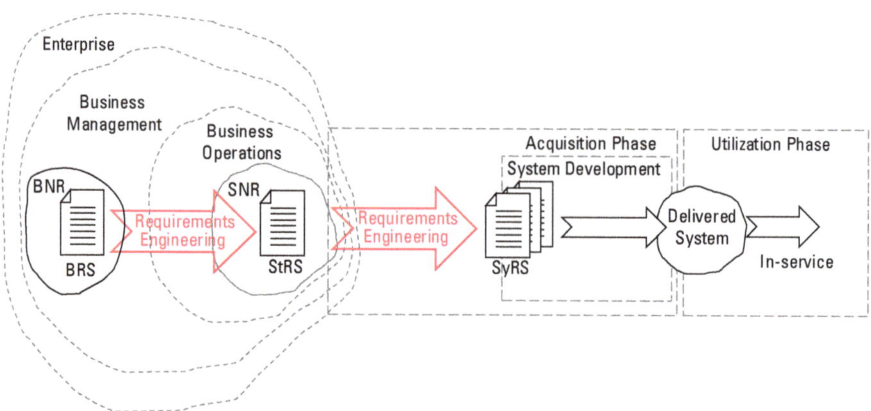

Figure 2.4 The role of requirements engineering in translating business needs and requirements (BNR) into stakeholder needs and requirements (SNR) and then into system requirements in one or more SyRS that define the capability system.

these definitions when we describe the supporting requirements engineering activities in Section 2.3.

2.2.1 Why We Need Requirements

It may not always be obvious why requirements are necessary in the first place. However, an agreed set of requirements is essential from a number of different perspectives.

From the point of view of the business management at the strategic level, the set of requirements provides a mechanism whereby they can agree that the organization's resources can be allocated to the project that has been established to bring the system into being. Unless the requirement set is well defined, the project has no firm base, and the resources allocated to it will be inappropriate. Worse still, without sufficient oversight, the business may be held to ransom by stakeholders insisting on unrealistic requirements that business management have not approved.

From the project manager's perspective, the set of requirements is an essential part of the definition of the scope of the project; it is from this scope that a suitable system will be developed. Scope management is one of the 10 major functions of a project manager [37], and a well-defined scope (set of requirements) is necessary to be able to justify any expenditure of funds or effort within the project, to be able to report adequately on the progress of work, and eventually to be able to determine when the project is complete. Clearly then, a well-defined, universally agreed set of requirements at the outset of the project phase, increment, iteration, spiral, or sprint.

From the systems engineering perspective, the requirements set instantiated in the SyRS represents the transition from the business world into the systems engineering and engineering worlds. In this way, the SyRS defines what an acceptable technical solution needs to be capable of and communicates this information to both technical and nontechnical stakeholders alike. It should also be emphasized that it is not possible to rectify a poor set of requirements with good design, good engineering, or good production/process control, particularly if that set of requirements is the basis of a fixed-price contract between the acquirer and the contractor. Consequently, a project can rarely recover from poor definition, regardless of how much good design work is performed subsequently.

2.2.2 Why We Need Requirements Engineering

An appropriate requirements engineering process allows us to have the best chance that we begin the project with a set of requirements that is complete, balanced, comprehensible, feasible, and consistent. A useful process ensures that all relevant business managers and stakeholders have had input, and the various points of view have been reconciled. To do that, we must therefore be able to trade off functionality for cost, which implies that we understand the required functionality, priorities, and costs. We also need to be able to use the process to manage changes in requirements for a wide variety of reasons.

Requirements engineering focuses on what is needed by the system, rather than how it is to be designed. The cost of developing requirements is estimated to be 10% to 15% of the total system cost; yet it is estimated that it costs some 200 times more to rectify an error in requirements during utilization than it does during early conceptual design [38]. As discussed in the previous section, poor requirements definition cannot be rectified by good design, so that it invariably follows that rigorous development of requirements is essential for the successful system; requirements engineering provides this rigor through the application of a formal methodology. When requirements engineering processes are ad hoc, the end result is almost always an unsatisfactory product or a canceled project [39]. Unfortunately, few organizations have a well-defined requirements engineering process and there are few, if any, standard methodologies that assist (see [40] for a useful conceptual design methodology that is consistent with this text).

2.3 Requirements Elicitation and Elaboration

Requirements elicitation and elaboration involve working with business managers and stakeholders to investigate the problem to be solved and to identify their needs and requirements. At the beginning of this chapter, we considered the three levels of requirement statements:

- Business needs and business requirements in the BRS;
- Stakeholder needs and stakeholder requirements in the StRS;
- System requirements in the SyRS.

We then discussed requirements engineering by referring to requirements in general terms. Before discussing in subsequent chapters how these requirements are part of systems engineering, it is useful here to discuss how the design moves through those three levels of requirement statement. To do so, we need to consider the hierarchical nature of the requirement statements and the processes of elicitation and elaboration [41].

- *Elicited requirements.* Elicited requirements can be attributed to the source (business manager or stakeholder), either directly or through negotiation. These requirements are normally gathered via interview or a structured workshop.
- *Elaborated requirements.* Elaboration is used to move between the levels in the requirements hierarchy; the process is also often referred to as requirements flowdown. Elaborated requirements are of two types:
  - *Decomposed requirements.* Elicited requirements tend to be at a high level because they were obtained from those directly involved in business management and business operations supported by the system of interest. Decomposition entails breaking a higher-level requirement into those lower-level requirements that are explicitly required by it. For example, if an aircraft needs to be able to land on a certain class of airfield, then it must be able to land within the appropriate runway length, adhere to strict mass-related constraints, be able to interface to the appropriate air traffic control systems, and so on. In decomposition, the need for the child requirements is obvious in the parent. An elicited requirement can be decomposed or derived into two or more subordinate requirements whose total meaning is equivalent to that of the original requirement. It should be noted that the original requirement normally becomes redundant and is replaced by the lower-level requirements. The essence of the upper-level requirement often becomes a heading or is noted in the rationale for the decomposed requirements, although it is good practice to retain the hierarchical structure of the RBS by abstracting the original requirement to a heading in the structure.
  - *Derived requirements.* In addition to being high level, elicited requirements also tend to be from the business management and operations perspectives and will not necessarily address all aspects of a systems design. Derivation entails drawing some inference from the higher-level requirements to obtain lower-level statements. That is, business management or the stakeholders did not state the requirement directly, but the derived requirement is a necessary part of the system design if one or more of the directly stated requirements are to be met. Like decomposed requirements, derived requirements will replace the original parent requirement.

Business, stakeholder, and system requirements are not collated randomly—requirements engineers undertake requirements analysis that involves, among others, elaboration of a statement of business and stakeholder needs through a systematic exposition of what the system must do to meet those needs [42]. Gathering, analyzing, and elaborating requirements are essential steps to ensure that the system, once built, actually meets business management and stakeholder needs [43]. High-level requirements are elaborated further into a number of lower-level requirements that are then grouped, along with other elaborated requirements, through the requirements analysis process [44] into the business-management-level descriptions in the BRS, the business operations-level descriptions in the StRS, and the system-level description in the SyRS.

The requirements in the SyRS are then further elaborated into lower-level requirements, which are then regrouped from the logical groupings of the SyRS into physical groupings of requirements (the subsystems). The subsystems (and their constituent CIs) are recorded in a number of subsystem specifications, which then define how the system will be implemented. The mapping of the logical groupings of requirements in the SyRS into the physical groupings in the subsystem specifications informs the definition of project deliverables.

For well-behaved systems, the requirements in each of the BRS, StRS, SyRS, and subsystem specifications are normally grouped in a hierarchical fashion (albeit grouped logically in the first two, and physically in the last). This hierarchy is useful because it allows requirements engineers to decompose the system into tasks of a complexity that can be more easily understood and communicated, as well as managed within the limitations of humans and our organizations.

A hierarchical description of a system is also useful in that it supports requirements traceability, which is an essential element of effective systems engineering as well as project management. Traceability is assisted by a hierarchical numbering of requirements. As illustrated in Figure 2.5, Requirements 1.1.1, 1.1.2, and 1.1.3 are child requirements of parent Requirement 1.1, which is a child of Requirement 1. For Requirements 1.1.1, 1.1.2, and 1.1.3 to be children of parent Requirement 1.1, they must be derived or decomposed from it; that is, Requirement 1.1 is a superset of Requirements 1.1.1, 1.1.2, and 1.1.3; and Requirement 1 is a superset of Requirements 1.1 and Requirements 1.2.

Elicitation involves engagement with the stakeholders in order to establish their requirements for the new system. There are a number of techniques for this engagement, such as facilitated structured workshops, brainstorming, interviews, surveys, questionnaires, process models, use cases and user scenarios, simulations, prototypes, observation of work studies (time and motion studies), participation in work activities, observation of the system's organizational and political environment, technical artifact review, benchmarking,

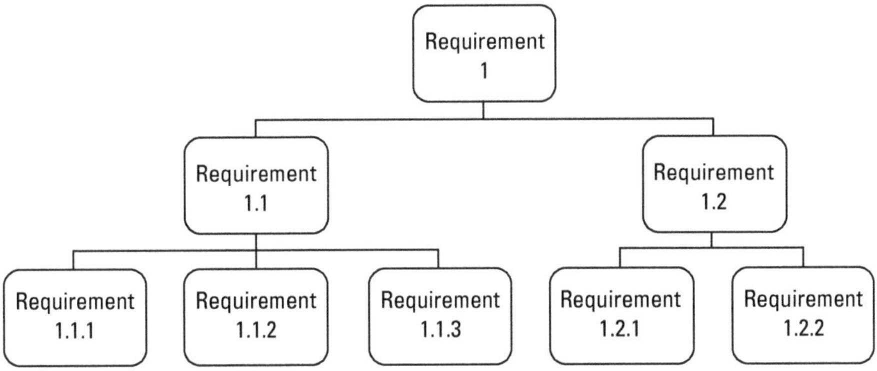

Figure 2.5 The hierarchical elaboration of requirements.

competitive system assessment, reverse engineering, and market analysis (see Section 2.3.2 for more detail).

2.3.1 Difficulties in Eliciting and Elaborating Requirements

Unfortunately, requirements elicitation is not a simple matter of transferring knowledge from stakeholders to requirements engineers, which means that the effective development of system-level requirements can be difficult and challenging. Stakeholders will probably not have given the new system much consideration and will not have had a chance to develop a clear view of their requirements; they may even be hearing about the system for the first time when contacted by requirements engineers requesting their input to the design. Even if they have an informed view, different stakeholders have conflicting requirements, they are not normally aware of the requirements engineering process and of the acquisition process in general, they are often not willing to compromise their expectations to meet any budget, and they are rarely aware of the limitations that technology has on their expectations. This section presents these and other difficulties.

Instead of focusing on the logical architecture of the system, stakeholders sometimes identify physical solutions that appear to meet their requirements. That is, instead of concentrating on what the system is required to do, there is an ever-present urge to describe how the system should be designed or built. While this is dangerous because it can preclude effective solutions from consideration, requirements engineers can use the predilection to some advantage because the stakeholder can at least visualize (and therefore hopefully describe) something that meets their perceived needs.

Further, stakeholders sometimes become confused between a need and a want. It is vital that system-level requirements are based on the true needs of the system and not merely a wish list of additional functionality. This reinforces the need for forward and backward traceability between the endorsed business needs and the SyRS. Additional (unendorsed) functionality may add complexity and expense onto the system without enhancing the system's ability to meet the true system-level requirements. This gold-plating of the requirements must be avoided.

Another difficulty is that stakeholders themselves do not always contribute positively to the requirements engineering process. They often do not really know what they want, have unrealistic expectations, and have no desire to compromise unrealistic demands when faced with cost and other constraints. Further, they do not understand the requirements engineering process and often do not participate in reviews when required.

Information on the application domain is not always collected in one place and may involve specialist terminology. It is highly unlikely in the development of large systems that any one user (or any single group of users, for that matter) can articulate all system-level requirements. Additionally, the language employed by users in most complex systems tends to be arcane to outsiders and even to other specialist users of the same system. Some considerable time may have to be invested by requirements engineers to be able to comprehend the domain knowledge of all users and, in the end, it may only be the requirements engineers that understand all aspects of the system requirements.

Requirements are often poorly stated because the real requirements are not known or understood. It cannot be assumed that the potential users of the system know what they need from the system, which is only to be expected since they have never seen it nor operated it. Requirements engineers must recognize this and try a number of different communication and elicitation techniques. As a corollary, stakeholders sometimes firmly believe that they know exactly what they need until they are presented with a system that has their specified functionality whereupon they realize that they do not require some of the features included and bemoan the lack of others. To overcome this, stakeholders must be presented with prototypes, simulations, and models as soon as possible to ensure that there is a strong correlation between the stakeholders' perceived requirements and the appropriate logical statements encapsulated in the requirement specifications. As mentioned earlier, the stakeholders' predilection for stating solutions in physical terms can be used to advantage by requirements engineers to confirm requirements through analogy by comparing stakeholders' stated desires with existing systems or prototypes that perform similar functions.

Conflicting requirements often result when two or more requirements are mutually exclusive such that satisfaction of one requirement will result in failing to meet another. Conflicts must be resolved early in the requirements engineering process, as they can be a source of ambiguity. Conflicts can also occur where the same requirement appears more than once, worded in slightly different ways, or requiring different levels of performance for the same function. Every requirement must be unique and stated only once to avoid confusion and conflict. There are also a number of organizational and political factors that are sources of conflicting requirements.

The final challenge is to know when enough is enough, as it is difficult to judge when the requirements analysis effort is complete. Although experienced personnel are best placed to make this judgment, no two system developments are likely to be the same, and the stopping criteria will require some consideration for each system.

2.3.2 Techniques for Eliciting/Generating Requirements

Requirements engineers select an appropriate requirements elicitation methodology to suit the system being developed. Most requirements engineering literature (see for example, Alexander and Stevens [45], Goguen and Linde [46], Hull [47], Katonya and Sommerville [48], Sommerville and Sawyer [49], Young [50, 51], and IEEE-STD-1233 [52]) have suggested such techniques as facilitated structured workshops, brainstorming, interviews, surveys, questionnaires, process models, use cases and user scenarios, simulations, prototypes, observation of work studies (time and motion studies), participation in work activities, observation of the system's organizational and political environment, technical artifact review, benchmarking, competitive system assessment, reverse engineering, and market analysis. The selection of technique is critical since inappropriate use can doom a project to failure [53]. Frameworks (see, for example, [54, 55]) have been proposed to assist in selecting an appropriate requirements engineering technique for a particular project.

The following sections describe each of the major requirements-elicitation/generation techniques in more detail.

2.3.2.1 Facilitated Structured Workshops

Perhaps the fastest (and most efficient) way to elicit requirements is through structured requirement workshops. It must be remembered that workshops are part of the iterative cycle of requirements development; the top-down approach greatly assists here in allowing participants to focus on appropriate levels of abstraction (level of detail). The iterative cycle does not start with a blank sheet of paper, however. A workshop should always start with a draft set of requirements (and other early conceptual design artifacts), particularly the visual artifacts that we describe at the end of this chapter. These visual representations (and other drawings and diagrams) are very useful as discussion and communication tools as well as requirements analysis tools. If possible, the set of draft upper-level artifacts should be sent out to participants before the workshop to allow them to prepare. Again, however, best results are achieved if workshop attendees have some basis for discussion; that is, it is essential that a first iteration of requirements engineering is conducted before the workshop is held. Conducting a workshop that is not informed by some well-focused preliminary artifacts normally leaves stakeholders with a poor opinion of the new system even before conceptual design is well under way. Another way to ensure that stakeholders make valuable contributions is to conduct a period of briefing/training in requirements elicitation before beginning the workshop.

Significant thought must be given to the administration and the mechanics of the conduct of the workshop. As mentioned earlier, it is essential that the workshop is structured and facilitated to avoid serious digression. The location, resources, and other logistical considerations must be considered to ensure that the workshop runs smoothly; the inability to run a simple workshop without problems is not likely to give stakeholders confidence that their new system is in good hands. A popular way of avoiding workplace distractions is to hold events at a convenient off-site location where stakeholders can devote their attention to articulating their needs. There must also be a scribe nominated to record the design as it is iterated.

The main disadvantages of workshops are the additional travel costs that may be incurred in gathering stakeholders in the one location and the difficulty associated with getting many busy people in the same place at the same time. Technologies that facilitate remote access (such as videoconferencing and modern virtual meeting applications) may assist in reducing costs.

2.3.2.2 Brainstorming Sessions

Brainstorming sessions have a similar format to structured workshops but are considerably less formal. It may well be worth conducting a brainstorming session with a small number of selected key stakeholders and designers before running a structured workshop as a means of obtaining the first iteration of the requirements engineering products. Brainstorming is particularly sensitive to the nature of the people involved as well as their number; the value of the exercise tends to diminish as the number of participants increases. As with a structured workshop, brainstorming sessions should be facilitated and can usefully be held off-site to allow participants to focus on the outcomes without the distractions of daily work.

2.3.2.3 Interviews, Surveys, and Questionnaires

Interviews, surveys, and questionnaires offer a formal process through which requirements may be garnered from individual stakeholders. However, their use should be considered carefully in that useful information is only provided in response to focused, well-balanced questions, which implies that a measure of requirements engineering has been conducted before the interviews are held or before the surveys and questionnaires are sent out. All three processes can also be very time-consuming, which again requires that careful thought is given to the questions to ensure that responses add to an understanding of the systems requirements, rather than add additional noise to the requirements definition process.

In some respects, interviews should be considered one-on-one structured workshops. Similar preparation is required; the interview should never be approached cold by either the requirements engineer or the stakeholder being interviewed. Interviews can be very effective but are very time-consuming and a structured workshop may be preferred if there are a large number of stakeholders. As an exercise in good stakeholder management, it is generally advisable to interview at least the key stakeholders.

While there is no satisfactory replacement for the personal contact possible in an interview, there are occasions when surveys and questionnaires are useful requirements elicitation tools. Their utility is generally limited to activities such as the collection of statistical data to reinforce or validate requirements collected by interview or workshop.

2.3.2.4 Use Cases and Operational Scenarios

One of the difficulties with asking users about their requirements for the new system is that users will probably have been trained on an existing system and may have difficulty in setting aside current operating procedures when considering requirements for the new system. Use cases and operational (or user) scenarios (sometimes also generically called storyboarding or user stories in Agile) are a simple but powerful way of assisting in providing users with a new frame of reference, particularly at the beginning of the requirements elicitation process where they may have difficulty in breaking free from existing practices.

Additionally, stakeholders will find it very difficult to understand requirements when they have been written as a series of simple "shall" statements. Use cases and scenarios are given in a narrative (and often visual) style to illustrate how the system will operate and how users will interact with it. Use cases and scenarios are therefore very useful because users often find it easiest to describe their view of the system in terms of examples of how it might be used.

Use cases therefore provide a good framework within which to elicit requirements, as well as serving throughout the project as upper-level descriptions of the required functionality. Use cases and scenarios also perform a very important function of defining the purpose of the system, which is useful during acceptance testing to ensure that the delivered system is fit for purpose; the use cases developed at the start of development to assist in eliciting requirements then effectively become test cases at the end of the development to assist in validation of the system.

To ensure a complete description of a system, it is important that the use cases and scenarios capture the exception conditions as well as the normal operational conditions. That is, users are often keen to describe the way in which the system works in terms of the positive aspects of its implementation, rather than describe what to do when the normal process is prevented from occurring due to such events as system failure, actions by a competitor or adversary, or simply operator error.

Use cases and use-case modeling are very popular in software requirements engineering and are described in a number of excellent references [56].

2.3.2.5 Process Models

Process models provide descriptions of the major activities involved in a particular process as well as their sequencing. Process models are normally provided at a number of different levels and from a number of different perspectives to allow for an understanding at different levels of abstraction from coarse to fine resolution and to ensure that all views of the process are considered.

Process models are particularly well suited to systems such as transaction-based systems where more formal methodologies such as entity-relation models can be used to define formal database structures. However, for requirements engineering during conceptual design for most other types of systems, process models are too low-level to be useful and use cases are preferred.

2.3.2.6 Prototypes

Requirements elicitation can be enhanced by the ability of requirements engineers to mock up a system, or part of a system, that meets the collected requirements and then present that to users for their consideration. Prototypes and simulations can be very complex, or they can be as simple as a series of role-playing exercises in which users are asked to describe how they would imagine the system operating under particular circumstances. At the highest level, use cases and operational scenarios could also be considered as a form of paper-based prototyping (a form of virtual walkthrough) in which stakeholders describe their requirements in a range of typical operations for the system.

Simulations and prototypes can also be used for requirements validation because they are a very useful (visual) way of showing stakeholders how the designers have captured and interpreted their requirements.

Prototypes have two main forms:

- Throw-away prototypes that are intended to elicit or validate requirements but not form part of the developed system;
- Evolutionary prototypes, that are intended to deliver a working system to the customer in incremental steps.

It is important to identify the purpose of the prototype. Although both forms are used to assist in eliciting and generating requirements, the nature of the prototype must be explained to the stakeholders when they are viewing it. If it is an evolutionary prototype, they can expect to see it, or elements of it, again. Care must be taken to explain that a throw-away prototype is just that, and they cannot expect to see it again. Some dangers with throw-away prototypes include:

- Stakeholders who are detail-oriented often pay more attention to the structure of the prototype than to its operation. This is particularly true of software prototypes where stakeholders are sometimes blinded by the (often rudimentary) graphical user interface and cannot see past it to the functionality being demonstrated.
- Having seen the prototype in software, stakeholders sometimes find it very hard to understand why it will take so long to develop a system that has the functionality they have just witnessed. They struggle to see the difference between a mocked-up prototype version and a properly tested, robust, operational version. Again, this is particularly true of software prototypes; while a pilot would never expect to take away and fly in a cardboard cut-out prototype used to illustrate cockpit layout, stakeholders often see the functionality in a software prototype and somehow extrapolate it into a working system.
- Developers of code for the prototype are reluctant to throw it away and feel compelled to use it in the final system, despite the fact that, again, the code is not well designed nor properly tested.

2.3.2.7 Modeling and Simulation

Mathematical modeling and simulation are widely used and are becoming increasingly popular as computers and software become more capable. As the name suggests, mathematical models provide a mathematical representation of a physical system or design. Simulation then uses the models to determine both static and dynamic characteristics of the system under varying conditions. An example of a static model would be one that represents a bridge's structural characteristics. Loads and stresses can be changed in magnitude, direction, and position to ensure that the structure will withstand the loads associated with its intended purpose. An example of a dynamic system is an aircraft's autopilot system, which is a classic example of a control system that can be used to determine characteristics such as stability and response times.

Models and simulations are excellent tools for performing what-if analyses and determining the limits of the design. Variables can be changed and conditions easily altered to allow designers to investigate the likely behavior of the design in varying conditions. Simulation is widely used for a variety of different synthesis tasks. Environmental design and human-machine interface design are two examples of synthesis tasks that can benefit from simulation. The fidelity of the simulations (and the models) must be defined and confirmed prior to their use as synthesis tools.

To build and use a mathematical model, designers need to define the problem and formulate the model to replicate the problem. The model can be investigated with pen and paper if desired, but the more common approach is to convert the model into a computer program for ease of use and speed. Once this is done, perhaps the most crucial step must be performed. The performance of the model must be verified to ensure that it is an accurate representation of the problem. It is pointless to collect data from a model if the model is incorrect (that is, it is not representative of the problem domain). Once the model has been built and verified, designers can experiment with it via simulation and collect important design data.

We discuss modeling further when we discuss model-based systems engineering (MBSE) in Chapter 4.

2.3.2.8 Analysis of Similar Systems

In a similar vein to the use of prototypes, it may be useful to assist stakeholders to investigate their requirements by asking them to describe what they like and do not like about an existing system that is similar in functionality (even if only in part) to the system under consideration. At the very least, requirements engineers will be able to develop an awareness of the stakeholders' level of understanding of the new system. Even if that is poor, requirements engineers should still gain an understanding of user interface issues and perhaps even of some required functionality in the new system. Additionally, requirements engineers may gain some insight into the new system by scanning problem reports from similar existing systems.

2.3.2.9 Observation of the Workplace

Although time-and-motion studies may not always be a suitable way of eliciting requirements (particularly at the systems level), an understanding of the work environment and the workplace context is essential to the development of user requirements. By observing users in their natural environment requirements engineers may often identify requirements that the stakeholders had not stated directly through more formal requirements elicitation techniques such as interviews or workshops. When describing their workflow, stakeholders often assume skills, processes, and tools and do not describe them explicitly. They are also not always consciously aware of all their work practices that are relevant to the new or modified system.

If the relevant portion of the workplace cannot be observed directly, key stakeholders can be selected to be permanent members of the requirement engineering team if it is considered that user input is continuously required. In some cases, user involvement may continue past the requirements stage and a user representative may be attached to the teams performing the lower-level design, to ensure that the workplace context is always taken into account.

2.3.2.10 Artifact Review

In addition to observing the workplace, requirements engineers may review relevant business and technical artifacts. Although few organizations have very many information-rich artifacts, many organizations have standard operating procedures (SOPs) that may describe in some detail how particular operations are conducted. Those artifacts may therefore be useful as alternate sources of information about the workplace. However, care must be taken because the artifacts may be good descriptions of the old system but may not be appropriate for the new one.

2.3.2.11 Market Analysis

Many systems do not have a representative group of users that is readily available. Consumer products such as vehicles, for example, have a wide range of users whose requirements are many and varied, often with some conflict. Market analysis techniques may be useful in identifying a common subset of achievable requirements that maximize the potential market for the system.

2.3.2.12 Competitive System Assessment and Trials

As with prototypes and simulations, competitive system assessment and trials are aimed at providing stakeholders with some tangible manifestation of early requirements to assist in requirements elicitation and validation. One simple, quick way of providing depth to early requirements is to identify a number of existing systems that seem suitable and ask stakeholders to evaluate them in the context of their needs.

2.4 Requirements Validation

Requirements must be validated to ensure that individual requirements are appropriate and that the requirements set is an adequate description of the desired system. This process is intended to detect any problems and gaps in the requirements before they are used as a basis for system development. In particular, as illustrated in Figure 2.6, the SyRS must be validated to show that it meets the stakeholders' original requirements in the StRS, which must be validated to show that it meets the business requirements in the BRS.

Requirements validation is very difficult; somehow, we must ensure that we have a complete set of requirements that, in their aggregation, represent the system required by the stakeholders and meets the business needs. This is difficult because there is no significant baseline to which we can refer for the validation. Later in the life cycle, we will look back to the previous baseline; for example, in the preliminary design, we look back to the SyRS; in detailed design and development, we look back to the development specifications; and

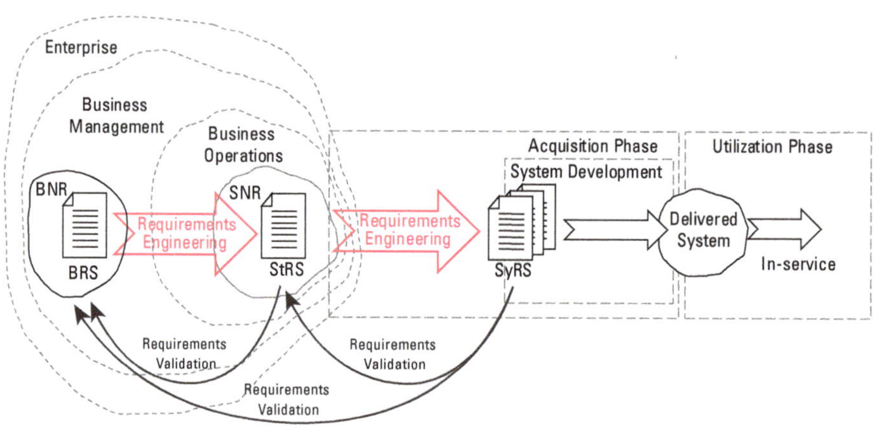

Figure 2.6 Validation of the requirements in the SyRS.

in construction and production, we look back to the product specifications. However, in conceptual design, we only have the high-level business descriptions as the starting point. It is therefore not a straightforward matter to validate that the StRS and then the SyRS represent the complete set of requirements associated with meeting those needs. Validation is normally conducted by a formal requirements review or a series of reviews.

2.5 Requirements Artifacts

Writing well-formed requirements can be achieved by using the following guidelines [3]:

- Ensure that each requirement is a necessary, short, definitive statement.
- Define the appropriate conditions (qualitative or quantitative measures) for each requirement.
- Ensure that each requirement is verifiable through the discipline associated with writing a verification statement for every requirement statement.
- Avoid overspecification, unnecessary constraints, and unbounded statements.
- Ensure readability by defining standard templates for describing requirements; using natural language simply, consistently, and concisely; using short sentences and paragraphs, as well as lists and tables; supplementing natural language with other descriptions where appropriate, such as equations if they are the most unambiguous way of describing requirements; and using diagrams where appropriate to show relationships between entities.

ID	Requirement	Attribute 1	Attribute 2	Attribute 3	Attribute 4	Attribute 5	
12.1.1	***Requirement statement***						
12.1.1P	Performance statement						
12.1.1V	Verification statement						
12.1.2	***Requirement statement***						
12.1.2P	Performance statement						
12.1.2V	Verification statement						
12.1.3	***Requirement statement***						

Figure 2.7 Example format for a simple paper-based specification.

Requirements must be placed into a structured format. An example of a very simple paper-based format for a specification is illustrated in Figure 2.7, where columns are included to contain the selected attributes required to be recorded along with the requirements.

Note that it is normally not possible to include all the information on a single sheet of paper in a manageable way. The difficulties with paper-based formats for specifications can be overcome with the use of requirements management tools, which are discussed in the following section.

2.6 Requirements Management

Requirements management provides the processes by which changes to requirements are managed throughout the system life cycle. Requirements management involves establishing and ensuring compliance with a formal procedure for the elicitation/generation, verification, and traceability of requirements. The many detailed requirements in lower-level specifications have been derived from a relatively simple statement made by a stakeholder; these lower-level requirements must be managed as they are developed to ensure that each of the lower-level requirements can be justified. Additionally, a formal procedure is required for the control of changes to requirements.

2.6.1 Requirements Change Management

Throughout the entire life cycle of the system of the system, we are very interested in managing the configuration of the system elements. In the earliest stages, we must manage the configuration of the requirements set; this process involves a formal change-management process to ensure that changes to the requirements set occur in a controlled way. Requirement change management is therefore a form of configuration management (which we discuss in Chapter 8). Later, we become interested in the configuration management of physical elements, which begins as soon as configuration items are defined at the end of preliminary design.

2.6.2 Change Management—Traceability

Requirements cannot be managed effectively without requirements traceability, which allows us to be able to identify where each requirement came from, what requirements are related to it, and what requirement or requirements stemmed from it (that is, which is its child, or are its children). If a requirement must change for some reason, good traceability allows us to identify all requirements affected by the change. It has been estimated that some 35% of requirements for the average project will have been added after the requirements engineering process is deemed to have ended [57]. Consequently, traceability is essential to support change management and supports the project management function of project scope management.

Broadly, there are two main sorts of traceability:

- *Forward traceability.* Forward traceability gives us confidence that each requirement has been addressed in the design. Through forward traceability, design decisions can be traced from any given system-level requirement (a parent requirement) down to a detailed design decision (a child requirement). For each requirement, we must be able to find at least one child in the subordinate designs. If there is more than one child, we must be able to identify all of them.
- *Backward traceability.* We must be able to trace from each requirement back to at least one requirement in the parent design. This backward traceability ensures that additional requirements (not formally endorsed by the customer) have not crept (through requirements creep) into the design. Any aspect of the design that cannot be traced back to a higher-level requirement is likely to represent unnecessary work for which the customer is most probably paying a premium. That is, an orphan requirement must, by definition in such a process, be out-of-scope.

Backward and forward traceability can mean that an enormous amount of information must be recorded. Consequently, one of the key decisions to be made early in the requirements engineering process is the degree to which traceability is implemented and the point at which traceability and change management start. If change management starts too early (with the mission, goals, and objectives, for example), the large number of changes in the early iterations is not important and tracking them leads to a huge amount of information that is relatively useless. It is generally better to allow the design to mature to a reasonable degree than it is to insert traceability information as soon as the requirements are entered into a database.

2.6.3 Requirements Management Tools

Due to the large number of requirements elicited/elaborated during the development of the SyRS, a requirements management tool is generally necessary to assist in the management of requirements. It is almost impossible to have requirements traceability without implementing the requirements in some automated context.

A simple internet search reveals that there are approximately two dozen requirements management tools commercially available [58]. Rather than attempt to review a subset of the tools, this section describes some of the most important features of a requirements management tool:

- *Supports requirements elicitation.* The selected requirements management tools must support the initial capture of requirements, which may involve importing requirements statements from existing requirements records or extracting individual requirements from within a larger, more general source record. Once the requirements have been captured, the management tool must support the editing and addition of requirements attributes. Because the requirements management tool will trace all changes and there is a very large rate of change of requirements during elicitation, there is some advantage in not populating a requirements management tool until the rate of requirements churn has reached an acceptably low level.
- *Allows readers to browse.* It is important for stakeholders to be able to relate to the requirements database in a natural way, so requirements management tools must allow the stakeholders to browse through the requirements in any order.
- *Allows specific requirements to be retrieved.* The tool must allow a specific requirement to be retrieved directly by its unique identifier, short title, or any part of its text.
- *Supports forward and backward traceability.* Traceability is critical to the systems engineering and requirements engineering processes. Requirements management tools must support and maintain traceability between the different levels of requirements (from business to stakeholder to system to subsystem to assembly to component, as well as in the reverse direction), and between logical design and physical architecture.
- *Supports the generation of good requirements.* Some requirements management tools assist with the aim of producing good requirements by providing automated functions to check for inconsistencies and ambiguities. These functions are valuable, but users should establish a degree of confidence in the process initially and always cross check that the requirements are well written.

- *Generates reports.* A good requirements management tool can generate reports of requirements based on selected criteria (a selected attribute, selected combinations of attributes, or some textual string) to assist in validation of the requirements set.
- *Allows import and export of requirements in various formats.* It is also essential that the tool allows import and export of requirements in various database, word processing, and spreadsheet formats, and to interface with other requirements management tools. This ensures that the overall requirements management process can be maintained even when stakeholders, contractors, and subcontractors use different tools.
- *Supports change control and change impact assessment.* Requirements change over time; performance levels may change, new requirements may be added, and old requirements may be deleted. The requirements management tool must support this configuration management function. Other configuration management functions such as impact analysis, change proposals and approvals, and change history should also be available. For example, support for some form of impact analysis is necessary to enable users to investigate the potential impact on the system of adding, deleting, or modifying requirements. As well as enabling changes, the tool should support the ability to restrict rights of users (to read-only, for example) in order to prevent unauthorized changes, and to keep track of which changes were made by whom.
- *Supports allocation and logical-to-physical translation.* During preliminary design, a physical architecture is established and logical requirements allocated to it. Some requirements management tools allow the physical architecture to be recorded within the tool and support the process of requirements allocation. This feature, combined with an ability to perform impact analysis, becomes extremely valuable as the system development passes through the more detailed design stages.
- *Does not enforce a particular requirements engineering process.* It is essential that the requirements engineers do not have to modify their preferred methodologies and processes because they have adopted a tool that instantiates a different set of processes. Although most tools are delivered with built-in specification templates based on popular specification standards, requirements engineers should also be able to select their own specification format, which will require an ability to perform some formatting of the information.

2.7 Requirements Engineering Tools

There are a large number of tools that may assist in requirements engineering, including the context diagram, functional flow block diagrams (FFBD), requirements breakdown structure (RBS), and N2 diagrams. Other tools include structured analysis, data flow diagrams (DFD), control flow diagrams (CFD), IDEF diagrams, behavior diagrams, action diagrams, state/mode diagrams, process flow diagrams, function hierarchy diagrams, state transition diagrams, entity relationship diagrams (ERD), structured analysis and design, object-oriented analysis (OOA), Unified Modeling Language (UML), structured systems analysis and design methodology (SSADM), and quality function deployment (QFD). Each of these techniques focuses on gathering requirements in a formal systematic way.

In this text, we focus on the philosophy of requirements engineering using the upper-level tools of the context diagram, FFBD, and RBS, leaving the reader to consult the many excellent references for the more detailed tools [59], particularly MBSE [60].

This section provides a brief introduction to the RBS and FFBD, which are used in the remaining chapters to illustrate the processes, activities, and steps necessary to implement the methodology presented. The use of a context diagram is discussed in more detail in Chapter 3.

2.7.1 RBS

The requirements engineering process can be assisted using a requirements framework around which the logical description of the system can be based. Here we call the requirements framework the RBS because, as illustrated in Figure 2.8, the requirements are shown in a tree-structured elaboration of the mission.

The words are deliberately chosen to differentiate this structure from the well-known project management artifact called the work breakdown structure (WBS) [61]; the RBS is grouped logically (functionally), the WBS is structured by physical work packages (including CIs) and contains other project-related work. The logical groupings of the RBS are then allocated to the physical groupings of the CIs in the WBS. The RBS is also called a functional hierarchy.

In addition to being a useful requirements analysis tool, the RBS can be used to capture the business requirements in the BRS, which can be further refined to reflect stakeholder requirements in the StRS, and then the system requirements in the SyRS. When completely populated, the RBS for a system provides an outline of the SyRS. In the first instance, the system is described

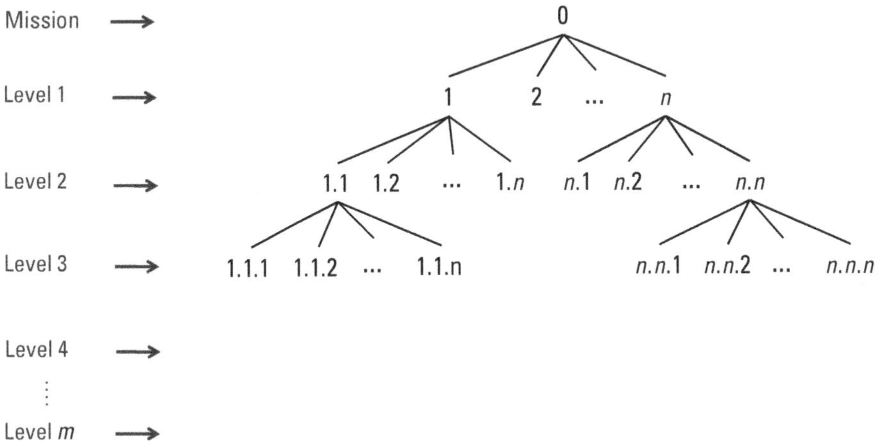

Figure 2.8 A simple RBS showing the hierarchical elaboration of requirements from the mission statement down to *m* levels.

by the words or short phrases of the RBS; in the SyRS, these short titles are elaborated into well-formed requirement statements articulating functional, performance, and verification requirements.

There are a number of advantages associated with the establishment of a suitable framework such as the RBS early in the requirements analysis activity. First, the framework will act as a reference source during requirements analysis to ensure that all aspects of the system requirements are addressed and that important areas are not omitted. The framework allows multiple people to work on the analysis simultaneously as it facilitates the effective allocation to individuals of responsibility for sections of requirements. The framework assists in avoiding duplication of requirements in different sections, which is undesirable as it raises the probability of conflict between requirements and often leads to ambiguity and confusion.

It should be noted that not all complex systems can be represented adequately through a hierarchical decomposition, and we must remain cognizant that we might need to extend the representation to describe fully the system to reflect its possible emergent properties. Still, almost all workable human-made systems are hierarchical, so the RBS is an adequate representation [62].

When developing the RBS, some general rules should be followed:

- The RBS is to be displayed hierarchically. The level of the RBS illustrated should be such that the portion displayed should be visible on a single A4 sheet of paper, at no less than 10-point font, with no lines leaving or entering the page.

- Naming RBS elements should be consistent so that the keyword phrases are either verb-based or noun-based.
- The length of the keyword phrases should be consistent (three or four words are normally sufficient).
- Each keyword phrase must be unique (since it must be fleshed out into a unique requirement).
- As illustrated in Figures 2.5 and 2.8, the RBS elements should be numbered with a hierarchical numbering system so that parent requirements can be traced to their children and vice versa.

2.7.2 FFBDs

The hierarchical representation of requirements in the RBS can be supported by FFBDs [63], a simple example of which is shown in Figure 2.9. While the RBS is very useful to show a hierarchical decomposition of the requirements, FFBDs are useful in showing the interrelationship of the functions to be performed by the system in accomplishing its mission. Some functions must be performed sequentially; some may be performed in parallel. Sometimes alternative paths may be taken based on certain conditions; the FFBDs can show these relationships much more easily than the RBS.

As with the RBS, FFBDs are developed top-down through hierarchical decomposition. Each top-level block can be expanded (decomposed) as necessary to ensure that the functionality of the system is adequately defined, as illustrated in Figure 2.9.

When developing FFBDs, some general rules should be followed:

- Each function is to be presented in a single box enclosed in a single solid line.
- As with the RBS, naming of the FFDB elements should be consistent so that the keyword phrases are unique, either verb-based or noun-based, and are no more than three or four words long. The same names should be used for the function as it is displayed in both the FFBD and the RBS.
- As shown in Figure 2.9, the FFBDs should be displayed as a hierarchical elaboration with the naming of the function appropriate for the level of abstraction.
- The level of the FFBD illustrated should be such that the portion displayed should be visible on a single A4 sheet of paper, at no less than 10-point font.

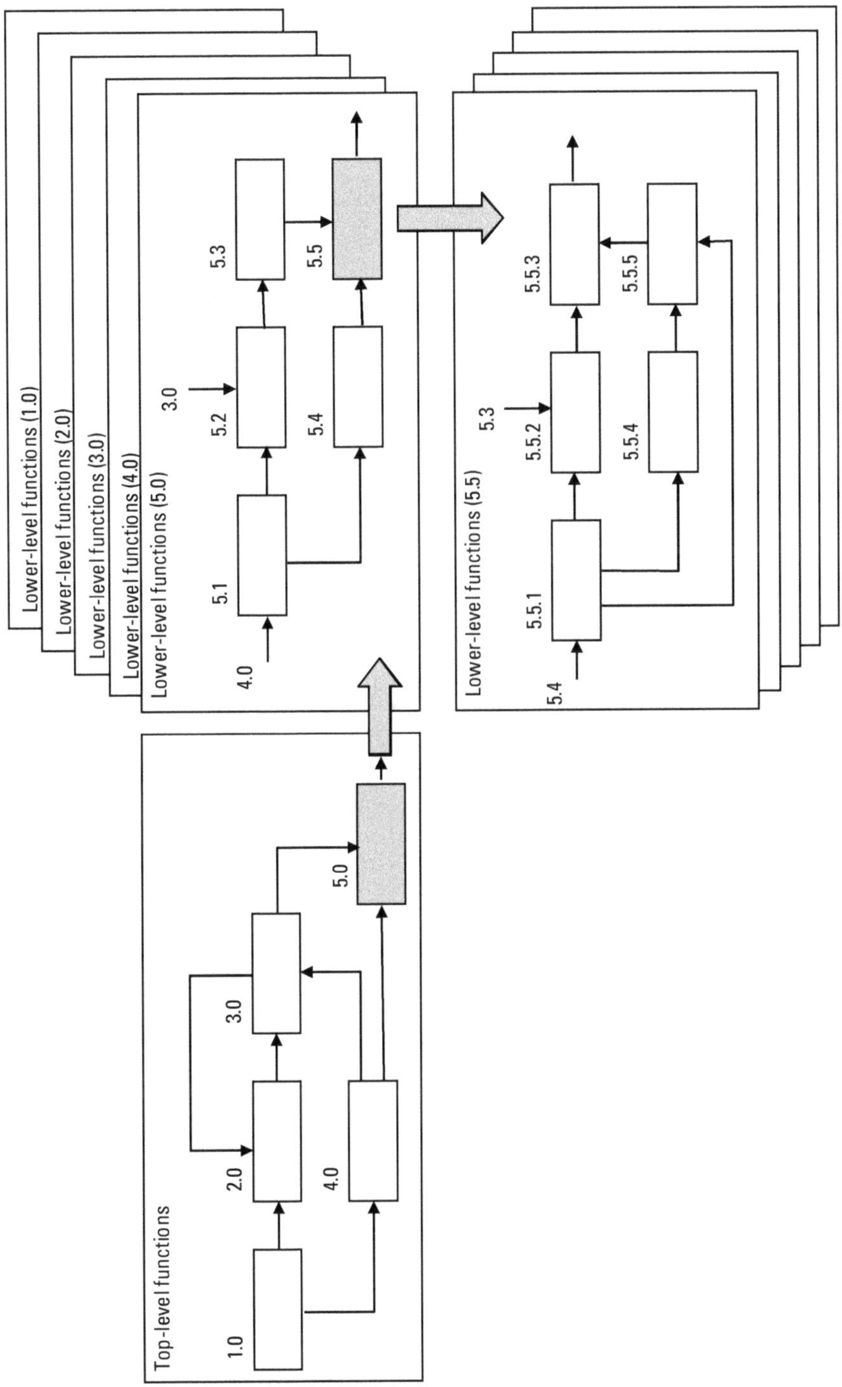

Figure 2.9 Example hierarchical decomposition of FFBDs.

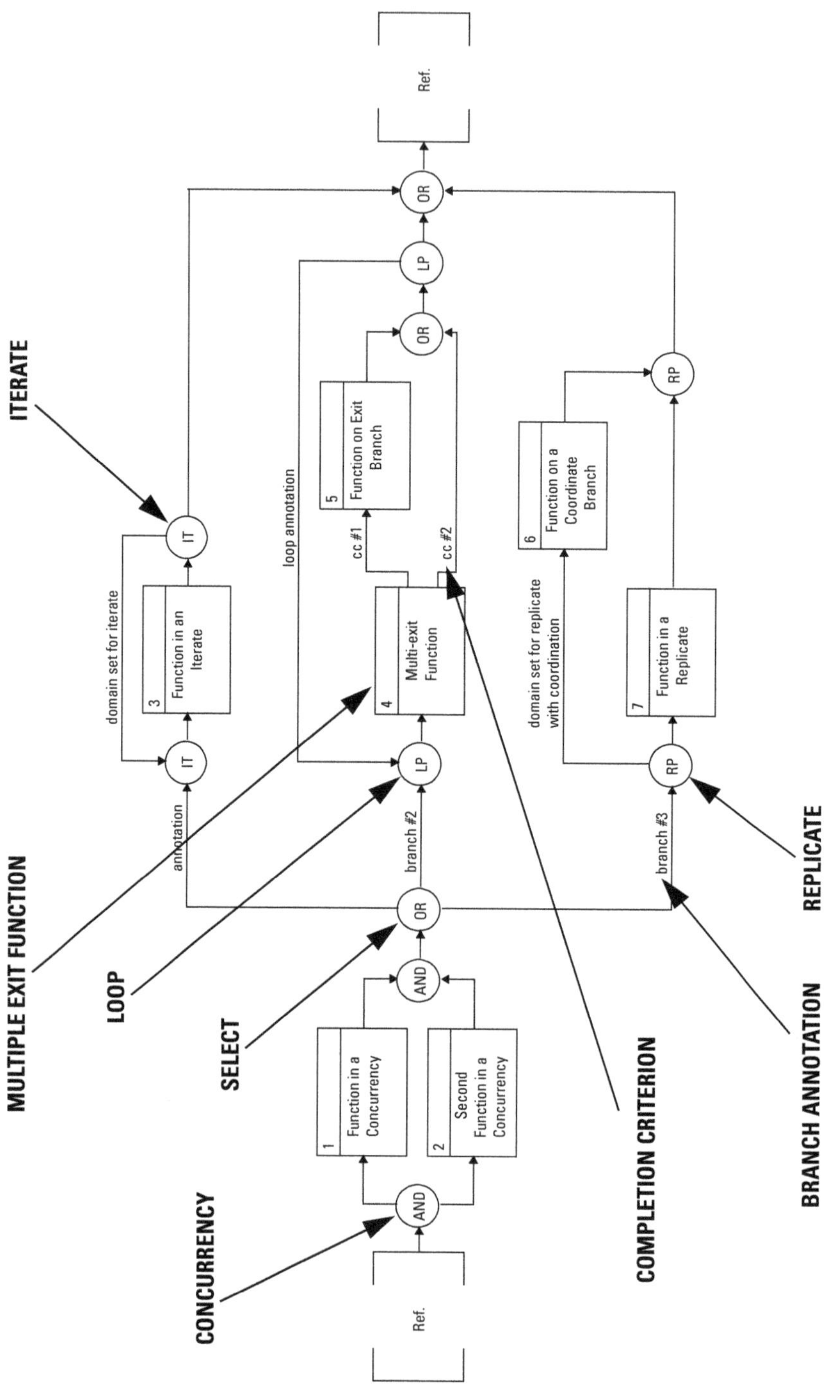

Figure 2.10 Example FFBD showing additional control constructs. (Reproduced with permission of Vitech Corporation.)

- As with the RBS, the FFBDs also make use of a hierarchical numbering system. The same numbering system is to be used in the RBS and the FFBDs with the allocated number being driven by the level of the function in the RBS.
- The lines that connect the blocks have arrows to show the flow of functions, not the elapsed time between them. Consistency in drawing is required; conventionally, the flow is from left to right of the FFBD.
- If useful, additional blocks may be used to show such additional control constructs as decision points, repetition, iteration, concurrent flow (logical AND), and alternative flow (logical OR).

If useful, additional blocks may be used to show such additional control constructs as decision points, repetition, iteration, concurrent flow (logical AND), and alternative flow (logical OR); an example is illustrated in Figure 2.10 [64].

2.8 Summary

This chapter has provided an overview of the extant requirements engineering body of knowledge because it is such an essential aspect of system development. With that background, the following chapters examine the life-cycle activities in more detail. Chapters 3 and 4 examine the issues associated with conceptual design and preliminary design. Emphasis is given to these early activities, as they have the greatest impact on the system development. Chapters 5, 6, and 7 deal with, respectively, detailed design and development, construction and/or production, and operational use and support.

Review Questions

1. Define what is meant by a concept, a need, and a requirement.
2. Briefly describe the role and content of the ConOps and the OpsCon.
3. Briefly describe the business needs and requirements definition process, including the roles of the PLC and the BRS.
4. Briefly describe the stakeholder needs and requirements definition process, including the roles of the LC and the StRS.
5. Briefly describe the system requirements definition process, including the roles of the SyRS.
6. Briefly describe what is meant by a requirement (functional, nonfunctional, and constraint).

7. Briefly describe the nine characteristics of a good requirement.
8. Briefly explain 10 of the 18 mandatory attributes that may be associated with each requirement.
9. Briefly explain why system-level requirements should describe what the system should do rather than how it should be done.
10. Briefly describe what is meant by the term emergent properties.
11. Briefly outline why we need requirements.
12. Briefly outline why we need requirements engineering.
13. Briefly describe what is meant by the terms elicitation, elaboration, decomposition, and derivation, in the context of requirements engineering.
14. Briefly describe the major difficulties in eliciting/elaborating requirements.
15. Briefly explain the major techniques for eliciting requirements.
16. Briefly describe the requirements engineering activity of requirements validation.
17. Briefly describe what is involved in requirements management.
18. Briefly describe why requirements traceability is important and describe what is meant by the terms backward traceability and forward traceability.
19. List 10 desirable features of a requirements management tool.
20. Briefly describe the RBS and its role as a requirements framework.
21. Briefly describe FFBDs and their role in defining functional requirements.

Endnotes

[1] In addition to information contained in the general systems engineering references listed at the end of Chapter 1, specific detail on requirements engineering can be found in:

Adzic, G., *Specification by Example*, Shelter Island, NY: Manning Publications Co., 2011.

Alexander, I. F., and L. Beus-Dukic, *Discovering Requirements: How to Specify Products and Services*, New York: John Wiley & Sons, 2007.

Alexander, I. F., and R. Stevens, *Writing Better Requirements*, Reading, MA: Addison Wesley, 2002.

Andriole, S., *Managing Systems Requirements: Methods, Tools and Cases*, New York: McGraw-Hill, 1996.

Aurum, A., and C. Wohlin, (eds.), *Engineering and Managing Software Requirements*, New York: Springer, 2005.

Berenbach, B., et al., *Software and Systems Requirements Engineering in Practice*, New York: McGraw-Hill, 2009.

Bray, I. K., *An Introduction to Requirements Engineering*, Reading, MA: Addison-Wesley, 2002.

Bron, J. -Y., *System Requirements Engineering: A SysML Supported Requirements Engineering Method*, New York: Wiley-ISTE, 2020.

Davis, A. M., *Just Enough Requirements Management: Where Software Development Meets Marketing*, New York: Dorset House Publishing, 2005.

Davis, A. M., *Software Requirements: Analysis and Specification*, Englewood Cliffs, NJ: Prentice Hall International, 1990.

Davis, A. M., *Software Requirements: Objects, Functions and States*, Englewood Cliffs, NJ: Prentice Hall International, 1993.

Dorfman, M., and R. H. Thayer, *System and Software Requirements Engineering*, New York: IEEE Computer Society Press, 1990.

Ferdinandi, P. L., *A Requirements Pattern: Succeeding in the Internet Economy*, Reading, MA: Addison-Wesley, 2001.

Fernandes, J., and R. Machado, *Requirements in Engineering Projects*, New York: Springer, 2016.

Fitchett, P., and J. Haslam, *Writing Engineering Specifications*, London, UK: Spon Press, 1988.

Gause, D., and G. Weinberg, *Exploring Requirements: Quality Before Design*, New York: Dorset House Publishing, 1989.

Gilb, T., *Competitive Engineering: A Handbook For Systems Engineering, Requirements Engineering, and Software Engineering Using Planguage*, Boston, MA: Butterworth-Heinemann, 2005.

Goldsmith, R. F., *Discovering Real Business Requirements for Software Project Success*, Norwood, MA: Artech House, 2004.

Gottesdiener, E., *Requirements by Collaboration: Workshops for Defining Needs*, Reading, MA: Addison Wesley, 2002.

Grady, J., *Systems Requirements Analysis*, New York: McGraw-Hill, 2006.

Hass, K. B., D. J. Wessels, and K. Brennan, *Getting It Right: Business Requirement Analysis Tools and Techniques*, Vienna, VA: Management Concepts, 2008.

Hatley, D. J., P. Hruschka, and I. A. Pirbhai, *Process for System Architecture and Requirements Engineering*, New York: Dorset House Publishing, 2000.

Hay, D. C., *Requirements Analysis: From Business Views to Architecture*, Upper Saddle River, NJ: Prentice Hall, 2003.

Hood, C., et al., *Requirements Management: The Interface Between Requirements Development and All Other Systems Engineering Processes*, New York: Springer, 2008.

Hooks, I. F., and K. A. Farry, *Customer-Centered Products: Creating Successful Products Through Smart Requirements Management*, New York: AMACOM, 2001.

Hossenlopp, R., and K. B. Hass, *Unearthing Business Requirements: Elicitation Tools and Techniques*, Vienna, VA: Management Concepts, 2008.

Hull, E., K. Jackson, and J. Dick, *Requirements Engineering*, New York: Springer, 2011.

International Institute for Business Analysis, *A Guide to the Business Analysis Body of Knowledge® (BABOK® Guide)*, Version 3, 2015.

ISO/IEC/IEEE 29148 *Systems and Software Engineering—Lifecycle Processes—Requirements Engineering*, 2018.

Jackson, M., *Software Requirements and Specifications: A Lexicon of Software Practice, Principles and Prejudices*, Reading, MA: Addison Wesley, 1995.

Jonasson, H., *Determining Project Requirements, Second Edition: Mastering the BABOK® and the CBAP® Exam*, Boca Raton, FL: Auerbach, CRC Press, 2012.

Kossmann, M., *Requirements Management: How to Ensure You Achieve What You Need From Your Projects*,: Gower, 2013.

Kotonya, G., and I. Sommerville, *Requirements Engineering: Processes and Techniques*, New York: John Wiley & Sons, 2000.

Kulak, D., and E. Guiney, *Use Cases: Requirements in Context*, Reading, MA: Addison-Wesley, 2004.

Laplante, P. A., *Requirements Engineering for Software and Systems*, Boca Raton, FL: CRC Press, 2014.

Larson, E., and R. Larson, *Practitioner's Guide to Requirements Engineering—Part 1: Requirements Planning*, Minneapolis, MN: Watermark Learning Publications, 2009.

Leffingwell, D., and D. Widrig, *Managing Software Requirements: A Unified Approach*, Reading, MA: Addison-Wesley, 2000.

Leffingwell, D., *Agile Software Requirements: Lean Requirements Practices for Teams, Programs, and the Enterprise*, Reading, MA: Addison-Wesley, 2011.

Loucopoulos, P., and V. Karakostas, *System Requirements Engineering*, New York: McGraw-Hill, 1995.

Lutowski, R., *Software Requirements: Encapsulation, Quality, and Reuse*, Boca Raton, FL: Taylor and Francis, 2005.

Macaulay, L., *Requirements Engineering*, New York: Springer, 1996.

Mate, J. L., and A. Silva, (eds.), *Requirements Engineering for Sociotechnical Systems*, Hershey, PA: Information Science Publishing, 2005.

Pohl, K., *Requirements Engineering: Fundamentals, Principles, and Techniques*, New York: Springer-Verlag, 2010.

Pohl, K., *Requirements Engineering Fundamentals*, Santa Barbara, CA: Rocky Nook, 2011.

Pohl, K., and C. Rupp, *Requirements Engineering: Fundamentals, Principles, and Techniques*, Heidelberg: Springer-Verlag, 2011.

Purdy, D., *A Guide to Writing Successful Engineering Specifications*, New York: McGraw-Hill, 1991.

Rinzler, B., *Telling Stories: A Short Path to Writing Better Software Requirements*, New York: Wiley, 2009.

Robertson, S., and J. Robertson, *Requirements-Led Project Management*, Reading, MA: Addison-Wesley, 2005.

Robertson, S., and J. Robertson, *Mastering the Requirements Process*, Reading, MA: Addison-Wesley, 2013.

Sommerville, I., and P. Sawyer, *Requirements Engineering*, New York, John Wiley & Sons, 1997.

Thayer, R. H., and M. Dorfman (eds.), *Software Requirements Engineering*, Los Alamitos, CA: IEEE Computer Society Press, 1997.

van Lamsweerde, A., *Requirements Engineering: From System Goals to UML Models to Software Specifications*, New York: John Wiley & Sons, 2009.

Wheatcraft, L. S., M. J. Ryan, and T. E. Katz, *Needs and Requirements Manual: Needs, Requirements, Verification, Validation Across the Lifecycle*, Hoboken, NJ: Wiley, 2024.

Wiegers, K. E., and J. Beatty, *Software Requirements*, Redmond, WA: Microsoft Press, 2013.

Wieringa, R., *Requirements Engineering: Frameworks for Understanding*, New York: Wiley, 1996.

Withall, S., *Software Requirements Patterns (Best Practices)*, Redmond, MA: Microsoft Press, 2007.

Young, R., *Effective Requirements Practices*, Reading, MA: Addison-Wesley, 2001.

Young, R., *The Requirements Engineering Handbook*, Norwood, MA: Artech House, 2004.

Young, R., *Project Requirements: A Guide to Best Practices*, Vienna, VA: Management Concepts, 2006.

[2] Ryan, M. J., and L. S. Wheatcraft, "On a Cohesive Set of Requirements Engineering Terms," *Systems Engineering*, Vol. 20, No. 2, 2017, pp. 118–130.

[3] Ryan, M. J., "An Improved Taxonomy for Major Needs and Requirements Artefacts," *INCOSE International Symposium IS2013*, Vol. 23, No. 1, June 2013, pp. 244–258.

[4] Ryan, M. J., and L. S. Wheatcraft, "On a Cohesive Set of Requirements Engineering Terms," *Systems Engineering*, Vol. 20, No. 2, 2017, pp. 118–130.

[5] Wheatcraft, L. S., M. J. Ryan, and T. E. Katz, *Needs and Requirements Manual: Needs, Requirements, Verification, Validation Across the Lifecycle*, Hoboken, NJ: Wiley, 2024.

[6] International Institute for Business Analysis, *A Guide to the Business Analysis Body of Knowledge® (BABOK® Guide)*, Version 3, 2015.

[7] ISO/IEC/IEEE 15288, *Systems and Software Engineering—System Lifecycle Processes*, 2023.

[8] ISO/IEC/IEEE 29148 *Systems and Software Engineering—Lifecycle Processes—Requirements Engineering*, 2018.

[9] ISO/IEC/IEEE 15288, *Systems and Software Engineering—System Lifecycle Processes*, 2023.

[10] ISO/IEC/IEEE 29148 *Systems and Software Engineering—Lifecycle Processes—Requirements Engineering*, 2018.

[11] ANSI/AIAA G-043B-2012, *Guide for the Preparation of Operational Concept Descriptions*, American National Standards Institute, American Institute of Aeronautics and Astronautics (sponsor), 2012.

[12] ISO/IEC/IEEE 29148 *Systems and Software Engineering—Lifecycle Processes—Requirements Engineering*, 2018.

[13] International Institute for Business Analysis, *A Guide to the Business Analysis Body of Knowledge® (BABOK® Guide)*, Version 3, 2015.

[14] ISO/IEC/IEEE 29148 *Systems and Software Engineering—Lifecycle Processes—Requirements Engineering*, 2018.

[15] ISO/IEC/IEEE 29148 *Systems and Software Engineering—Lifecycle Processes—Requirements Engineering*, 2018.

[16] International Institute for Business Analysis, *A Guide to the Business Analysis Body of Knowledge® (BABOK® Guide)*, Version 3, 2015.

[17] MIL-STD-499B, *Military Standard—Systems Engineering—Draft*, Washington, DC: US Department of Defense, 1994.

[18] Wheatcraft, L. S., M. J. Ryan, and T. E. Katz, *Needs and Requirements Manual: Needs, Requirements, Verification, Validation Across the Lifecycle*, Hoboken, NJ: Wiley, 2024.

[19] Wheatcraft, L. S., M. J. Ryan, and T. E. Katz, *Needs and Requirements Manual: Needs, Requirements, Verification, Validation Across the Lifecycle*, Hoboken, NJ: Wiley, 2024.

[20] ANSI/AIAA G-043B-2012, *Guide for the Preparation of Operational Concept Descriptions*, American National Standards Institute, American Institute of Aeronautics and Astronautics (sponsor), 2012.

[21] ISO/IEC/IEEE 29148, *Systems and Software Engineering—Lifecycle Processes—Requirements Engineering*, 2018.

[22] ISO/IEC/IEEE 29148, *Systems and Software Engineering—Lifecycle Processes—Requirements Engineering*, 2018.

[23] International Council on Systems Engineering (INCOSE), INCOSE-TP-2010 -004-04, *Guide to Writing Requirements*, 2024.

[24] ISO/IEC/IEEE 29148, *Systems and Software Engineering—Lifecycle Processes—Requirements Engineering*, 2018.

[25] International Council on Systems Engineering (INCOSE), INCOSE-TP-2010-004-04, *Guide to Writing Requirements*, 2024.

[26] Ryan, M. J., and L. S. Wheatcraft, "On a Cohesive Set of Requirements Engineering Terms," *Systems Engineering*, Vol. 20, No. 2, 2017, pp. 118–130.

[27] Ryan, M. J., and L. S. Wheatcraft, "On a Cohesive Set of Requirements Engineering Terms," *Systems Engineering*, Vol. 20, No. 2, 2017, pp. 118–130.

[28] Wheatcraft, L. S., M. J. Ryan, and J. Dick, "On the Use of Attributes to Manage Requirements," *Systems Engineering*, Vol. 19, No. 5, September 2016, pp. 448–458.

[29] IEEE-STD-1233, *IEEE Guide for Developing System Requirements Specifications*, New York: The Institute of Electrical and Electronic Engineers (IEEE), 1998. Note that a similar list is in IEEE 29148.

[30] Wheatcraft, L. S., M. J. Ryan, and T. E. Katz, *Needs and Requirements Manual: Needs, Requirements, Verification, Validation Across the Lifecycle*, Hoboken, NJ: Wiley, 2024.

[31] Wheatcraft, L. S., M. J. Ryan, and T. E. Katz, *Needs and Requirements Manual: Needs, Requirements, Verification, Validation Across the Lifecycle*, Hoboken, NJ: Wiley, 2024.

[32] Verner, J., et al., "Requirements Engineering and Software Project Success: An Industrial Survey in Australian and the U.S.," *Australasian Journal of Information Systems*, Vol. 13, No. 1, 2005, pp. 225–238.

[33] Andriole, S., "The Politics of Requirements Management," *IEEE Software*, November/December 1998, pp. 82–84.

[34] Dorfman, M., and R. H. Thayer, *System and Software Requirements Engineering*, IEEE Computer Society Press, 1990.

[35] Kotonya, G., and I. Sommerville, *Requirements Engineering*, New York: John Wiley & Sons, 2000.

[36] ISO/IEC/IEEE 29148, *Systems and Software Engineering—Lifecycle Processes—Requirements Engineering*, 2018.

[37] Project Management Institute Standards Committee, *A Guide to the Project Management Body of Knowledge*, Upper Darby, PA: Project Management Institute, Version 6, 2017.

[38] Davis, A. M., *Software Requirements: Objects, Functions and States*, Englewood Cliffs, NJ: Prentice Hall, 1993.

[39] Verner, J., et al., "Requirements Engineering and Software Project Success: an Industrial Survey in Australia and the U.S.," *Australasian Journal of Information Systems*, Adelaide, Vol. 13, No. 1, September 2005.

[40] Ryan, M. J., *Requirements Practice in Conceptual Design: A Practical Guide*, Norwood, MA: Artech House, 2026.

[41] Hitchins, D., *Systems Engineering: A 21st Century Systems Methodology*, New York: John Wiley & Sons, 2007.

[42] Grady, J. O., *System Requirements Analysis*, Burlington, MA: Academic Press, 2006.

[43] Daniels, J., and T. Bahill, "The Hybrid Process That Combines Traditional Requirements and Use Cases," *Systems Engineering*, Vol. 7, No. 4, 2004.

[44] IEEE-STD-1220, *IEEE Standard for Application and Management of the Systems Engineering Process*, New York: IEEE Computer Society, 2005.

[45] Alexander, I. F., and R. Stevens, *Writing Better Requirements*, Reading, MA: Addison-Wesley, 2002.

[46] Goguen, J., and C. Linde, "Techniques for Requirements Elicitation," *International Symposium on Requirements Engineering*, San Diego, CA: IEEE Computer Society Press, 1993, pp. 152–164.

[47] Hull, M. E. C., et al., *Requirements Engineering*, New York: Springer, 2002.

[48] Katonya, G., and I. Sommerville, *Requirements Engineering: Processes and Techniques*, New York: John Wiley & Sons, 2000.

[49] Sommerville, I., and P. Sawyer, *Requirements Engineering: A Good Practice Guide*, New York: John Wiley & Sons, 1997.

[50] Young, R. R., *Effective Requirements Practices*, Reading, MA: Addison-Wesley, 2001.

[51] Young, R. R., *The Requirements Engineering Handbook*, Norwood, MA: Artech House, 2004.

[52] IEEE-STD-1233, *IEEE Guide for Developing System Requirements Specifications*, New York: The Institute of Electrical and Electronic Engineers (IEEE), 1998.

[53] Verner, J., et al., "Requirements Engineering and Software Project Success: An Industrial Survey in Australia and the U.S.," *Australasian Journal of Information Systems*, Adelaide, Vol. 13, No. 1, 2005, pp. 225–238.

[54] Coulin, C., D. Zowghi, and A. E. K. Sahraoui, "A Situational Method Engineering Approach to Requirements Elicitation Workshops in the Software Development Process," *Software Process Improvement and Practice*, Vol. 11, 2006, pp. 451–464.

[55] Tsumaki, T., and T. Tamai, "Framework for Matching Requirements Elicitation Techniques to Project Characteristics, *Software Process Improvement and Practice*, Vol. 11, 2006, pp. 505–519.

[56] Specific details on use-case modeling can be found in:

Bittner, K., and I. Spence, *Use Case Modeling*, Reading, MA: Addison-Wesley, 2003.

Callahan, S., *Putting Stories to Work*, Melbourne, Australia: Pepperberg Press, 2016.

Cockburn, A., *Writing Effective Use Cases*, Reading, MA: Addison-Wesley, 2002.

Kulak, D., and E, Guiney, *Use Cases: Requirements in Context*, Reading, MA: Addison-Wesley, 2004.

Rinzler, B., *Telling Stories: A Short Path to Writing Better Software Requirements*, New York: Wiley, 2009.

[57] Jones, C., *Software Quality—Analysis and Guidelines for Success*, International Thomson, 1997.

[58] See also Andriole, S., *Managing Systems Requirements: Methods, Tools and Cases*, New York: McGraw-Hill, 1996.

[59] Specific details on analysis tools can be found in:

Blanchard, B., and W. Fabrycky, *Systems Engineering and Analysis*, Upper Saddle River, NJ: Prentice Hall, 2011.

Coad, P., and E. Yourdon, *Object Oriented Analysis*, Upper Saddle River, NJ: Prentice Hall, 1990.

Defense Systems Management College (DSMC), *Systems Engineering Management Guide*, Washington, DC: US Government Printing Office, 1990.

Eisner, H., *Essentials of Project and Systems Engineering Management*, New York: John Wiley & Sons, 2008.

Gause, D., and G. Weinberg, *Exploring Requirements: Quality Before Design*, New York: Dorset House Publishing, 1989.

Goodland, M., and C. Slater, *SSADM: A Practical Approach*, New York: McGraw-Hill, 1995.

Grady, J. O., *Systems Requirements Analysis*, Burlington, MA, Academic Press, 2006.

Heap, G., J. Stanway, and A. Windsor, *A Structured Approach to Systems Development*, New York: McGraw-Hill, 1992.

Hitchins, D., *Putting Systems to Work*, New York: John Wiley & Sons, 1992.

Kotonya, G., and I. Sommerville, *Requirements Engineering: Processes and Techniques*, New York: John Wiley & Sons, 2000.

Lacy, J., *Systems Engineering Management: Achieving Total Quality*, New York: McGraw-Hill, 1992.

Leffingwell, D., and D. Widrig, *Managing Software Requirements: A Unified Approach*, Reading, MA: Addison-Wesley, 2000.

Macaulay, L., *Requirements Engineering*, London, England: Springer, 1996.

Maier, M., and E. Rechtin, *The Art of Systems Architecting*, Boca Raton, FL: CRC Press, 2009.

Martin, J., *Principles of Object-Oriented Analysis and Design*, Upper Saddle River, NJ: Prentice Hall, 1993.

Martin, J., *Systems Engineering Guidebook: A Process for Developing Systems and Products*, Boca Raton, FL: CRC Press, 1997.

Martin, J., and J. Odell, *Object-Oriented Analysis and Design*, Upper Saddle River, NJ: Prentice Hall, 1992.

Rumbaugh, J., et al., *Object-Oriented Modeling and Design*, Upper Saddle River, NJ: Prentice Hall, 1991.

Sage, A. and J. Armstrong, *Introduction to Systems Engineering*, New York: John Wiley & Sons, 2000.

Shlaer, S. and S. Mellor, *Object-Oriented Systems Analysis*, Upper Saddle River, NJ: Prentice Hall, 1988.

Stevens, R., et al., *Systems Engineering: Coping with Complexity*, Hertfordshire, England: Prentice Hall Europe, 1998.

Wheatcraft, L. S., M. J. Ryan, and T. E. Katz, *Needs and Requirements Manual: Needs, Requirements, Verification, Validation Across the Lifecycle*, Hoboken, NJ: Wiley, 2024.

Young, R., *Effective Requirements Practices*, Reading, MA: Addison-Wesley, 2001.

Westerman, H., *Systems Engineering Principles and Practice*, Norwood, MA: Artech House, 2001.

[60] Excellent coverage of MBSE and its application can be found in a number of texts:

Buede, D. M., and W. D. Miller, *The Engineering Design of Systems: Models and Methods*, New York: Wiley, 2006.

Delligatti, L., *SysML Distilled: A Brief Guide to the Systems Modeling Language*, Boston, MA: Pearson Education, 2010.

Dori, D., *Model-Based Systems Engineering with OPM and SysML*, Springer, 2016.

Friedenhal, S., A. Moore, and P. Stein, *A Practical Guide to SysML: The Systems Modeling Language*, Waltham, MA: The MK/OMG Press, 2015.

Hetherington, D., *Simple SySML for Beginners: Using Sparx Enterprise Architect*, Austin, TX: Asatte Press, 2019.

Holt, J., *Systems Engineering Demystified: A Practitioner's Handbook for Developing Complex Systems Using a Model-Based Approach*, Birmingham, UK: Packt, 2021.

Holt, J., and S. Perry, *SysML for Systems Engineering: A Model-Based Approach*, London: Institution of Engineering and Technology, 2013.

Micouin, P., *Model-Based Systems Engineering: Fundamentals and Methods*, New York: Wiley, 2014.

Vitech Corporation, *A Primer for Model-Based Systems Engineering*, 2nd Edition, Vitech Corporation, 2011.

[61] Project Management Institute (PMI), *Guide to the Project Management Body of Knowledge*, Newtown Square, PA: PMI, Version 6, 2017.

[62] Simon, H., *The New Science of Management Decision*, Englewood Cliffs, NJ: Prentice Hall, 1977.

[63] Blanchard and Fabrycky, (*Systems Engineering and Analysis*, Prentice-Hall, 1998) noted that these generic FFBDs can be prepared by any one of the graphical methods including the Integrated DEFinition (IDEF) modeling method, the Behavioral Diagram Method, and the N^2 Charting Method, which are compared and discussed in Blanchard, B., D. Verna, and E. Peterson, *Maintainability: A Key to Effective Serviceability and Maintenance Management*, New York: John Wiley & Sons, 1995. Defense Systems Management College (DSMC), *Systems Engineering Management Guide*, Washington, DC, 1990, p. 6.3–6.5) explains how FFBDs can be used during function identification. See also J. Long, *Relationship Between Common Graphical Representations in Systems Engineering*, Vitech Corporation, 2002.

[64] Long, J., "Relationship Between Common Graphical Representations Used in Systems Engineering," *National Council on Systems Engineering (NCOSE) Symposium*, 1995.

3

Conceptual Design

3.1 Introduction

The systems engineering processes begin in earnest with the first activity in the acquisition phase: conceptual design (or logical design or problem-domain design), which aims to articulate the mission; to analyze, integrate, and record the business and stakeholders needs flowing from the mission for that system; to transform those into an integrated set of needs; and to complete a functional design of the system as represented in the SyRS. The major product of conceptual design, the initial functional baseline (FBL), provides, among other things, a system-level logical architecture that is the basis for subsequent lower-level (physical) design in subsequent stages, beginning with preliminary design.

Conceptual design is about the problem domain and is therefore normally the domain of the customer, acquirer, or tasking activity, who is responsible for and heavily involved in this activity. While other organizations may be involved, it is the customer who must determine what the system needs to do and how well it needs to do it; after all, it is their business from which the needs for the system arise.

Conceptual design is perhaps the most critical of the life-cycle activities because:

- Conceptual design is responsible for the expansion of the system definition from a small number of relatively brief business needs into a logical set of system-level requirements that may be hundreds of pages long.

- All subsequent aspects of the system design will be traced back to the FBL that ends this activity; any errors here will flow down to the remainder of the acquisition activities.
- Conceptual design is concerned with the transition from the problem domain into the solution domain. This phase also often ends with the transition from the acquirer to a developer; that is, as we will see later, the SyRS defined in conceptual design may form the basis of an agreement between the customer (the acquirer) and a developer. It is therefore essential that the output of conceptual design adequately represents the business and stakeholder needs and requirements.

Figure 3.1 illustrates that five major processes are conducted during conceptual design, including:

- Definition of the BNR in the PLC and the BRS;
- Definition of SNR in the LC and StRS;
- Definition of draft system requirements in the draft SyRS;
- System synthesis to finalize the SyRS with the preferred solution in mind;
- The system design review (SDR), at which the initial FBL is established.

Remember, as discussed in Chapter 2, the BNR and SNR described in ISO/IEC/IEEE 15288 are combined in the INCOSE NRM's integrated set of needs and the SyRS described in ISO/IEC/IEEE 29148 is called the design-input specification in the NRM. Figure 3.1 shows both taxonomies to help to visualize the relationship as well as mapping the conceptual design processes that we consider here onto the ISO/IEC/IEEE 15288 processes.

Note that, as with all systems engineering effort, the processes are iterative, even at the highest level of abstraction; at each stage, the improved understanding of the system is used to revisit the previous process or processes if necessary. These conceptual design processes are described in more detail in this chapter.

3.2 C1. Define Business Needs and Requirements

Before any work can commence on developing the system, the BNR must be articulated clearly and completely by business management. It would appear obvious that an understanding of what is to be achieved must be clearly articulated by the business before any further work is undertaken. Yet a surprising

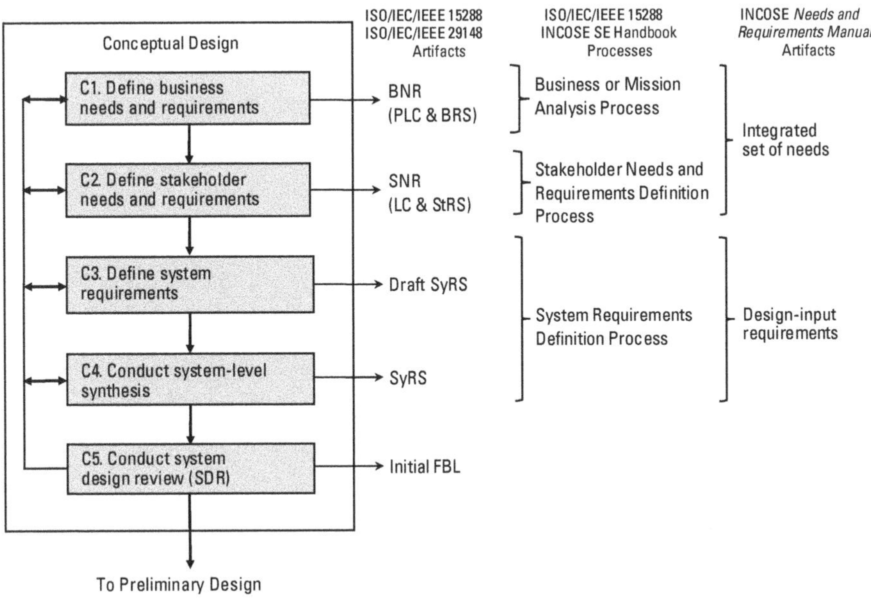

Figure 3.1 The five major processes performed iteratively during conceptual design.

number of system developments commence without a clear and complete understanding of the fundamental business needs and requirements. Not surprisingly, most of these developments founder or, at the very least, enter utilization with poor levels of user acceptance.

Figure 3.2 summarizes the activities and steps involved with the definition of BNR resulting in the development of the PLC and the BRS. Note once again that the process is iterative, allowing current understanding to improve the quality of previously generated considerations. The activities of Figure 3.2 are described in more detail in the following sections.

3.2.1 C1.1 Identify Major Stakeholders and Constraints

At the strategic level of the organization, business management starts the definition of BNR with the identification of the major stakeholders who are to be engaged to elicit business needs and requirements and to verify design artifacts as the design progresses. They also must identify the business, project, external, and design constraints that provide important context within which to begin consideration of the system requirements, as well as being the source of many requirements.

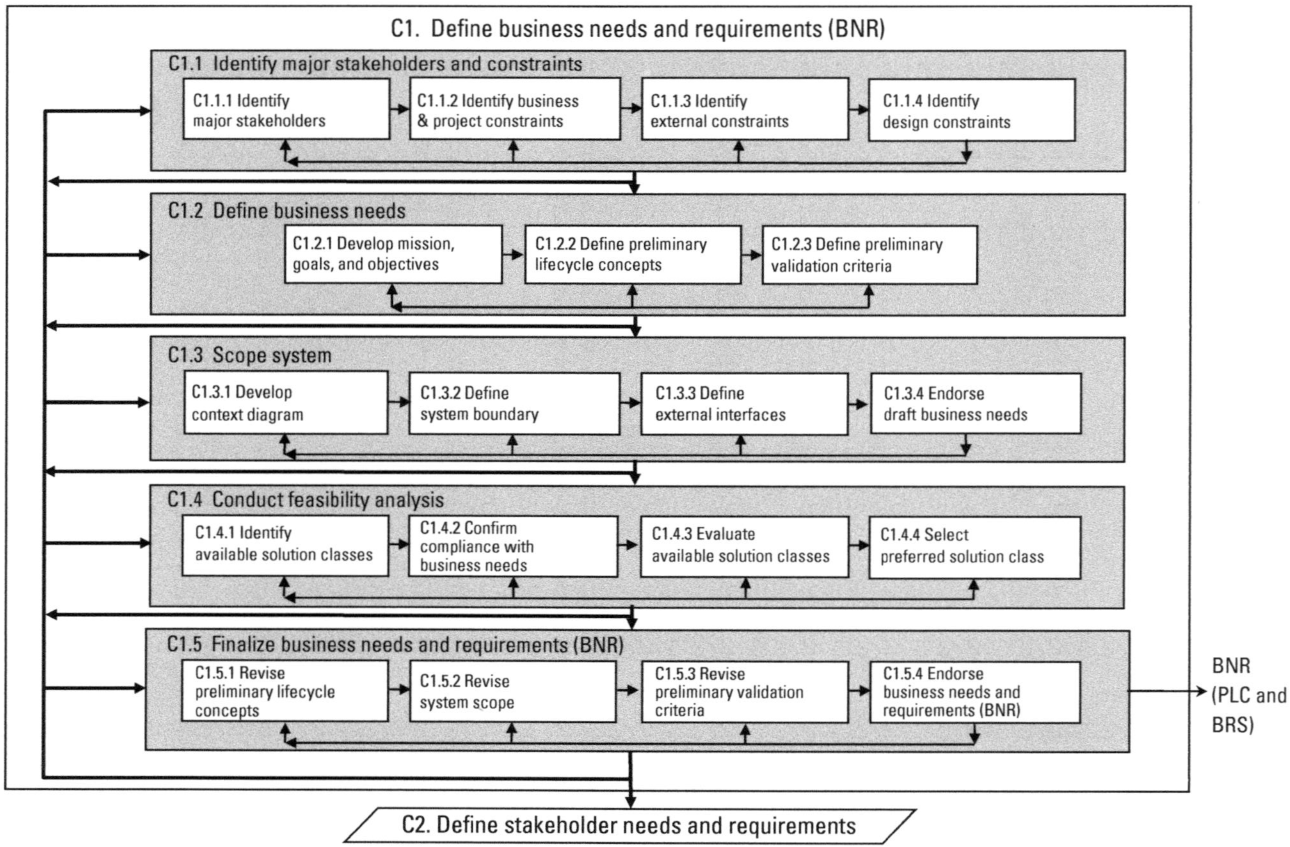

Figure 3.2 A summary of C1. Define business needs and requirements.

3.2.1.1 C1.1.1 Identify Major Stakeholders

The first step in developing the BNR (Figure 3.2, step C1.1.1) requires the identification of the major stakeholders for all aspects of the new system including those that will be the source of any constraints that the business wishes to place on the system and its development. These stakeholders are the individuals and groups from the business units at the organization's operational level that business management identifies as those who will be responsible for writing and agreeing to the BNR prior to their endorsement. These major stakeholders also have a significant role during later activities to assist with the requirements analysis effort; that is, they will most likely lead the development of the SNR.

Since their input into the system life cycle is crucial, careful selection of appropriate stakeholders is fundamental to the success of the project. It follows that, if a different set of stakeholders is nominated, the process will most likely end with a different set of system requirements.

A stakeholder is commonly defined as someone who has a stake in the project; that is, someone who is affected by the system in some way or can affect the system in some way. However, the identification of stakeholders cannot involve simply listing those who have, or perceive themselves to have, a stake in the project. In most systems, this is not a useful definition because it is often difficult to find someone who is not affected by the system in some way. Even in a simple system such as an automatic teller machine (ATM) network for a bank, there may be millions of stakeholders by such a definition. This is also true in committee-based organizations such as public service organizations, where the number of potential stakeholders is almost limitless.

Conceptual design cannot progress with such a definition of stakeholders. First, if there are so many, there is not sufficient time to engage with everyone who will have an involvement with the new system. Second, it assumes that anyone affected by the system is a source of requirements; the businesses competitors will be affected, for example, but we are not going to ask them what they would like to see in the new systems (although we would always take them into account, albeit as adversaries). Additionally, it would mean that anyone who wants to have a say should be able to influence requirements; conceptual design would therefore be at the mercy of the stakeholders who want to contribute requirements, whether they are useful or not. Finally, the new system will have disparate effects on stakeholders; they will not be affected equally, and they should not therefore have equal rights in expressing requirements. No complex system can be optimum to all parties concerned, nor can all functions be optimized [1].

The first step of identifying stakeholders is therefore much more complicated than simply listing all those who could be considered to have a stake in the new system. Business management must restrict stakeholders to the owners of the problem and of the solution presented by the new system. While the definition of needs and requirements must consider anyone who is affected by the new system, just because they are affected does not necessarily mean that they are stakeholders. More usefully, a stakeholder could be defined as some individual (or some group or organization) who has a right to influence the outcome of the system, rather than someone who is simply affected by the system. However, note that a stakeholder is also invariably affected by the system.

This first act of stakeholder identification is therefore crucial; business management must decide which of the potentially very large number of actors who are candidate stakeholders (those that are affected by or can affect the system) are to be considered as stakeholders (those that are to have the right to define the system) for the system of interest. They must also decide on what leadership approaches are to be taken to address individuals or groups that may well be disenfranchised by the implementation of the new system (such as the bank tellers in our ATM example); such actors will certainly be affected by the system but will not be nominated as stakeholders.

Example 3.1: Potential Stakeholders for an ATM System

For example, when considering the design of an ATM system, the armed guards who distribute the cash around the system are clearly important actors within the system; however, they are not stakeholders, even though their requirements place considerable constraints on the design of a number of elements of the system. Similarly, there are a number of parties (such as the tellers whose employment may be threatened by the proliferation of ATMs and other competitor banks), who are clearly affected by the system, but they are not necessarily stakeholders.

All these actors must be accounted for by the system, but, other than the right to a safe workplace and fair working conditions, they have limited right to mandate any other aspect of the system. Consequently, the actors must be considered and consulted but their input to the requirements for the system must be moderated by the business, not taken down and implemented without scrutiny.

Example 3.2: Potential Stakeholders for the Aircraft System

There are a number of potential stakeholders within ACME Air who are a major source of information relating to the functional and performance needs of the aircraft system. Corporately, the business owners of the airline have a set of more commercial needs for the aircraft, such as fuel efficiency and passenger and freight capacity. Users, including air crew and ground crew, personnel who will eventually operate the aircraft, will have a significant input. Marketing people articulate the features necessary in the aircraft system in order to make the aircraft marketable to passengers. Maintainers influence the design to ensure that the aircraft is maintainable within the time and cost constraints set by the business case. Accessibility, reliability, and maintainability will be high on the priority list of the maintenance representatives. Logistics and supportability issues such as the provision of spare parts are also likely to influence design decisions.

In the first instance, these stakeholders will be represented by their business units for the development of BNR. Lower-level stakeholders make their major contribution during the development of SNR.

3.2.1.2 C1.1.2 Identify Business and Project Constraints

Constraints are requirements that are imposed on the system by circumstance, force, or compulsion. Therefore, they absolutely limit the options open to a designer by imposing immovable boundaries and limits [2]. Before focusing on the details of the desired system, it is therefore essential to identify the business and project constraints (Figure 3.2, step C1.1.2) that are relevant to the system and its acquisition; a system will fail if it cannot operate under the constraints imposed upon it. This analysis provides essential information about the development environment for the system and begins the top-down approach to system development.

Business constraints include management guidance, organizational policies, procedures, standards, or guidelines that guide system development and procurement. These constraints can include partnering relationships with other companies, use of established life-cycle processes, contracting policies, human resource limitations, budget restrictions, and specific management guidance to the project. Such constraints must be articulated early because it has a significant effect on the acquisition and the design of the remainder of the system.

Project constraints include budget and schedule constraints but also include the resource allocations within the project as well as any externally imposed deliverables and acquisition timeframes. Many companies have business-wide standards for processes such as quality assurance and systems engineering and these methodologies guide the way that projects can operate. Additionally, projects might be constrained to conform to specified engineering and technical standards; mandated toolsets; metrics; documentation sets, artifacts, and plan templates; technology use; and control and reporting mechanisms.

Example 3.3: Example Constraints for the Aircraft System

An example of a business constraint from ACME Air may be a policy of only buying engines from a particular manufacturer to ensure that through-life support issues are simplified across the entire fleet. This constraint must be articulated early, as it has a significant effect on the procurement.

Project constraints will include budget and schedule but will also include cash-flow constraints (availability of funding), location(s) of the work, preferred partnering arrangements, project methodology, project team size and location, and many more.

Additionally, bounds such as all-up mass may be a stated design constraint for an aircraft system if it is to meet weight-related restrictions at certain classes of airfield.

3.2.1.3 C1.1.3 Identify External Constraints

In addition to enterprise-imposed constraints, there are wider external constraints (Figure 3.2, step C1.1.3) on system development that arise from the requirement for conformance to national and international laws and regulations, compliance with industry-wide standards, and ethical and legal considerations. Other external constraints include the requirement for interoperability and the capabilities required for interfacing with other systems. Additionally, the capability of competitors, as well as the availability of human resources, specific skill sets, technologies, and tools might provide external constraints. Again, an important aspect of top-down design is to understand these constraints while considering needs and requirements.

3.2.1.4 C1.1.4 Identify Design Constraints

Design constraints (Figure 3.2, step C1.1.4) include those factors that directly affect how the system is designed and the way the design is conducted. Typical constraints include the state of the art of relevant technologies as well as extant methodologies and tools to assist in the design, development, construction, and production of the system. Such issues must be considered by business management who will address the risk associated with embracing new technologies, the impact of upgrading construction facilities, workforce reskilling issues, and so on.

3.2.2 C1.2 Elicit Business Needs

Having established the context by identifying major stakeholders and constraints, the next activity is to elicit business needs. The major activities are the development of the mission, goals, and objectives; preliminary operational scenarios; preliminary validation criteria; and the preliminary concepts to be encapsulated in the PLC.

3.2.2.1 C1.2.1 Develop Mission, Goals, and Objectives

It is axiomatic that every project should begin with a concise statement of the mission of the system, which is elaborated by statements of the upper-level goals and objectives for the system (Figure 3.2, step C1.2.1). The mission statement should be stated in a single sentence, although it must have a word or phrase for every important aspect of the system. While business management often finds it difficult to state the mission in a single, short sentence, the project is invariably doomed to failure if the owners of the mission cannot articulate it succinctly at the outset. Interestingly, it will often take a small number of participants several hours to craft an agreed mission statement; therefore, the process is useful not only to start the design process, but also to serve as a mechanism for developing among stakeholders a common understanding of the system. That it could take several hours to agree to a mission statement should also sound a warning that the project is doomed to considerable confusion if the stakeholders continued, each with a differing view of what the system is to be.

The discipline of limiting the mission statement to a single sentence is important because an inability to express the mission in a single sentence tends to translate into an inability to constrain the scope of the project sufficiently to enable subsequent requirements analysis. The urge to write two sentences in English is normally driven by two separate, or at least loosely coupled, thoughts (which probably relate to two separate, or at least loosely coupled, solutions). For similar reasons, conjunctions should also not be used in the mission statement (except to connect adjectives and adjectival phrases). As well as being contained in a single sentence, the mission should be short (the sentence should be read in a single breath) because any convoluted concepts in the mission will invariably translate into a convoluted set of ambiguous requirements.

Once drafted, the mission statement is then expanded and qualified by short declarative statements of the system goals and objectives. Goals are normally relatively broad statements, each of which spawns a number of more specific objectives (although in business planning, these are sometimes treated in reverse order and objectives are considered to lead to goals).

We saw in Chapter 2 that requirements are represented in a hierarchy, first in a logical hierarchy in the BRS, the StRS, and the SyRS and then in a physical hierarchy in the subsystems and CIs in the development specifications. We also saw that traceability is assisted by a hierarchical numbering of requirements.

Backward traceability from children to parent requirements leads to a useful observation related to heredity, in that the highest-level requirements also ought to be traceable to some source. Therefore, it follows that there is a single parent statement that serves as the highest-level requirement statement, back to which all subsequent requirement statements should be able to be traced. Then this is the principal role of the mission statement in systems/requirements engineering; the mission is the parent statement that provides the seminal point for requirements engineering and is the ultimate system requirement [3] from which all other requirements flow [4].

Once again, it is clear that the mission statement performs an essential role as its contents must be sufficient to start the system design. While it is elaborated further by the high-level requirements that are articulated subsequently, the mission statement cannot be at odds with them, nor can it contain information that is not then elaborated on by at least one subsequent requirement.

That is, in the hierarchy, the mission statement provides the base statement from which the high-level needs and then requirements are drawn. Therefore, the mission statement must be a complete description of the system and contain something pointing to every function that is to be performed by the system.

To extend the hereditary relationship, it follows that the parent mission statement should have the same desirable properties as those requirement statements that are ultimately its children. We saw in Chapter 2 that requirement statements should be necessary, singular, correct, unambiguous, feasible, appropriate to level, complete, conforming, and verifiable. Therefore, it follows that the parent statement, the system mission statement, should have similar properties if they are to be inherited.

Of these desirable properties for a system mission statement, the properties of necessary, correct, feasible, and verifiable are related to the content of the statement and must be considered carefully during its drafting (addressed in the next section). The properties of singular, unambiguous, appropriate to level, complete, and conforming are similarly a matter of the content, but are also assisted by a focus on an appropriate format for the statement. The attainment of these latter properties is greatly assisted if the following guidelines are used for the format of the system mission statement:

- *A single sentence.* To ensure that the system mission statement is singular, it must be expressed in a single sentence. In good written English, a sentence may be defined as encapsulating a single thought; if we want to express two separate thoughts, a second sentence is normally required. The discipline of writing a single sentence to describe the system mission therefore keeps us at the level of abstraction that defines the system as a single thought, which is the level of abstraction we are seeking at this stage in the project in order to define a unique endeavor. If we feel the need to start a second sentence, we run the risk of adding a second "thought," which may well be describing a second project.

- *No conjunctions.* In addition to being expressed in a single sentence, the system mission statement should not contain any conjunctions if it is to be singular. The statement cannot state, for example, that the purpose of the system is to "do this. . . and . . . do that. . . ." The presence of the "and" in the statement implies that there are two purposes and therefore two systems. Having said that, we can have conjunctions in phrases and adjectival clauses. For example, we can say that a clause in a mission statement proposing a system that is ". . . comfortable and secure . . .," but we cannot have a system that provides "a medical service and riding training. . ." because two (very different) systems are implied. Note that we could have a system mission statement that is for "GP services and dental services and physiotherapy services . . .," but only because we can (and therefore probably should) logically group those functions into a mission statement for a "medical services." Generally, therefore, we should not have any conjunctions because, if elements of the same type are joined by "and," there could be a group noun or adjective that can replace them.
- *Contain no more than 5 to 7 concepts.* The system mission statement cannot contain any more than 5 to 7 key elements if it is to remain within the intuition of the reader [5] and to remain unambiguous.
- *Include an "in order to" clause.* The system justification, or rationale, can be made explicit by the inclusion of an ". . . in order to . . ." clause at the end of the system mission statement. This clause then ties the system mission statement back to the business case for the system.
- *Avoid physical terms.* To ensure that the subsequent system mission statement remains in the problem domain and is independent of the physical implementation means, the elements of the mission should always be couched in logical, not physical, terms. The system mission statement must not imply any particular physical solution.
- *Rely on iteration.* It does not follow that a perfect mission statement can be written in a single pass of the following process. The understanding that stakeholders and designers have of the system must increase as the design process progresses; regardless of when it occurs,

this increased understanding must be able to flow back into previous artifacts. A temporary lack of detailed understanding should not be allowed to halt the design process. If unsure of any element, continue and come back to the stumbling block when a better understanding has been developed.

Example 3.4: Mission Statement for the Aircraft System

For the ACME Air example, the mission statement for our aircraft might be: ". . . to provide a medium-sized aircraft that can provide class-leading comfort to passengers between Class X airfields on domestic and international routes."

In this example, the goals and objectives would then need to elaborate on such matters as what is meant by "class-leading" and "medium-range," how many passengers are to be carried, and operational issues such as crewing and maintenance.

Having satisfied ourselves that we have a draft mission statement that is a suitable start point for the design (recognizing that we will invariably return later and modify the mission statement as we understand the system of interest better), we can now begin the process of identifying the requirements that are explicit in the draft mission statement (that is, the mission statement can be decomposed into those requirements) and those requirements that can be inferred (that is, derived) from it. We have now begun the elaboration process that continues throughout the remainder of the design.

Once we have decomposed or derived the required goal elements from the draft mission statement, we can begin to group these elements into goals. Note that this is the start of the allocation and grouping process. We cannot have a large number of goals, or else the design will be unnecessarily complicated at this very high level; it is desirable to have approximately 5 to 7 goals so that the design at this level stays within our intuition.

Once we have established the goal statements and formally numbered them, we revisit them to confirm (at our present level of understanding) that the goals are balanced and complete. The properties of a good goal statement and the principles for its construction are similar to those discussed previously for the mission statement. We adjust it again if necessary and revisit the draft mission statement to confirm that it is still appropriate (and adjust it if necessary). We may again choose to confirm the draft mission and goals by communicating them to the broader community of stakeholders, not just those immediately involved in their development.

Our next task is to elaborate the objectives for each goal by continuing the decomposition and derivation process. Again, a bottom-up process can be used within a top-down framework. Starting with Goal 1, we decompose or derive from it the objective elements required to be performed to achieve that goal. We then continue the hierarchical elaboration for Goal 2 and the subsequent goals.

Once we have developed a first draft of the objectives for each goal, we need to group and allocate them to provide a balanced structure of objectives. Any redundancy or overlap here will result in redundancy and overlap in lower-level requirements if we continue to elaborate those requirements without first grouping and allocating objectives here. The objective elements are not discarded once we have the goals because they form the next level in the elaboration process. The hierarchical numbering of the goals and objectives continues (and starts the hierarchical numbering system that will end in the SyRS requirements at the end of the process).

The decomposition, derivation, grouping, and allocation of objectives that we have discussed so far are only the beginning of the process, which will continue until a complete set of balanced objectives has been derived. Should any element of the design require more detailed consideration, trade studies should be commissioned to inform any design decisions and to provide subsequent rationale for the requirements (we discuss trade studies in more detail in Chapter 4).

Again, the objective statements are initially brief (three-word, verb-based phrases) to allow them to be readily understood; once iteration is complete, each should be written as a full sentence. The properties of a good objective statement and the principles for its construction are similar to those discussed previously for mission and goal statements.

Once we have established the objectives, we revisit the goal statements to confirm (at our present level of understanding, at least) that the goals are balanced and complete. If necessary, we adjust them and also revisit the draft mission statement to confirm that it is still appropriate (and adjust it if necessary). See [6] for a more detailed description and some examples of mission statement development.

We may again choose to confirm the draft mission, goals, and objectives by communicating them to the broader community of stakeholders, not just those immediately involved in their development. Once we have confirmed the artifacts with stakeholders, we finalize the mission, goals, and objectives into a formal set of statements that provide a complete, balanced description of the system, covering all aspects of operation, support, and maintenance (albeit at a high level of abstraction). The set is the starting point for the system design back to which every subsequent requirement must be traceable.

Example 3.5: Mission and Goals for the Domestic Security Alarm System

For the ACME Domestic Security Alarm, the mission statement might be: ". . . to provide a market-leading alarm system that can deter, detect, classify, and report unauthorized entry to a residence in order that the resident is aware of the residence's state of security," which could then be supported by the following goals that can be decomposed from the mission:

- *Deter unauthorized entry.*
- *Detect unauthorized entry.*
- *Classify unauthorized entry.*
- *Report unauthorized entry.*
- *Provide market-leading features.*

3.2.2.2 C1.2.2 Define Preliminary Life-Cycle Concepts

Having defined a preliminary operational concept (OpsCon), business management must give some guidance on the other life-cycle concepts (Figure 3.2, step C1.2.4) related to the acquisition, deployment, operation, support, and retirement of the system. While the systems engineering procedures that follow will ensure a life-cycle focus, it is important that there is an early business focus on the major cost drivers that will impact on all aspects of the system throughout its life. Additionally, there are a number of life cycle-related trade-offs at a business-case level that need to be concluded to inform subsequent considerations (perhaps providing constraints).

It should be noted that the life-cycle concepts must be considered in the context of achievement of the capability required to meet operational scenarios (which is why the preliminary operational scenarios and concepts are developed before the other preliminary life-cycle concepts). For example, the support aspects of a system must not be optimized in their own right without consideration of the operational use of the system. For example, the lowest-cost through-life support for an aircraft is achieved by minimizing fuel consumption and reducing wear and tear that requires maintenance, in short, by not operating it all. Therefore, clearly, through-life support must be optimized based on the desirable operational scenarios, not independent of them. Importantly, however, the business must also recognize that through-life support optimized for the operational scenarios may not necessarily be optimal for any other scenarios, should they change their operational context in the future.

Example 3.6: Life Cycle-Related Trade-Offs

Preliminary life-cycle concepts are continuously tested and adjusted, if necessary, throughout subsequent design as the system is understood in more detail. Consider the example of a need for an inexpensive, dashboard-mounted Global Positioning System (GPS) for a large trucking company. Since the item is to be procured in large numbers at a low cost, the business case may originally deem that it would not be cost-effective to implement a repair system for defective items, which could be discarded when broken. That preliminary concept may well be adjusted during subsequent, more detailed consideration that there would still be a need to inspect "broken" items before being discarded to ensure that they were indeed broken, not just subject to problems such as finger-faults or sticking buttons that could be rectified simply. If technical resources are to be applied to inspection, perhaps the additional cost of slightly more detailed training may allow some items to be repaired rather than discarded, resulting in a reduction in TCO.

Before looking briefly at the content of each of the PLC, it should be noted that those concepts are indeed preliminary and will be adjusted over time as more is understood about the system and the life-cycle issues. However, the value of the preliminary investigations is to understand the options available in each concept so that they can be iterated and adjusted to optimized as a set.

Preliminary OpsCon

There are a number of excellent references that provide guidance on process, content, and layout of an OpsCon. Perhaps the principal guidance is in ANSI/AIAA G-043A-2012 [7] and ISO/IEC/IEEE 29148 [8], which suggest that the OpsCon contains the following sections:

- *Scope.* This section provides an overview section describing in broad terms the organizational context, the proposed capability system, and the system's intended purpose within that context.
- *Glossary and referenced artifacts.* A glossary is essential to ensure that terms are used consistently throughout the OpsCon (and other LC). It is also critical to keep a record of the source of information in the OpsCon in case of conflicting and ambiguous requirements or some disagreement between stakeholders.
- *Proposed system operational overview.* This section identifies the needs (business and then stakeholder needs) for the system, including the major mission(s), how these will be performed, techniques used, and

any constraints. Each mission is investigated to determine common needs in order to develop an RBS based on the system mission, goals, and objectives. The section addresses proposed operational environments as well as any operational policies and constraints on the system. The section also includes personnel requirements derived from the types of personnel involved in the system, as well as consideration of operational support issues (in concert with the support concept).

- *Operational scenarios.* A description of possible operational scenarios should be provided to help describe the likely events facing the system (these are often summarized in the body of the OpsCon and outlined in detail in an annex).
- *Existing system(s).* This section outlines the existing system(s) to the extent that they are relevant to the acquisition of the new system, including the current major mission(s), how these are performed, major stakeholders, and any current constraints.
- *Nature of (and justification for) proposed changes.* The existing system(s) (to the extent relevant to the acquisition of the new system) are examined to identify the shortcomings that lead to the needs for the proposed system. There are many types of justification for a proposed system: there may not be a current system in service, the current system may be beyond economical repair, there may be new operational needs for which the current system was not designed, and so on.
- *System scope.* This section includes a description of the systems context diagram, external interfaces, significant constraints, system modes and states, and any architectural considerations that need to be addressed.

The preliminary OpsCon (as with all PLC) is written by business management and contains as much detail as they wish to communicate regarding (and sometimes constrain) the solution, which is to be refined by stakeholders at the business operations level.

Operational Scenarios

Once the mission, goals, and objectives have been articulated, business management in conjunction with stakeholders from business operations identify the operational scenarios (Figure 3.2, step C1.2.2) relevant to the system. Scenarios help stakeholders to explain how the system will be used. These scenarios, associated with what are called use cases or vignettes or use stories, provide valuable guidance to the system designers and also form the basis of major validation events in the acquisition phase such as operational testing of the system as it is introduced into service. Despite any more detailed technical verification and validation procedures, the system's fitness for purpose is fundamentally related to its ability to perform in accordance with the operational scenarios defined at this stage.

In many cases, it is also useful to define the various modes of operation for the system products under development. Designers need to understand that, if the system exists in different modes, even if it is as simple as the difference between the fully operational mode and the training mode. Complex systems may have their needs and requirements stated in a number of modes. For example, a modern fighter aircraft may have modes defined for air-to-air combat, ground attack, reconnaissance, naval operations, nontactical flights, air-show deployments, and so on. Each mode must be associated with the particular conditions (such as mission, operational, environmental, and configurational) that define it.

In our aircraft example, a number of modes may be defined for international and domestic operation including taxi, take-off, cruise, approach, landing, turn-around, and so on. Modes may also be defined for maintenance and for administrative movement of the aircraft.

Operational Scenario Development

Operational scenarios are developed using the following general guidelines.

- First, the general operational environment for the system is described to identify all of the environmental factors that may influence the operation of the system (these may or may not have been identified in the development of the mission, goals, and objectives). This scenario is often called "A day in the life of . . ." because it describes a typical day or cycle of system use by all involved. This general scenario is then used as the context for more detailed scenarios.
- It is not possible to develop all possible scenarios from all possible perspectives for all possible stakeholders, so consideration must be given to the manner in which a suitable (minimal) set of scenarios is to be developed. While the smallest number of scenarios is sought, all of the system requirements must be covered, and all relevant actors must be described. Additional details on the development of a minimal set of operational scenarios were shown earlier.
- Specific operational scenarios are then described to depict the full range of circumstances under which the system is required to operate. It is not necessary to describe every possible scenario, but all types of operation must be represented. There must be at least one scenario that represents the perspective of each major stakeholder.

- Where necessary, the exception conditions of each scenario are outlined. For example, in a scenario that describes the use of a movie theater by patrons, care must be taken to identify how the scenario would vary in exceptional conditions such as an emergency in the theater (such as a fire or power blackout) or a disruption such as a faulty projector.
- There must be as little duplication between scenarios as possible. It is often of assistance in minimizing repetition to group scenarios in a hierarchy where the lower-level scenarios inherit the properties of those at the level above.
- Scenarios also tend to be more useful and more easily interpreted if a standard structure is used as a broad template for each scenario.
- Scenarios must be consistent in the use of terms for functions and actors; this consistency is greatly assisted by the development of a supporting glossary.

At this stage in the process, the operational scenarios are preliminary scenarios in that they are relatively high level and provide broad guidance from the business management perspective in terms of how the system is to fit into the business and the organizational context. The details of the preliminary scenarios are to be fleshed out by the stakeholders at the business operational (business unit) level when the SNR are developed.

Functional Flow Block Diagrams (FFBD)

A possible way to define, or at least support, operational scenarios is through the use of FFBD as described in Chapter 2. FFBD are developed for the main purpose of structuring the scenarios into a series of logical flows that both serve as descriptions of system use and provide excellent communications tools to allow all involved to understand readily how the system is to be employed. The FFBD at this level identify the major system needs in accomplishing its mission. Each top-level block is expanded as necessary to ensure that the functionality of the system is adequately defined. This may require the functional analysis to break the system down further than the major system-level functions even at this early stage. The scenarios should be elaborated (decomposed or derived) as far down as necessary to achieve this aim. Subsequently, the FFBD will help the designers to evaluate each function adequately and will also assist in the definition of requirements (in the BRS, StRS, and SyRS).

Developing a Minimal Set of Operational Scenarios

As observed earlier, it is not possible to develop all possible scenarios for all possible perspectives for all possible stakeholders; there would be too many that would take too long to capture, and there would simply not be enough time in the project to complete such an exhaustive writing or modeling task. Not only that, but there is no guarantee that such a large volume of scenarios is even useful in determining needs and requirements. Yet we need sufficient scenarios to elaborate the sufficient set of concepts from which we can transform the appropriate needs and requirements.

So consideration must be given to the selection of the smallest set (the minimal set) of scenarios that is sufficient to cover all actors and all system requirements. Development of the set of scenarios must therefore start with an analysis of which scenarios constitute the minimum set. One way to do so is to look in advance for potential repetition and overlap as well as commonalities in scenarios. For example, rather than writing a scenario for each flight ACME Air intends for the new aircraft, they might develop a scenario for a "normal flight" to describe the operation of the aircraft as it moves between airfields, covering all activities for aircrew and passengers between the shutting of the aircraft doors at one airfield and opening them at another. They might also develop a "normal turn-around" scenario to describe the actions taken by air crew, ground crew, and passengers between doors opening and closing again. They would then develop normal scenarios to describe maintenance, refueling, support, training, and so on.

Then, having developed a number of general normal scenarios, ACME Air may develop a number of exception scenarios for situations that deviate from the normal scenarios. For example, it could be a normal deviation such as a slightly different procedure at one airfield, or it could be an exception as a result to an incident during flight or an emergency on the ground or in the air. Exception scenarios are normally written so that that can be inserted into, and inherit the process flow from, normal scenarios.

The development of operational scenarios is assisted by the presence of standard operating procedures (SOP) or doctrine that describes the organization's operational procedures and processes. Should such SOP exist (and be desired in the new system), the operational scenarios do not have to be so elaborate or detailed but simply have to provide the context and rely on the SOP to provide the detail.

Preliminary Acquisition Concept

Based on the operational needs for the proposed system, business management outlines the proposed acquisition concept. Any business needs for acquisition are articulated, such as budget and schedule, preferred (and perhaps prohibited) supply sources, types of contracts desired, relevant existing contractual arrangements, any relevant existing acquisition support arrangements, program/project management considerations, reporting issues, and the relationship of the system of interest to other system acquisitions within the organization. This preliminary concept is continually refined: first, after the feasibility analysis undertaken in Figure 3.2, activity C1.4, and then after stakeholders have begun to consider their needs and requirements; and then, at every major review as the design continues.

Preliminary Deployment Concept

In this concept, business management outlines the issues surrounding the deployment of the system (to the extent that they are concerned with the introduction into service of this system). For example, business management may dictate the nature of transition arrangements between an existing and replacement system, since business continuity may well be at risk during this period. Additionally, the preliminary deployment concept may address issues such as transition between facilities, training of operational personnel (and possibly engaging new staff and retiring others), and transition support arrangements.

Preliminary Support Concept

The preliminary support concept will capture the needs of business management for the support of the system throughout its life cycle. Issues addressed would include relevant support policies and procedures, any relevant existing support arrangements, the support environment, desired maintenance levels and cycles, and the anticipated impact of the proposed system on support facilities, equipment, personnel, and training.

A number of support services are provided (by what is sometimes called the support system) during the utilization phase. The constituent capabilities that provide those support services are acquired initially during the acquisition phase and then refreshed during utilization.

- *Operating support capability.* This capability encompasses the support elements of operating facilities system operators, support equipment, operator manuals and technical data, operating support procedures, and operating support information systems.
- *Engineering support capability.* Engineering support encompasses the support elements of engineering facilities, engineering personnel, engineering support and test equipment, engineering technical data, engineering processes, engineering information management system, and software support.
- *Maintenance support capability.* Maintenance support is the capability that includes the support elements necessary to develop, establish, and

integrate a maintenance support system capable of sustaining a system throughout its life. The capability comprises the elements of maintenance facilities, maintenance personnel, maintenance support and test equipment, maintenance technical data, maintenance processes, and maintenance information management system.

- *Supply support capability.* Supply support encompasses the following support elements: supply facilities, supply personnel, supply support equipment, supply technical data, supply processes, supply information management system, spares, and packaging.
- *Training support capability.* The training support capability includes the support elements: training facilities training personnel, training equipment, training materials and other technical data, training processes, and training information management system.

It should be noted that the life-cycle concepts must be considered in the context of the capability required to meet operational scenarios (which is why the preliminary operational scenarios are developed first). In particular, the support aspects of a system must not be optimized in their own right without consideration of the operational use of the system. For example, the lowest-cost through-life support for an aircraft is achieved by minimizing fuel consumption and reducing wear and tear that require maintenance, in short, by not operating it all. Clearly, therefore, through-life support must be optimized based on the desirable operational scenarios, not independently of them. Customers must also recognize that through-life support optimized for the contracted operational scenarios may not necessarily be optimal for any other scenarios, should they change their operational context in the future.

Preliminary Retirement Concept

The preliminary retirement concept addresses the retirement phase, which is the final stage in the system life cycle. Functions associated with phase-out and disposal include transportation and handling, decomposition, and processing of the retiring system. If considered early, disposal and phase-out issues will form some of the criteria against which the system is designed. However, it is important that system designers focus on retirement, rather than the more limiting issues of disposal; planning for disposal is important, but a system may retire from a number of life cycles before its ultimate disposal.

Section 7.3 provides more detail on the considerations of the retirement concept.

3.2.2.3 C1.2.3 Define Preliminary Validation Criteria

An important top-level activity is the identification and definition of preliminary validation criteria (Figure 3.2, step C1.2.3). Broadly, validation criteria encompass any mechanism by which the acquirer will measure satisfaction with the products of the acquisition phase. Key criteria include performance in each operational scenario, safety, reliability, supportability, maintainability, ease of use, and time and cost to train, among others.

There are a number of issues that need to be considered when identifying validation criteria:

- *Measurability.* The selected measure clearly has to be measurable to be useful. That does not always mean that there has to be some quantitative measurement possible; there may be a qualitative assessment. What has to be defined is the way in which the measurement is to be conducted. A quantitative measurement is normally straightforward to define (such as the measurement of length and mass or the aggregation of events). A qualitative measure might require a questionnaire providing responses on some scale (such as 1 to 10) to indicate degree.
- *Baseline.* For the measurement to be meaningful, it is often useful to be able to identify the change in the measurement, rather than the absolute value at any particular time. Most measurements therefore must be made at the start of the project and then at selected times throughout the project so that some change can be calculated or inferred; that is, we would prefer to derive measurements that reflect our progress as well as our current position.
- *Relevance.* A measure must be able to be related to at least one or more requirements. If that is not possible, the measure is seeking to observe a function that is not part of the design; either the measure is irrelevant or a function is missing.

For a measure to be relevant, it must be related to one or more requirements. One common way of doing this is to develop a matrix that relates measures to requirements. The use of relationship matrices is very common in systems engineering; we have traceability matrices that show inheritance of requirements, we have verification matrices to show how verification means relate to individual requirements or groups of requirements, and we could have matrices to relate measures to requirements.

The use of matrices is common because it is often perceived that the entities related by the matrix are orthogonal; for example, there is a hierarchy of requirements, and there is a hierarchy of measures. However, orthogonality in engineering and mathematics implies independence of variables, which is not always the case in the matrices that we develop in systems engineering. In the case of requirements, we have developed performance statements and verification statements for each requirement. If we are looking for a measure of effectiveness or performance, we already have the basis in the information related to each requirement. Rather than developing the requirements database and then stepping sideways to develop independently the measures (and then have to develop a matrix to relate them), it would make more sense to use the same functional, performance, and verification requirements structure to develop the measures.

In doing so, as illustrated in Figure 3.3, it is common to develop a hierarchy of measures, such as: critical strategic issues (CSI) that relate to the measurement of system mission and goals, critical operational issues (COI) that relate to the measurement of objectives, measures of effectiveness (MOE) that relate to the next level of the requirements hierarchy, measures of performance (MOP) at the next level, and down to verification statements at the lowest level.

- *Critical strategic issue (CSI).* A CSI is a measure of an aspect of the system that is of primary importance to business management and must be met before the system is allowed to proceed, either to the next development phase or into service. CSIs are commonly phrased as questions that must be answered in the affirmative.

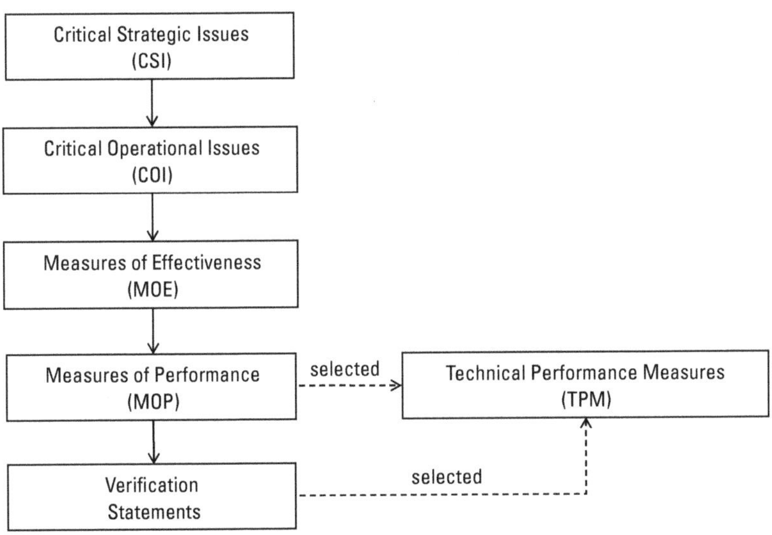

Figure 3.3 A typical measures hierarchy.

- *Critical operational issue (COI).* A COI is a measure of an issue that must be examined to determine whether a system meets its mission; COIs are selected subsets of CSIs. COIs are also commonly phrased as questions that must be answered in the affirmative.
- *Measure of effectiveness (MOE).* An MOE is a measure (quantitative or qualitative) of the degree to which a system objective is met under specified conditions. MOEs address the effectiveness of the system in terms of both how well it accomplishes its mission and how well it integrates into the organization and operational environment (that is, how well it meets its interfaces). MOE are supported by sibling measures of measures of suitability (MOS), which measure subjective suitability of the system to the organization. The difference is often clouded, and MOE can be qualitative, so here we avoid the complication and include such measures as MOE.
- *Measure of performance (MOP).* An MOP is a measure of performance that is one of a subset of measures that combine to form an MOE. Examples of MOP include performance measures for range, velocity, frequency range, and all-up mass.
- *Verification statement.* It is good practice to ensure that, every time a requirement is articulated, a corresponding verification statement is made. While the common practice of attachment of a letter to illustrate the verification methods is a useful indication of the customer's intent, it is much better practice to write a verification statement to make the verification methods more specific. That is particularly true for the analysis and test verification means, because the use of a single letter begs a number of important questions that really should be answered at the time the requirement is written.
- *Technical performance measure (TPM).* It is normally not possible for an acquirer and developer to jointly track each verification statement or even each MOP. Therefore, selected MOPs and verification statements are nominated as TPMs that are used to assess conformance throughout the development process. Commonly, TPMs are reported across the contractual boundary from the developer to the acquirer as a means of providing confidence to the acquirer that the contract is on track.

As discussed in Section 3.4.4, selected MOPs and verification statements are nominated as TPMs that are used to assess conformance throughout the development process.

Note that MOEs are related to their parent COI and grandparent CSI by a hierarchical system (just as objectives are related to goals). As illustrated in Figure 3.4, there then should be a direct relationship between the requirements hierarchy and the measures hierarchy.

At this stage in the process, the criteria being developed are upper-level, preliminary criteria (probably at the level of CSI and COI) that will be refined further during the development of SNR (which will address MOE, MOP, verification criteria, and TPM); see Section 3.3.2.4.

3.2.3 C1.3 Scope System

The next phase in understanding business needs and requirements involves the development of an understanding of the scope of the system development effort (called scoping). This scoping activity helps to establish a clear understanding of what the system is expected to do by defining the system context, the system boundary, and any external interfaces.

3.2.3.1 C1.3.1 Develop Context Diagram

To assist with the scoping process, a tool called a context diagram (Figure 3.2, step C1.3.1) may be used to illustrate the interrelationship among related systems, relevant regulatory environments, major stakeholders, external systems, external interfaces, and so on. The context diagram is a very useful tool

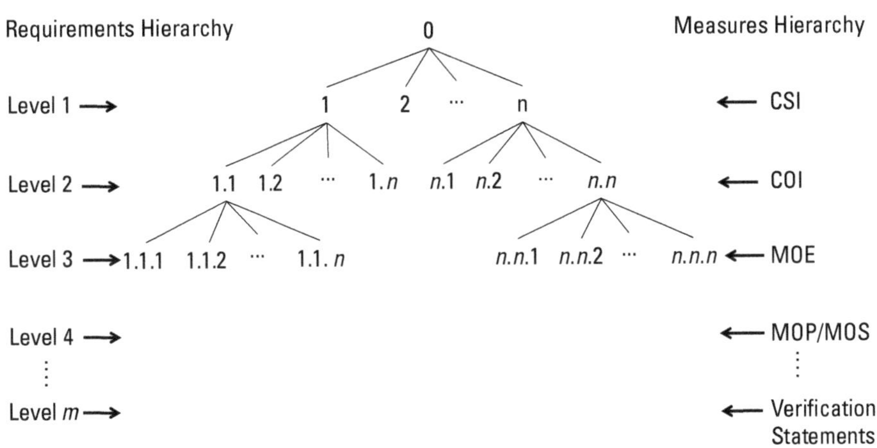

Figure 3.4 The relationship between the requirements hierarchy and the measures hierarchy.

because, in this case, a picture is definitely worth a thousand words. In systems engineering, there is a common heuristic that, if you cannot draw a picture of the system, or you cannot explain it in 5 minutes, it is either broken or you do not understand it [9].

The exercise of developing the context diagram is also very useful; it is not uncommon at the start of a requirements workshop to take a day of discussion to complete a suitable context diagram. Although the task seems simple enough at the outset to most stakeholders, it quickly becomes apparent that very few of them have a comprehensive overview of the system, its interfaces to the external environment, or its boundary.

There are a number of different forms of context diagram in use as part of the various design methodologies; in this book, we propose a context diagram that is fairly free-form, with a format depending on the nature of the system under investigation. The context diagram is best developed bottom-up within a top-down construct:

- The process begins with a simple list of all the possible candidate elements for the context diagram. This can be produced as a textual list in a word processor, as a list on a whiteboard, or perhaps as separate pieces of sticky paper on a board or wall.
- Next, similar elements are grouped in accordance with their relationship with the system, that is, in terms of their impact on it, or their interface with it.
- The groupings of external elements are then drawn as goose eggs. The system is shown as a goose egg in the center of the diagram; nothing inside the system is shown because the system's internal elements have not yet been defined (and will not be until system elements are selected in preliminary design). In a simple context diagram:
  - Two overlapping goose eggs represent an interface. An interface is invariably where something crosses the system boundary; the overlap shows that the element is sometimes outside the system boundary (an external element) and, at other times, is inside the boundary (an internal element). As we discuss in more detail in Section 3.2.3.2, this is an important first step in the identification of the system boundary (Figure 3.2, step C1.3.2) and the formal process of external interface definition (Figure 3.2, step C1.3.3) and the identification of system-level interface requirements.
  - An arrow between two goose eggs illustrates an influence of one element over the other (in the direction of the arrow) or a constraint placed on one element by another. Just as the overlap of

elements prompts interface definition, the identification of these influences and constraints confirms the enterprise, project, external, and design constraints considered in Figure 3.2, steps C1.1.2, C1.1.3, and C1.1.4.

- Using this draft diagram, stakeholders move their focus around the boundary of the system verifying each relationship as it is depicted. The diagram is adjusted as necessary, and the process continues until the diagram is complete.

Again, at this stage, we still have a preliminary understanding of the system and attendant life-cycle issues. The context diagram is also therefore preliminary but is a great vehicle for capturing the current view of how the system works and then to ensure that view is common across, and accommodates, the views of all relevant stakeholders. The diagram, as with other aspects of the PLC, will mature as understanding matures.

If the context diagram becomes complicated, it may be useful to separate the components out into a number of context diagrams illustrating more than one view of the system. For example, there could be a context diagram for:

- *Operational view.* In this view, we would expect to see all external elements that relate to the system as it goes through its operational day. In particular, we would expect to see any actors involved as well as any regulations and constraints. The OpsCon (and particularly the operational scenarios in Figure 3.2, step C1.2.2) should be reflected in this view.
- *Acquisition view.* Before the system comes into operation, the acquisition view shows the relationships of the acquirer, the project office, the contractor, and subcontractors, as well as any facilities to be used in acquisition such as test facilities. The acquisition concept from Figure 3.2, step C1.2.2 should be reflected in this view, which will reflect the statement of work (SOW) requirements in the contract.
- *Support/sustainment view.* This view would include training, support contractors, maintenance, supplies, and facilities required to support the system. The support concept from Figure 3.2, step C1.2.2 should be reflected in this view.

If the project is to have multiple phases during acquisition, there could even be one context diagram for each phase. In some cases, the system exists quite differently in different phases of operation, so very different context diagrams can result.

Example 3.7: Simple Context Diagram for Our Security Alarm Example

Figure 3.5 illustrates a simple context diagram for the operational view of our domestic security alarm example; the acquisition and support (particularly maintenance) views would be shown on a separate diagram. In reality, the diagram would be a little more detailed, but this example shows the main elements of a useful diagram. Additionally, even this operational diagram would best be split to show the normal state of operation (when the intruder is not present) and the exception case (when the resident is absent, but the intruder is present).

- Alarm. *It is obvious from the context diagram that the alarm is entirely contained within the larger house system since one goose egg is a subset of the other. It is also obvious from the way that the alarm subset is drawn that the designers accept the prospect that the alarm might not cover 100% of the area covered by the house system (since the boundaries of the two are not coincident). So, in this case, the context diagram shows a relationship between two elements; one is totally subsumed by the other.*

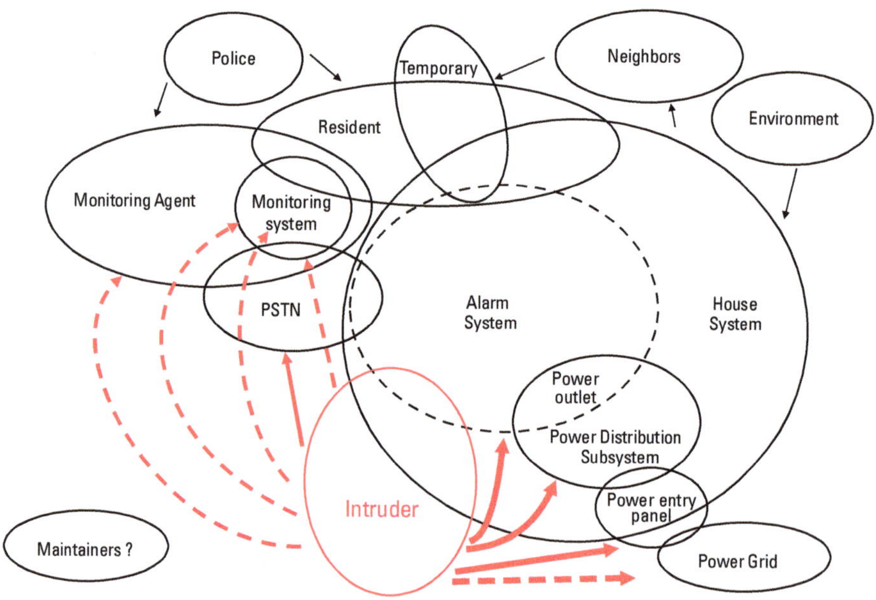

Figure 3.5 A simple context diagram for the operational view of our domestic security alarm example.

- Resident. *The resident has a number of interfaces evident on the context diagram, shown as overlapping goose eggs. There are the physical interfaces in which the resident first enters the house and then enters the sphere of influence of the alarm. In this case, the context diagram shows an interface of the resident with the house and the interface of the resident with the alarm. However, more than that, it shows that the resident's interface with the alarm occurs after their interface with the house; that is, the resident has no relationship or interface with the alarm until they are inside the house. Also evident is the interface between the resident and the monitoring agent, which is shown as being independent of the alarm system. Further, the diagram shows the notion that there might be temporary residents (guests of the residents, for example) who will interface with the alarm (in the same way as a resident) for the period during which there are guests.*
- Intruder. *The intruder has a physical interface that is very similar to the resident (both are humans who may enter the house and then the alarm boundary by a number of means), so it can be concluded that the alarm must be able to distinguish between the two in some manner. Note that, in addition to the physical interface, the diagram also shows a number of influences that the intruder may have on the alarm and house systems; the solid lines show the influences of most interest in this design (disabling alarm sensor functions or their power source, disconnecting house power, disconnecting the phone line) and those that the threat analysis considered less likely for a domestic burglary (disabling the suburban power grid, logging across the phone line to mimic the alarm, hacking into the monitoring system, paying an accomplice inside the monitoring agent to disable or ignore the alarm, or destroying the monitoring agent). These latter potential influences (dotted lines) are considered out of scope for this system.*
- Monitoring agent. *Note that there is an expectation in this design that there is a monitoring agent (one does not necessarily have to exist in an alarm system) and that the mandated interface is via the Public Switched Telephone Network (PSTN), rather than a more modern version of alarm where a cellular data service might be incorporated. In this case, the context diagram assumes the presence of an element (the monitoring agent) and the use of a technology (the PSTN); both assumptions are clearly obvious on the diagram, which makes the context diagram a very useful communications device since such assumptions should be surfaced and tested as early as possible. We test this assumption in the next example.*
- Power. *The diagram shows the expected interface of the alarm to the house power subsystem; note that, because a power outlet is identified explicitly,*

we can infer that the alarm is not hardwired into the power subsystem (perhaps by a licensed electrician), but instead is able to be plugged in by the resident so that the alarm can be transported between houses.

- Police. *The context diagram illustrates that the police do not have any direct interface with this system; for example, they are not expected to respond when the alarm is triggered but may respond to any incident as a function of their normal policing duties.*
- Neighbors. *Again, neighbors are shown to have no particular role in the alarm system but are affected by it (perhaps by the noise it may make) and may influence the resident (perhaps to complain about the noise).*
- Maintainers. *The diagram shows that there has not yet been a decision regarding the way in which maintenance is to be conducted. Such an arrangement, once known, will be able to be reflected in the diagram, or perhaps in a separate support context diagram since operations and maintenance are unlikely to be undertaken at the same time.*

The context diagram assists in capturing stakeholder needs and requirements, but it is also useful to assist in performing the necessary feasibility analyses and upper-level trade-offs at the business level. That is, the context diagram is a useful visual design artifact that allows stakeholders to visualize the system at an abstract level to allow them to make decisions at that level that will then drive (constrain) the subsequent considerations.

Example 3.8: Example Upper-Level Trade-Offs Using the Context Diagram for Our Domestic Security Alarm Example

Even at this early stage, there are a number of upper-level trade-offs that can be conducted for our security alarm example. When developing the context diagram, the stakeholders included a monitoring agent connected to the alarm via the wired telephone system (the PSTN). Before progressing and transforming those implied needs into system requirements, there are some decisions to be made. For example, is the PSTN still relevant; after all, how many households have a fixed phone line these days? Would an internet connection or a mobile phone interface not be more relevant? If so, which one? Or both? Or should we remain technology agnostic at this early stage?

Similarly, why is there a monitoring agent? Although the business originally requested a back-to-base option that was an essential element of such an alarm in the past, but, in a modern system, the alarm could report directly to the resident when they are anywhere in the world (on the internet, by message via mobile phone, or both). Would the business like to reconsider its original need now that there are more details available?

The value of the context diagram is that it brings such issues to the fore to ensure that they are considered sufficiently at a high level of abstraction before they are simply accepted as given.

3.2.3.2 C1.3.2 Define System Boundary

Definition of the system boundary (Figure 3.2, step C1.3.2) is essential early in the acquisition phase so that it is clear which elements are included in the system and which are outside. This activity also identifies which aspects are inside the systems acquisition and which are not, which is also particularly important to the project manager who is principally responsible for defining project scope; that is, what is to be included in the project as well as what is to be excluded. In that sense, the context diagram is an essential tool in the project management activity of scope management.

The boundary of a system is normally straightforward to describe in physical terms (such as a fence line or external building walls), but it is often necessary to describe the boundary in conceptual or logical terms as well. Additionally, although we traditionally describe the boundary in terms of what is inside the boundary, that is, what is included inside the system, it is often useful to describe what is not inside the boundary. This is particularly useful when there are a number of elements that may commonly be assumed to be part of the system but are not in this particular case.

In Figure 3.5, the boundary of the alarm is shown by the dotted line. Note, in this example, that the boundary shows that the alarm system is totally contained within the house system. Additionally, the two goose eggs are drawn such that there does not seem to be any intention that the alarm covers every part of the house; the extent of coverage will be defined in subsequent analysis.

The boundary is also important in the next activity, which is to define the external interfaces, that is, those interfaces between the system of interest and to those other external systems to which it is interconnected.

3.2.3.3 C1.3.3 Define External Interfaces

External interfaces are between the system of interest and each of the other existing or future external systems (Figure 3.2, step C1.3.3) to which it is interconnected. The interfaces describe the inputs and the outputs of the system, the interfaces to its external environment. The specification, development, test, and management of these external interfaces will place considerable requirements on the system. While the external systems are not directly related to the current SOI or the scope of the project that will bring it into being, the success of the fielded system is often dictated by its ability to interface to those systems in its external environment.

The definition of an interface requires considerably more detail than simply identifying and naming the interface. Broadly, there are three main steps in interface definition:

- *Interface description.* The interface is given a name, a short title, and a unique identifier. The nature of the interface is described in terms of who, what, when, where, why, and how.
- *Interface impact analysis.* The interface is analyzed in terms of its impact on the system of interest. In particular, any constraints imposed by or on the system are identified. Additionally, a risk analysis is conducted to determine the impact of the interface on the operation and design of the system.
- *Interface control analysis.* Each external interface must be analyzed to determine the extent to which it can be controlled so that designers and operators of the system are not at the mercy of its external interfaces (noting the requirement to obtain agreement from the owners of systems on the other sides of external interfaces).

Figure 3.6 shows how the external interfaces might be enumerated on the context diagram for our domestic security alarm example. The nomenclature will be defined by the organization, but the example shows one simple method of listing each type of interface, from E01 to E10. Each of those interfaces becomes the source of external interface requirements for the system.

Figure 3.6 also shows examples of secondary interfaces: E08, E09, and E10. These are not explicitly interfaces to the alarm because they are outside the alarm's boundary. However, it is essential that the alarm is aware of the actions of the intruder in interfering with the phone line (and hence affecting E05 and E06) or the external power connection (and hence E08).

3.2.3.4 C1.3.4 Endorse Draft Business Needs

At this point, business management should pause to review and endorse the draft business needs as articulated in the artifacts produced so far (predominantly in the PLC) (Figure 3.2, step C1.3.4). Because the next major activities are related to the transformation of those business needs into formal requirements, it is important that the business is satisfied that their stated needs will result in satisfaction of the business case for the system. This endorsement is most likely undertaken through a review conducted by the business owner in the presence of business management and the major stakeholders from business operations.

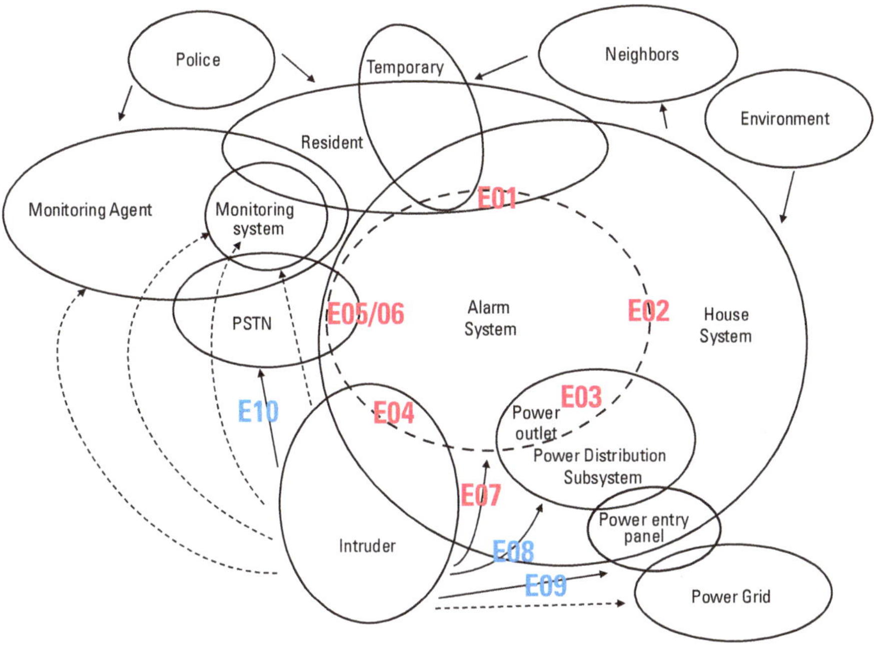

Figure 3.6 External interfaces enumerated on the context diagram for our domestic security alarm example.

3.2.4 C1.4 Conduct Feasibility Analysis

Once business management has defined their needs, the next major activities are related to the transformation of those needs into formal requirements, perhaps in the form of the BRS in accordance with ISO/IEC/IEEE 29148. This last step is really only useful if the business intends to use those requirements to engage a design partner to complete the design, especially if the engagement is via a formal contract. Otherwise, the process can continue with the business needs, which can then be combined with the stakeholder needs in what the INCOSE *Needs and Requirements Manual* [10] refers to the integrated set of needs (as an alternative to the development of the BNR/SNR and BRS/StRS).

At this stage of conceptual design, business management have stated their needs in logical terms, that is, in terms of the problem domain (the management and operational environment of the business). However, for any problem domain, there is normally a range of potential classes of solutions within the solution domain. Each alternative solution class may represent a completely different type of project; therefore, the feasibility analysis is an essential step to narrowing the solution domain so that the subsequent project can be managed effectively. If we continue without narrowing down the design into one of those classes, we will have great difficulty coping with the breadth of the design. Consequently, we next undertake a feasibility analysis, the aim of which is to select the desired solution class, which is then to be elaborated further by the stakeholders at the business operations level.

Business requirements are generally couched in "business" or "capability" terms and are elicited first so that we can understand what is required by the business of the deployed system. Requirements analysis elaborates these BNR into SNR and then goes on to transform SNR into system requirements. The idea of a feasibility analysis is to pause after identifying BNR to determine what possible system-level alternatives are available before going on to conduct detailed requirements analysis and eventually to establish the system requirements. In other words, there is most likely a variety of broad classes of system that might meet the upper-level BNR and it is often very useful to choose one of these before proceeding to determine stakeholder needs and requirements, or else the SyRS that would result would be far too general to be useful (either for subsequent design or, more particularly, as a centerpiece of the SOW for a contract). Not only that, but the choice among solution classes is invariably a business choice and should therefore be made before time is wasted in elaborating classes of solution that the business would not wish to embrace.

Once the BNR have been articulated, the feasible system-level alternatives available to meet those needs can be investigated. Note that does not mean that the business must choose an alternative after feasibility analysis; they may not be able to. It is just that, if a feasible alternative can be chosen, the remainder of the process will be much more concise and focused.

Feasible alternatives are considered in terms of available resources such as money, time, personnel, and materials. As illustrated in Figure 3.2, the following steps are completed during feasibility analysis:

- *Step C1.4.1.* Identify the alternative system-level solution classes capable of satisfying the BNR.
- *Step C1.4.2.* Confirm compliance of each solution class with the BNR and, if not completely compliant, note the likely level of achievement.
- *Step C1.4.3.* Evaluate the alternative solution classes in terms of feasibility, performance, effectiveness, technical and project risk, and other selected measures.
- *Step C1.4.4.* Select the best of the alternative solution classes, ensuring that the options are narrowed down as much as is practicable at this stage.

A great deal of care should be taken during this stage of the project, as the feasibility analysis sets the direction for the rest of the system life cycle. To that end, the feasibility analysis should address the entire life cycle, not just the operational aspects of the system.

There may be an opportunity to examine the marketplace to identify existing systems that may be assessed to assist with feasibility analysis. Assessing operational systems may also assist with the refinement of the mission, goals, and objectives of the new system based on a realistic understanding of what is currently available. The feasibility of using existing systems to meet emerging needs may be assessed as a possible way of reducing the schedule and technical risk associated with the system development. The process is sometimes called concept demonstration or technical demonstration.

Example 3.9: Solution Class Selection—Family Vehicle(s)

Consider a family of four who are also avid hobby farmers on a small block of land at the edge of town. They have a family car, but they also need a way to transport significant amounts of waste to the rubbish dump as well as gardening supplies such as fertilizer, garden soil, turf, bricks, and so on.

Briefly addressing the steps of a feasibility analysis:

- *Step C1.4.1. Identify alternative system-level solution classes:*
  - *Fit a towbar the family vehicle to tow a trailer (buy or hire when needed).*
  - *An additional utility vehicle (buy or hire when needed).*
  - *Replace family car with a large utility vehicle that has a dual cab to accommodate four people and a sufficient tray to carry garden waste and supplies.*
- *Step C1.4.2. Confirm compliance with BNR. Each of the candidate solutions is suitable in terms of meeting the need for transporting humans and material. Note that we will need to consider all the BNR before we can make a choice: is there sufficient parking space available for the two-vehicle option, what is the total cost of ownership of each option through its life cycle (or perhaps compared over 10 years), are sufficient funds available for an outright purchase of a second vehicle or is a loan required, and many more. Note that, at this level, the decision is a business decision: what is best for the family business (not which vehicle one family member might prefer to drive).*
- *Step C1.4.3. Evaluate alternative solution classes. Criteria would be established and each solution class would be evaluated against those criteria (noting that the first two solution classes have two ways of being implemented and may be considered as separate solution classes).*
- *Step C1.4.4. Select the best of the alternative solution classes.*

Example 3.10: Feasibility Analysis—Solution Class for the Aircraft System

ACME wants to service selected short to medium domestic and international routes but does not have the airframes to do it. Based on this simple statement of business needs, the feasible alternative solution classes might be:

- *Leasing or buying an existing aircraft;*
- *Leasing or buying and modify an existing aircraft;*
- *Contracting to have a new aircraft developed;*
- *Outsourcing the operation to another airline;*
- *Not servicing these medium routes.*

Therefore, there is a wide range of possible solution classes for ACME in replacing its existing aircraft. Each of the alternatives is a completely different type of project; therefore, feasibility analysis is an essential step to narrowing the solution domain so that we can manage the subsequent project. It is also an important step for the business to confirm for itself which type of solution it prefers.

Then, for each approach, there is still a range of subordinate alternative system solutions. It may be desirable to establish another feasibility analysis to further narrow down the solution space.

The feasibility analysis narrows the solution domain in order to reduce the scope of the resultant requirements that are recorded in the BRS. It should also be noted that the narrowing of the solution classes to one preferred class is a business management decision and not one that should normally be made by stakeholders at the business operations level, nor by project management, nor by systems engineering.

It should be noted that one result of the feasibility analysis could be that no preferred system solution exists (or exists at a sufficient level of maturity or within the organization's budget or schedule) and the remaining process may not continue because the business case cannot be met at the current time.

The feasibility analysis and selection of the resultant desired solution class may require the draft business needs to be revisited in light of the more detailed investigations of the feasibility analysis. The hierarchical representation of business needs (mission, goals, objectives) is now further elaborated (decomposed or derived) and formalized into a balanced set of business requirements, which are recorded in the BRS.

The development of the business requirements in the BRS can be assisted by a requirements framework around which the logical description of the system can be based. Here we call the requirements framework the requirements breakdown structure (RBS). The words are deliberately chosen to differentiate this structure from the well-known project management artifact called the work breakdown structure (WBS); the RBS is grouped by function, the WBS is structured by physical work packages, including the configuration items in the systems breakdown structure (SBS) and contains other project-related work. We demonstrate the relationship between RBS and WBS in Chapter 4.

In addition to being a useful requirements analysis tool, the RBS can be further refined to reflect stakeholder requirements in the StRS and then the system requirements in the SyRS. When completely populated, the RBS for a system provides an outline of the SyRS.

Example 3.11: Example RBS for the Domestic Security Alarm Example

> *A representative requirements framework for our domestic security alarm example is shown in Figure 3.7 showing the detail completed at the BNR/SNR (integrated set of needs) level.*

There are a number of advantages associated with the establishment of a suitable framework such as the RBS early in the requirements analysis activity. First, the framework will act as a reference source during requirements analysis to ensure that all aspects of the system requirements are addressed and that important areas are not omitted. The framework is ideally suited to the

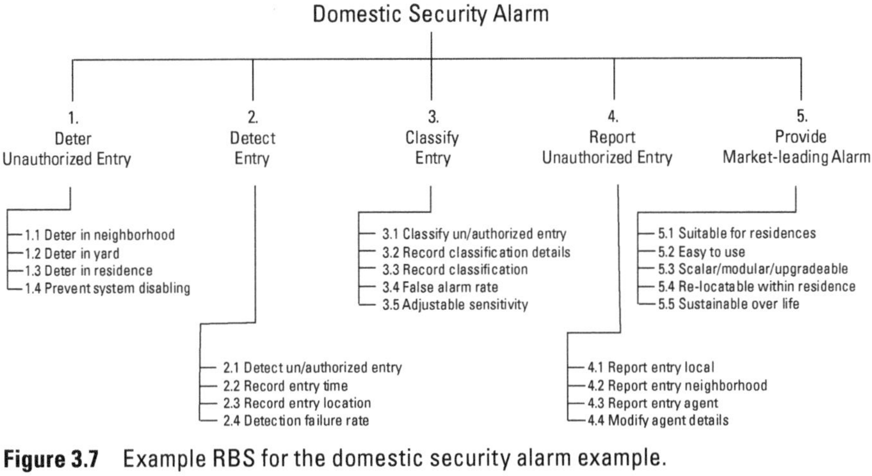

Figure 3.7 Example RBS for the domestic security alarm example.

top-down approach and provides confidence at any level that the needs and requirements at that level are complete, unambiguous, and nonredundant. The framework allows multiple people to work on the analysis at the same time as it allows the effective allocation of responsibility to individuals for sections of requirements. By assisting in a hierarchical decomposition and derivation of requirements, the RBS also assists in avoiding duplication of requirements in different sections. Duplication is undesirable because, at best, it raises the probability of conflict between requirements and often leads to ambiguity and confusion and, at worst, duplicate requirements may end up being instantiated in a duplicated manner when inadvertently allocated to different system elements during preliminary design.

3.2.5 C1.5 Finalize Business Needs and Requirements (BNR)

At this stage, the business now understands the system much better than it did when it first drafted the mission statement; not only has much more detail been elicited, but also a solution class has been identified, which provides a much more concrete view of the system and its operation. Consequently, the BNR can be finalized by revisiting earlier artifacts and revising them to incorporate the increased understanding.

The preliminary life-cycle concepts are revised (Figure 3.2, step C1.5.1), as is the solution scope in terms of the mission, goals, objectives, context diagram, boundary, and external interfaces (Figure 3.2, step C1.5.2) and the preliminary validation criteria (Figure 3.2, step C1.5.3). Finally, the BNR are endorsed (Figure 3.2, step C1.5.4) by business management, most likely at a major review.

3.3 C2. Define Stakeholder Needs and Requirements

The endorsement of the BNR signifies that business managements are confident that they have defined the problem domain in sufficient detail to communicate their needs and requirements to the stakeholders at the business unit (operations) level. These stakeholders then go on to develop their needs and requirements (the SNR) within the context of the needs and requirements and constraints of the BNR.

Figure 3.8 illustrates the activities involved with the process of definition of SNR resulting in the development of the life cycle and the StRS. Note once again that the process is iterative, allowing current understanding to improve the quality of previously generated artifacts. The activities of Figure 3.8 are described in more detail in the following sections.

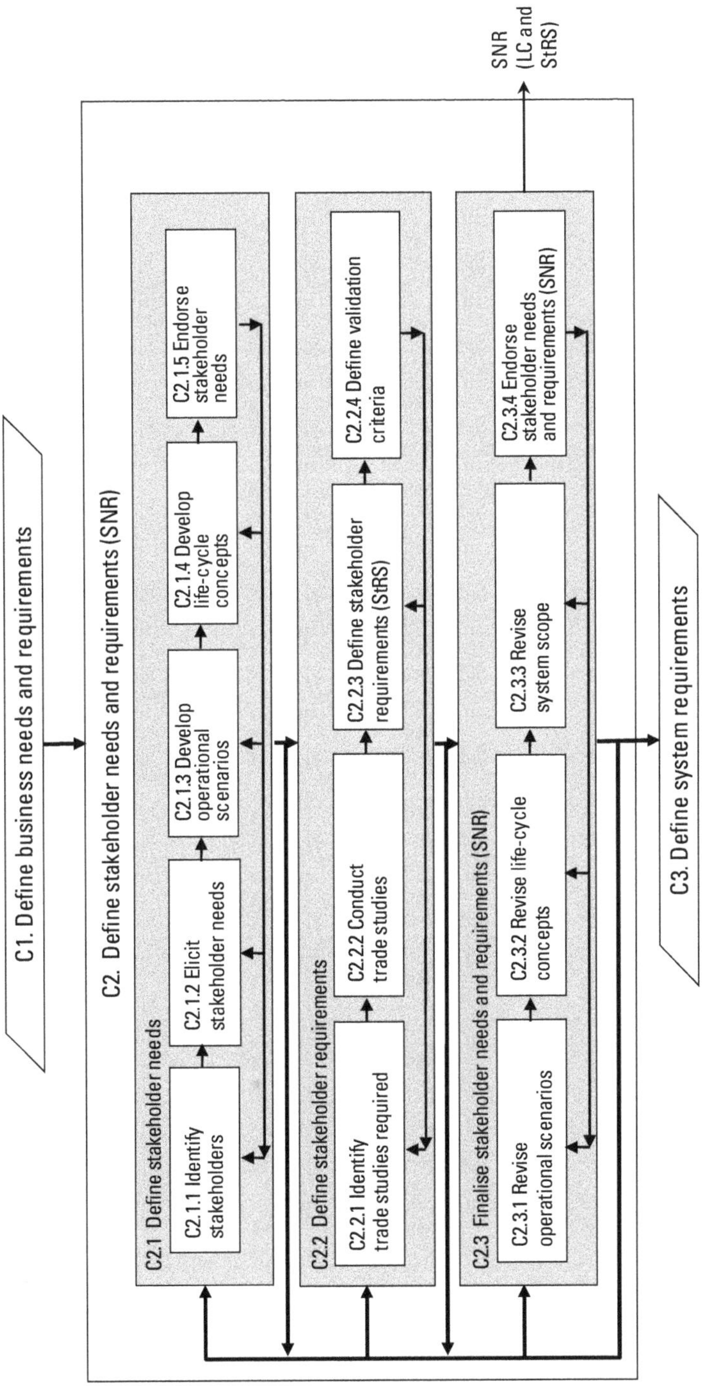

Figure 3.8 A summary of C2. Define stakeholder needs and requirements.

3.3.1 C2.1 Define Stakeholder Needs

Once the BNR have been confirmed within the preferred solution class, the nominated stakeholders at the business operations level can begin to articulate their needs and requirements (Figure 3.8, step C2.1). Although the major stakeholders were nominated by business management during the definition of BNR, the first act in the process of defining SNR is to refine those stakeholders by naming individuals within the major stakeholder groupings and adding lower-level stakeholders at the operations level (Figure 3.8, step C2.1.1).

Within the context set by the BNR, stakeholders at the business operations level articulate their needs through the refinement of the mission, goals, and objectives (Figure 3.8, step C2.1.2). The preliminary operational scenarios defined by business management are now defined in more detail, and other lower-level scenarios (often involving use cases or vignettes) are developed as the design progresses (Figure 3.8, step C2.1.3). Similarly, the other PLC are developed further and fleshed out with details from the various stakeholders to be more complete concepts in the life cycle (Figure 3.8, step C2.1.4).

Stakeholder needs are endorsed first by the stakeholders at the business operations level and then by business management (Figure 3.8, step C2.1.5). Again, endorsement is most likely achieved through a major review.

3.3.2 C2.2 Define Stakeholders Requirements

Once the stakeholder needs have been endorsed, the process of transforming their needs into formal stakeholder requirements is undertaken. A hierarchical structure is a useful framework for refining the BRS into the detail of the StRS while also assisting in the maintenance of traceability.

3.3.2.1 C2.2.1 Identify Trade Studies Required

In a similar manner to the feasibility analysis that business management used to narrow down the alternative solution classes, stakeholders at the operations level will have a number of alternative ways in which their needs can be met. The selection between various options is best undertaken by a formal trade (trade-off) study. The first activity (Figure 3.8, step C2.2.1) lists those trade studies that are required, groups related trade studies for efficiency, prioritizes studies, and allocates appropriate resources.

3.3.2.2 C2.2.2 Conduct Trade Studies

As noted in the previous section, the conduct of trade studies did not begin here (the feasibility analysis conducted earlier can be considered to be a system-level trade study) and will not end here as trade studies are an important design tool and are used throughout system development and the remainder of the life cycle. Trade studies and their conduct are described in more detail in Chapter 4.

3.3.2.3 C2.2.3 Define Stakeholder Requirements (StRS)

The development of the StRS is a first step towards successful system development and is given a number of titles, including simply stakeholder requirements or user requirements. A reader of the StRS should be able to understand completely the likely applications or missions for which the system is intended, the major operational characteristics to be exhibited by the system, the operational constraints that limit the design and development of the system, the external systems and interfaces with which the system under development must operate, the operational and support environment within which the system must exist, and the support concept to be employed to support the system and enable it to continue performing in accordance with customer expectations.

Although the StRS is a formal requirements artifact, it must be written in the language of the stakeholders to ensure that the artifact is accessible to all stakeholders. However, as it is critical that the StRS is comprehensive, readable, unambiguous, and subsequently easily traceable, the language must still be formal. In many cases, the difference between the StRS and the SyRS is marked because the user is operationally focused and uses very little technical language. In other cases, such as our aircraft system, the stakeholders are groups such as air crew, ground crew, maintainers, and engineers, all of whom will use relatively technical language in their StRS statements. However, the difference is always that StRS statements are high-level and are made in more natural language, which is generally much more easily read and understood than the language of formal specifications.

Example 3.12: StRS Language

In our aircraft development example, the StRS will contain ACME Air's intentions in the language of the stakeholders. For example, there may be such statements as:

- *The Aircraft shall operate from a Class X airfield.*
- *The Aircraft shall provide class-leading comfort for passengers.*
- *The Aircraft shall be turned around to its next flight within 30 minutes.*

Note that the StRS requirement statements are similar in construction and language to those that will follow in the SyRS, but they do not have to be as precise and the verification requirements can be less stringent. That is, for example, class-leading comfort in our aircraft example is not directly verifiable, but it is sufficiently bounded for the next level of design (conducted by the system designers) to understand what quantitative statements must be made that, when implemented, will lead to "class-leading" comfort.

The StRS also refers to the whole aircraft capability system, that is, the aircraft, air crew, ground crew, simulators, support systems, training systems, and so on. As noted illustrated in Figure 3.9, each element of the capability will most likely be procured separately and even at a different time, so, while the StRS covers the capability system, it may well lead to several SyRS, one for each input to the capability.

The potential audience for the StRS includes everyone who has an interest in the delivered system, from the eventual users, operators, and support personnel to the systems engineers and designers responsible for designing, developing, and manufacturing the system as well as the test personnel responsible for confirming that the system is fit for its intended purpose. Each audience set provides a different view of the requirements.

While there may be many stakeholders who have an interest in the development of the StRS, the population of the StRS is generally best conducted by a small team of experienced stakeholder representatives. However, it is important that all stakeholders contribute to the preparation of the artifact, and elicitation or generation techniques such as structured workshops, surveys, interviews, and prototyping are necessary to ensure that all relevant information is considered.

Again, as for the business requirements, the StRS (as required by ISO/IEC/IEEE 29148) does not necessarily have to be prepared, and the BNR/SNR can be consolidated in the integrated set of needs (as suggested by the INCOSE *Needs and Requirements Manual*). The business may wish to engage

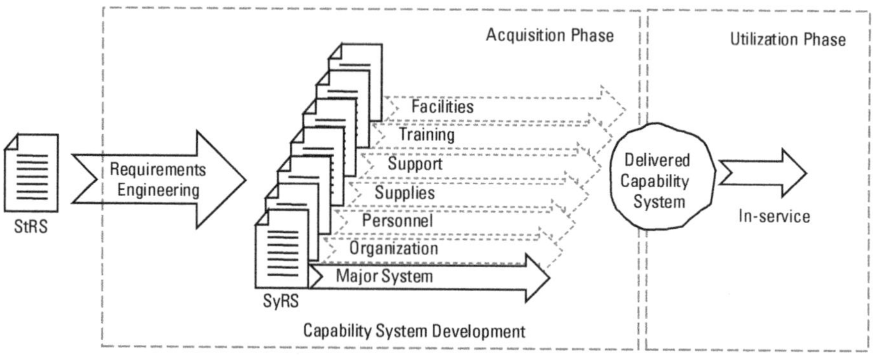

Figure 3.9 An example in which one SyRS is developed for each of the elements of a capability system.

a design partner at this point, in which case they may formalize the StRS to be used as the basis of a partnering agreement.

3.3.2.4 C2.2.4 Define Validation Criteria

Each requirement in the StRS must be accompanied by a validation statement; that is, the stakeholders must not only say what it is that they require, but also how it is they will know that the system has met that requirement. In Figure 3.8, step 1.2.3 (see Section 3.2.2.4), upper-level, preliminary criteria (probably at the level of CSI and COI) were developed and now need to be refined further to develop MOE, MOP, verification criteria, and TPM.

3.3.3 C2.3 Finalize SNR

At this stage, the stakeholders now understand the system much better than they did at the beginning of this process; the broad guidance of the BNR has been significantly refined and much more detail has been added as a result of the additional needs elicited from stakeholders and the decisions made as a result of a number of trade studies. Consequently, the SNR can be finalized by revisiting earlier artifacts and revising them to incorporate the increased understanding. The operational scenarios are revised (Figure 3.8, step C2.3.1), as are the other life cycles (Figure 3.8, step C2.3.2), and the solution scope in terms of the mission, goals, objectives, context diagram, boundary, and external interfaces (Figure 3.8, step C2.3.3).

Finally, the SNR are endorsed (Figure 3.8, step C2.3.4) by stakeholders and by business management, most likely at a major review. Endorsement of the SNR is a critical step in developing the system in order to provide confidence that all of the fundamental stakeholder requirements have been captured. In particular, endorsement of the StRS is not to be taken lightly. For requirements engineers, the StRS is the basis of further development of the requirements to populate the SyRS that forms the centerpiece of the FBL of the system. From a project management perspective, the StRS assists in bounding the scope of the project from which early cost and schedule estimates can be developed.

3.4 C3. Define System Requirements

The next process in conceptual design is the definition of system requirements, which starts with the endorsed SNR and then decomposes or derives them into system requirements. Figure 3.10 summarizes the process involved with system requirements definition.

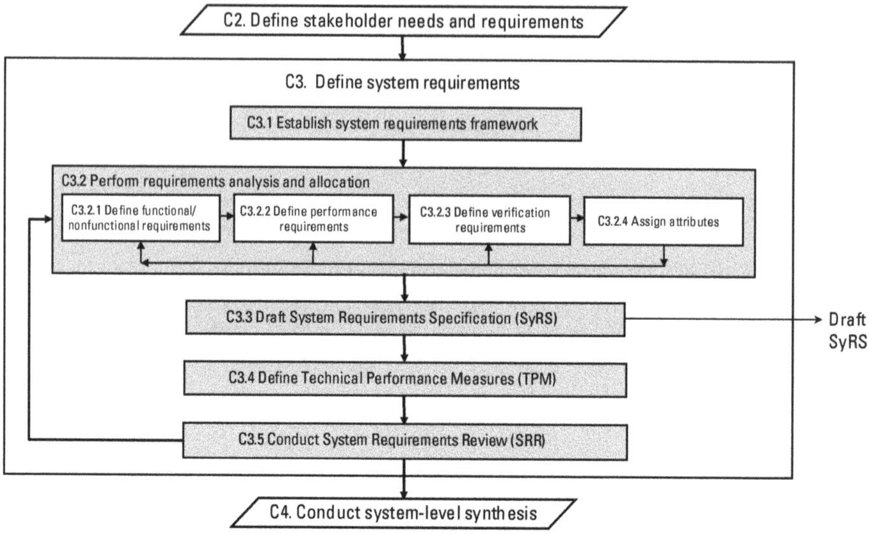

Figure 3.10 A summary of C3. Define system requirements.

The aim of defining system requirements is to determine what the system must do in order to meet the business and stakeholder needs and requirements; that is, to describe the requirements at the system level and to be able to relate the functional design back to the BNR/SNR, particularly the StRS. Therefore, system requirements definition uses elaboration to bridge from stakeholder requirements to system requirements.

In the system requirements definition, it is important to concentrate on what is required rather than how to do it. For example, a system-level performance requirement of the aircraft system might be a specified cruising speed. The physical subsystems that would be needed to perform this function (such as engines, airframes, and flight controls) may not be considered at this stage. The latter stages of the systems engineering processes focus on the physical subsystems (and lower) and how system-level requirements will be met best through a disciplined design approach.

Specifying how to meet the requirements at this embryonic stage may result in ill-informed design decisions leading to suboptimal designs. Because the customer is most likely to be performing the conceptual design tasks, specifying at this stage how requirements are to be met will also have an undesirable effect of the acquirer taking on some responsibility and risk for achieving system performance; such a risk should be taken by the developer.

System requirements must be complete and must meet all the needs of the major stakeholders. The requirements should be objective, designable, measurable, verifiable (or testable), and traceable. In contrast, system requirements are often ambiguous and incomplete, conflict with other requirements, and are not traceable to any endorsed customer requirement. Stakeholders sometimes prefer ambiguous and nonspecific system-level requirements in the mistaken belief that it does not lock them into specific requirements early in the project, or alternatively stakeholders are not sure of their real requirements. Untraceable requirements sometimes appear in an SyRS because the "needed" system described in the StRS is not the same as the "wanted" system that is described in the SyRS. Regardless of why the SyRS contains poor requirements, ambiguous and loose statements of requirement and untraceable requirements have an adverse effect on the remainder of the project and the system that is eventually produced. Time invested in ensuring that the requirements are "correct" will significantly reduce technical, schedule, and financial risks associated with the system acquisition by basing project estimates on a complete requirement set.

In Chapter 1, we saw that, because a system is often seen as an operational capability, it is commonly referred to as a capability system or an asset. We also saw that a capability system (asset) comprises constituent elements such as the major equipment solution (hardware and software), organization, personnel, collective training, facilities, data, support system (including supplies), and operating procedures and organizational policies. At this point in the design, the business owners and stakeholders must agree on how the elements of the capability will be acquired. There could be a single SyRS developed for the whole capability system or as illustrated in Figure 3.9, there could be one SyRS developed for each element of the capability system or for any combination of elements.

As outlined in Chapter 1, for the majority of the remaining chapters of this book, we focus on the acquisition of the major equipment element (often called the materiel system or the combination of mission system and support system) of the capability system. However, we must always keep in mind that all the elements must be brought back together prior to introduction into service to field an operational capability.

3.4.1 C3.1 Establish Requirements Framework

The first step in defining system requirements is to establish a requirements framework (Figure 3.10, activity C3.1) around which the logical design of the system is to be based. As noted earlier, the RBS is a useful framework for refining the StRS into the detail of the SyRS while also assisting in traceability. Since we have started with a hierarchical decomposition of mission, goals, and objectives, we can use that information to begin our development of a logical hierarchy, which can be captured in an RBS.

3.4.2 C3.2 Perform Requirements Analysis and Allocation

This activity involves the elaboration of major requirements for the system (with associated performance, verification, and rationale) including operations, maintenance, support, and other necessary functions. When we use the word system from now on, we are using the systems paradigm because, as noted earlier, there will be more than one SyRS (design-to specification) produced for each StRS (integrated set of needs) for the capability system: one for the aircraft, one for the personnel, one for training, and so on. As noted earlier, the process of requirements analysis is the bridge between stakeholder requirements in the StRS to the system requirements in each of the SyRS.

3.4.2.1 C3.2.1 Define Functional/Nonfunctional Requirements

The central step in requirements definition (and in conceptual design, for that matter) is the definition of the requirements (functional, nonfunctional, constraints, and interfaces) (Figure 3.10, step C3.2.1) for the system. While users are understandably focused on operational requirements, maintenance and support requirements also need to be defined during conceptual design.

Defining requirements (Figure 3.10, step C3.2.1) is an important part of the overall system requirements analysis that, as the name suggests, identifies the many different functions that the system needs to perform in order to meet the requirements contained in the endorsed StRS. The analysis identifies all functions including maintenance functions that are necessary to maintain or return the system to operational use. At this stage in the development, the concentration is still on what needs to be achieved, not how it should be done. FFBDs are useful tools to assist.

Operational needs and requirements are normally given the highest priority during the early stages of a project because operational aspects tend to be the most tangible and most easily understood. The life-cycle approach emphasized in systems engineering also acknowledges that design decisions made early in the acquisition phase can have a large impact on the maintenance and support of the system. To that end, maintenance and support requirements need to be determined and specified during conceptual design, so that the design of the system ensures efficient and effective support during the utilization phase.

Human factors considerations also produce a number of additional requirements. Requirements for the system are obtained from an analysis of physical space limitations, climatic limits, eye movement, operator reach, occupational health and safety issues, ergonomics, cognitive limits, usability, and so on. The assessed skill levels of known users also help to constrain the design.

3.4.2.2 C3.2.2 Define Performance Requirements

Once the functions have been identified and grouped according to the agreed RBS, performance-related parameters must be considered (Figure 3.10, step C3.2.2) that the new system must achieve. Having decided what the system must do, the designer must now determine how well the system is to perform each of those functions. A good practice is to ensure that each time a functional statement is articulated, one or more corresponding performance statements are made.

Most of the operational functions will have obvious performance parameters associated with them such as speed, accuracy, endurance, and acceleration. Support and other functions also require parametric definition to define completely the requirement. Examples of these parametric requirements include environmental considerations, quality factors, support requirements, and utilization requirements.

3.4.2.3 C3.2.3 Define Verification Requirements

The definition of requirements is not complete until verification requirements (Figure 3.10, step C3.2.3) are also included. It is good practice to ensure that each time a requirement is articulated, a corresponding verification statement is made. It is often difficult to write verification requirements for system-level functions, but the discipline of doing so is important since there is little point in stating a requirement for a function or constraint without consideration of how the function is to be verified. An additional benefit of adhering to the discipline at this stage is that test plans become much easier to write because the tests required at any level can be considered within a framework provided by an aggregation of the verification requirements at that level.

3.4.2.4 C3.2.4 Assign Attributes

As outlined in Section 2.1.3.2, an attribute is additional information associated with a need or requirements that is used to aid in its definition and management. Attributes also aid in achieving many of the characteristics of needs and requirements defined in the INCOSE *Guide to Writing Requirements* [11]. Need and requirement statements are therefore supported by associated attributes (Figure 3.10, step C3.2.4) that aid in the definition and management of needs and requirements and of sets of need and requirements [12].

For example, it is useful to record the rationale behind each requirement, which will further assist with the removal of ambiguity but, more importantly, will assist in situations where more than one person is involved in the requirements analysis over the course of the project. The rationale explains why each requirement is necessary as well as the logic behind performance levels assigned to the requirement. A requirement should not be accepted unless the rationale is well understood. This rationale remains attached to a requirement throughout the acquisition phase, providing guidance to lower-level designers as to why and how the requirement has been decomposed or derived and why the performance and verification requirements were selected.

Example 3.13: Utility of Rationale

> *An example commonly used in systems engineering circles involves the design of a back windshield of a passenger car, which had a user requirement to be able to withstand the air pressure associated with the car moving at 60 mph. During the design, it was established that the car's maximum speed in reverse gear would only be 10 mph, so designers reduced the requirement for the rear windshield to withstand air pressure from 60 mph to 10 mph. Such a decision seemed to make sense, particularly since there was likely to be a subsequent cost saving in the back windshield component. However, when the new vehicles were reversed onto a car transporter for delivery from the factory to retail outlets, the rationale for the original user requirement became readily apparent when the rear windshields were subjected to 60 mph on the transporter.*

3.4.2.5 Notes on Analysis and Allocation

The requirements articulated by the stakeholder are generally at a high level and often tend to be more qualitative than quantitative. For example, our stakeholders require that the new aircraft can operate from any Class X airfield. While stakeholders will articulate some additional requirements during requirements elicitation or generation, these broader functions need to be analyzed to identify the lower-level requirements necessary to achieve the parent requirement. As discussed in Chapter 2, these lower-level requirements are often called decomposed or derived requirements because they are not stated directly by a stakeholder but are an artifact of the requirements analysis process. For example, if the parent requirement is to be able to operate from any Class X airport, designers would decompose or derive requirements and constraints on the aircraft associated with operation from that class of airfield, which may include requirements such as minimum runway lengths, runway surfaces, and maximum allowable aircraft mass.

Requirements analysis also involves investigation of the relationships between requirements. In addition to the derivation of requirements during requirements analysis and allocation, system designers also group and allocate requirements. Allocation has three aspects. First, requirements are allocated to the appropriate group of the RBS. Second, higher-level requirements flow down to lower-level requirements. Finally, there is the logical-to-physical allocation in preliminary design as the system-level functions are allocated to the physical configuration items.

The RBS as it is developed is a very useful tool to support the design as well as to communicate with stakeholders. However, the RBS simply states the function required of the system, and it is sometimes difficult to determine how those functions relate to each other. The flow of functions is ideally illustrated by FFBDs, which are also very useful tools at this stage, both as analysis tools and as additional mechanisms for communication with stakeholders.

In performing the system requirements analysis, the following tasks are performed:

- Determine the major functions required of the new system (including maintenance, support, operational, and any other necessary functions).
- Assign the necessary performance-related parameters or requirements to each of the functions, as applicable.
- Articulate any other system-level requirements, such as those in the requirement categories shown in the example RBS in Figure 3.7.
- Determine the TPMs to be applied during the systems engineering effort.

Many tools exist to assist with the process of requirements analysis. For example, IEEE-STD-1220 [13] is a systems engineering standard that suggests a procedure for performing requirements analysis. IEEE-STD-1220 and other standards are briefly described in Chapter 6.

Example 3.14: Moving Toward System-Level Requirements in the SyRS

Three possible StRS statements relating to our aircraft example are listed briefly below, along with examples of decomposed or derived system-level statements developing the lower-level of requirements moving toward the system-level requirements in the SyRS.

1. *Operate from Class X Airfield*
    - 1.1 *Operate with a specified maximum takeoff distance*
    - 1.2 *Operate from specified runway surface*
    - 1.3 *Operate with specified maximum allowable aircraft mass*
    - 1.4 *Operate with specified essential navigation aids*
    - 1.5 *Operate with specified essential automatic landing systems*
    - 1.6 *Operate with specified essential communications systems*
    - . . .

2. *Provide class-leading comfort to passengers*
   2.1 *Provide class-leading seating*
   2.2 *Provide class-leading entertainment systems*
   2.3 *Provide class-leading bathroom facilities*
   2.4 *Provide class-leading catering services*
   . . .
3. *Turn the aircraft around in less than 30 minutes*
   3.1 *Load and unload passengers, cargo, and catering in less than 30 minutes*
   3.2 *Conduct refueling operations in less than 30 minutes*
   3.3 *Conduct operational maintainability in less than 30 minutes*
   . . .

Naturally, there are many requirements needed to define fully the logical design of the aircraft system. We obviously do not have space here to include them all and only list those that result from elaboration of just three StRS requirements in order to illustrate the concepts associated with conceptual design and preliminary design. Note that the numbering system has been provided to support traceability throughout the process. For example, the requirements associated with 3.3 Conduct Refueling Operations in Less Than 30 Minutes relates directly to (is derived from, or is a child of) StRS statement 3.0 Turn the Aircraft Around in Less Than 30 Minutes.

Note that the functions in the above example are offered to illustrate the decomposition and derivation processes. The functions at which level they are shown are not low enough in level to be adequate system requirements; that is, requirements that refer specifically to the system of interest and nothing else. For example, as was noted earlier when referring to the FFBD, the function 3.3 Conduct Refueling will require a number of actors and other equipment that will be outside the system boundary. The decomposition or derivation process must continue until the system requirements in the SyRS refer only to the aircraft system, not any entity outside the system.

Example 3.15: Use of FFBDs During Conceptual Design

During conceptual design, designers use FFBDs to identify system-level functions and to assist in requirements analysis. An FFBD incorporating the turn-around operation of the aircraft may take the form shown in Figure 3.11. "Turn around" involves processing the passengers, luggage, freight, and so on between flights and preparing the aircraft for its next flight. Note also that many such FFBDs are required to cover the other functions required of the aircraft system.

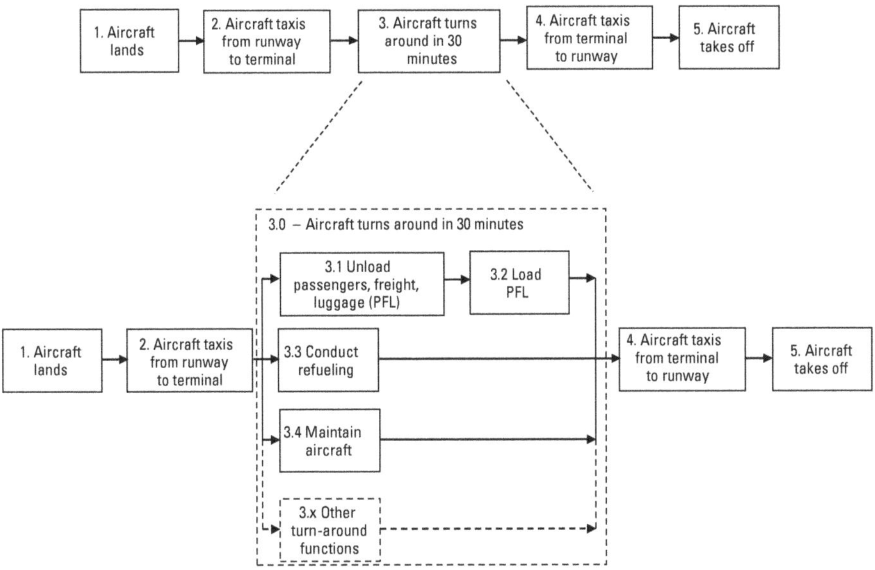

Figure 3.11 An FFBD describing the functions involved in aircraft turn-around.

The StRS requirement 3.0 to turn the aircraft around results in a number of system-level functions, including 3.3 Conduct Refueling, which we must take no more than 30 minutes to complete. This is an example of a functional requirement (refueling) and an associated performance parameter (less than or equal to 30 minutes).

Note that, in this example, we have not resolved differences in the terms used in the list of requirement statements (as they would appear in the RBS) and the functions detailed in the above FFBD. These two tools have different views and are often used simultaneously. We have left the two artifacts in their initial states to illustrate that the two views must be reconciled. For example, the RBS requirement statements refer to passengers, cargo, and catering whereas the FFBD refers to passengers, freight, and luggage. These are similar but have additional information in each, so the designers must reconcile them in order to identify all the functions required (as well as to ensure that, in the end, all functions are referred to in only one way).

In these last two examples, we consider two of the three aspects of allocation introduced earlier. First, requirements are allocated to the appropriate group of the RBS. Second, higher-level requirements flow down to lower-level requirements. For example, the aircraft must be able to be turned around within 30 minutes. Figure 3.11 shows that this requirement consists of the lower-level functions of unloading passengers, conducting refueling, and maintaining aircraft, which can all be conducted in parallel within the timeframe. Therefore, we can allocate the performance requirement to lower functions and can allocate the performance measure so that we state that unloading passengers, refueling, and maintenance must each be completed within 30 minutes.

3.4.3 C3.3 Draft System Requirement Specification (SyRS)

The complete and populated RBS forms the framework for the SyRS, which, when approved, becomes the centerpiece of the system's FBL. However, at this stage, the SyRS is only in draft form (Figure 3.10, activity C3.3) because the requirements collected and analyzed so far have yet to be synthesized into an architectural solution. Therefore, there will be some degree of latitude allowed in certain requirements and priority labels such as mandatory, important, and desirable leave some flexibility for subsequent design.

The SyRS may take any one of many forms [14] and the most appropriate form depends on the application. The SyRS is perhaps the most important of all systems engineering artifacts because it becomes the source of reference for all the subordinate specifications that are produced at later stages through the design process. A solid SyRS provides an effective foundation for the remainder of the design and development effort. Errors or omissions in the SyRS flow into the remaining design effort. The later these errors are discovered, the more expensive and time-consuming the rectification becomes.

The endorsed stakeholder requirements in the StRS are expanded in the SyRS to be a set of system-level requirements that, when developed, will result in a system that meets the StRS requirements. Although the SyRS is much more detailed and has transformed the StRS statements in the stakeholders' language into more formal requirements that relate solely to the system. That is, the SyRS only includes statements that relate to the system of interest that is the subject of that SyRS: the aircraft, the facility, the simulator, and so on.

Example 3.16: StRS Versus SyRS Language

> *The language utilized in the StRS is intended for easy reading for a larger audience than the specification language, which tends to be more precise. For example, the StRS requirement statement: "The Aircraft shall operate from Class X airfields" will spawn, among many other requirements, the more detailed SyRS statement: "The Aircraft shall exert less than Y kN/m^2 of pressure on the tarmac when the aircraft is stationary and at maximum all-up mass."*

Note that the aircraft operations may be defined in terms of modes and states. In that case, there may be a Parked_State in which the aircraft is fully loaded and stationary, so that the requirement might then say: "When in the Parked_State, the Aircraft shall exert no more than Y kN/m^2 of pressure on the tarmac when the aircraft is stationary and at maximum all-up mass."

3.4.4 C3.4 Define TPMs

Some of the requirements from the set determined in the preceding activities may be considered as useful to be key indicators of system performance (or may represent particular risk), a sort of "health check." These selected requirements are called TPMs (Figure 3.10, activity C3.4).

IEEE-STD-1220 states that appropriately selected TPMs can be used to: assess conformance to requirements, assess conformance to levels of technical risk, trigger development of recovery plans for identified deficiencies, and examine marginal cost benefits of performance in excess of requirements [15]. TPMs are identified early in the system development effort and are continually monitored and tracked throughout the system development as a means of managing the progress and risks associated with system development.

The first step in identifying TPMs is to identify the quantitative parameters that require tracking throughout the project. Once the parameters have been identified, they should be prioritized in terms of their importance as viewed by the customer. A second list of TPMs may be established, prioritized, and maintained by the contractor. As is often the case with subjective issues such as risk, a second perspective is always valuable. Both lists will imply priority to potential suppliers of the system during its design and development.

Throughout the design process, a list of TPMs and associated metrics should be maintained along with the priority, the "benchmark" objective, and the current level of achievement and projected or estimated performance. It is normal for some TPMs to move down in the priority order or drop from the list entirely as the design proceeds due to the fact that their ultimate achievement is no longer uncertain. It is also normal for requirements that were not considered to be at risk during early stages of design to be added or moved up in priority due to previously unforeseen complications. This assists both the acquirer and the supplier in the identification and mitigation of technical risks during the acquisition phases.

Bands of acceptable variation from the expected value will also be determined and agreed upon so that unnecessary risk management actions are not instigated at the first sign of a very minor variation. The variation bands will be typically large in the early stages of the design but will become increasingly narrow as the design matures.

An example may clarify the role of TPMs in the acquisition phase of the systems engineering effort and emphasize the uses of TPMs as described by IEEE-STD-1220.

Example 3.17: TPM Example for Our Aircraft Example

Our aircraft system must be capable of operating from Class X airfields, achieving a certain rate of climb, achieving and maintaining a certain cruising speed, and achieving certain fuel economies (relating to aircraft range and flying costs). All these requirements rely on the aircraft being delivered at or below a certain maximum all-up mass.

The designers of the aircraft system may identify mass to be a major design challenge. They can estimate the likely mass of an aircraft system of this type using past experience and modeling. They will also know the mass of all the components on the aircraft as they design it using tools such as CAD. Their estimates will not be confirmed until the aircraft is developed, integrated, and produced, at which point it is too late to make substantial changes to the aircraft's all-up mass.

All-up mass may be considered to be a TPM because it is a key indicator of likely performance. By monitoring the changes in aircraft all-up mass as the aircraft is developed, the designers may be able to identify promptly situations where this is a trend or growth of actual results away from expected results. This helps to identify technical risks as they are occurring with a view to initiating action on those risks in a timely fashion.

The designers will have a development strategy in mind that may involve traditional design and development, followed by the use of a prototype to finalize the design. The strategy may involve the following phases:

- Phase 1. *In the early stages of the system development, the aircraft design will exist "on paper" and will be enhanced by computer models. Based on this paper design, the designers can calculate an initial estimate of aircraft mass to be made.*
- Phase 2. *The design process continues as the subsystems are designed and incorporated into the overall aircraft design. Each subsystem will add mass to the aircraft (except if the subsystem is entirely made up of computer software). To that end, as the subsystem design matures, the initial estimate of all-up mass may be adjusted.*
- Phase 3. *During the later stages of design (before production), an initial prototype aircraft may be produced and may undergo a test flight. The*

plan may be for the initial prototype aircraft to fly in a very basic configuration, without full loads (fuel, freight, and passengers) and without some of the nonflight-critical components such as bathrooms, galleys, seats, and in-flight entertainment systems. In this basic configuration, the prototype will be (artificially) light.

- Phase 4. *The configuration of the prototype will continue to mature as increased functionality is integrated into the prototype and is tested. As this happens, designers expect the prototype to increase in mass.*
- Phase 5. *Eventually, the prototype will represent the production-ready design. It may fly with dummy mass loads in it to simulate a full passenger load, and it may fly with full fuel tanks and all the food and water required to support its intended applications.*
- Phase 6. *Designers may use the production system to produce a number of pre-production aircraft to confirm that the production system is operating correctly. Each of these pre-production aircraft will be tested and slight adjustments will be made to the product baseline. In this way, each pre-production aircraft gets closer to the final design.*
- Phase 7. *Eventually, the designers will be satisfied that the design is complete, and the aircraft is ready to move to production. The production baseline will be finalized, and the production system will be used to produce the aircraft to be delivered to the customer. As each aircraft is produced and tested, the designer and customer will be able to confirm the as delivered all-up mass of the aircraft.*

At each step in this process, the designers will have an expectation of where the aircraft's all-up mass should be. The "actual" all-up mass at each point can be measured and compared with the expected mass at that point. Major departures from expected values or major changes in mass trend can then be identified and discussed.

To monitor "actual" versus "expected" for this critical TPM, the aircraft all-up mass will be estimated, investigated, and reported upon at discrete stages throughout the acquisition phase. Design reviews are a convenient forum for such activities, but TPMs may also be reported on more frequently in artifacts such as status or progress reports. These reports will be of interest to both the designers and the customers of the aircraft system because a failure of the aircraft system to meet this requirement will prevent the aircraft from meeting its original (and endorsed) needs.

During Phases 1 and 2, the designers will be able to provide information on how they expect aircraft mass to change over the subsequent phases. They estimate that the aircraft all-up mass will be below the maximum of

130,000 lbs and they expect the first prototype aircraft to fly at approximately 65,000 lbs. Their estimates and expectations may be presented at a design review or progress meeting as in Figure 3.12.

The designers will then be able to maintain this representation as the aircraft system passes through the subsequent development phases. For example, as we move through Phase 4, the designers will have been collecting mass data on the prototypes and will be able to present Figure 3.13.

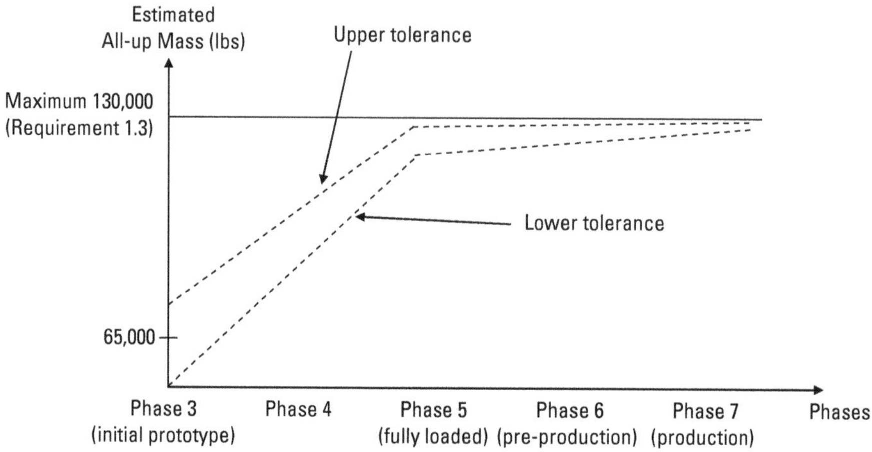

Figure 3.12 Expectations for all-up mass at each of the phases of development.

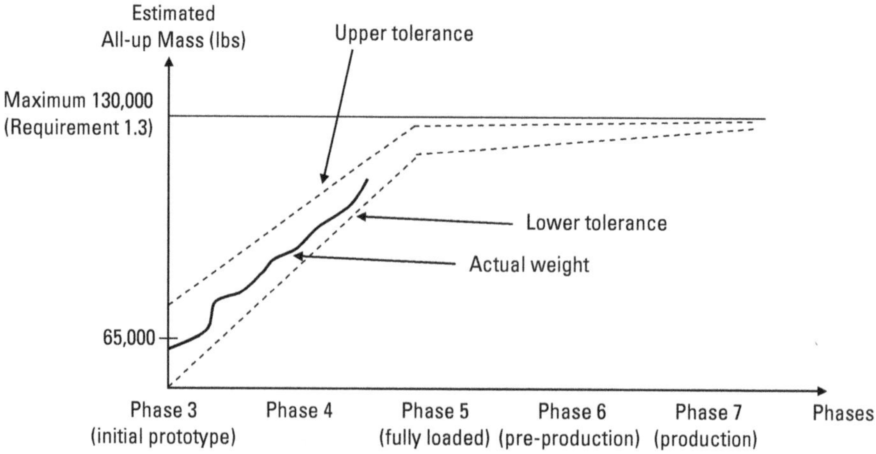

Figure 3.13 Example of data collected and compared with expectations for all-up mass at each of the phases of development.

From this graph, all parties can gain confidence that the mass of the prototypes is under control and is behaving as expected (within the tolerances set by the designers).

If things are not proceeding as well as expected, the following graph may be presented (showing that the mass of the prototypes is growing beyond what was expected and is now moving beyond the upper tolerance band). As soon as the prototype mass starts to move beyond the upper band, the designers will need to explain what is going wrong. That is, they will need to explain why this has occurred and what steps they are taking to regain control over the aircraft mass, noting that the mass is still below the maximum all-up mass of 130,000 lbs. By detecting this problem early, the designers stand the best chance of being able to address it before finalizing design and moving into production (see Figure 3.14).

In some cases, aircraft mass might appear to be performing better than expected, such as in Figure 3.15.

These results could be great news for everyone. The prototypes appear to be coming in at masses much lower than expected, which could provide all sorts of opportunities (perhaps to ease mass constraints on one of the critical subsystems, for example); the earlier these are exposed to the customer, the better. There is a famous adage about news that appears too good to be true. The situation illustrated above could be caused by progress problems where subsystems that were meant to have been installed and tested are running late, meaning that the prototype aircraft is lighter than expected. The delay in these subsystems being integrated could be an early warning of delivery

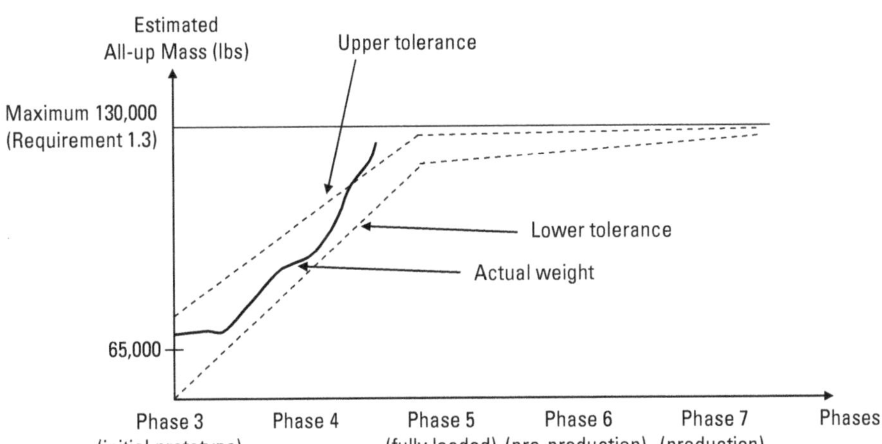

Figure 3.14 Example of data collected showing all-up mass exceeding expectations at a particular phase of development.

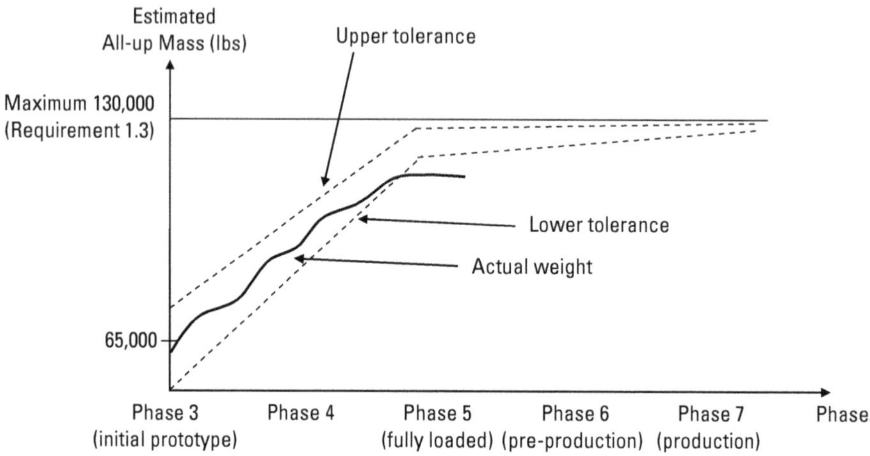

Figure 3.15 Example of data collected showing all-up mass being lower than expectations at a particular phase of development.

> *delays and cost overruns. For these reasons, if the mass TPM fell below the lower tolerance, the designers would need to provide some explanation of this observation.*

TPM tracking should occur throughout the acquisition phase, and the parameters should be assessed periodically during technical reviews or in status or progress reports. The establishment of TPM management (including identification, prioritization, and setting of values) is therefore an important part of conceptual design (and subsequent stages).

3.4.5 C3.5 Conduct System Requirements Reviews

System requirements reviews (SRR) may be conducted periodically throughout conceptual design to verify and approve versions of system-level requirements (Figure 3.10, activity C3.5). The aim of the SRRs is to progressively monitor and approve the system-level requirements that are developed on the way to the initial FBL. Progressive reviews allow the requirements analysis effort to continue to lower levels in the logical hierarchy in the RBS by providing validation of the higher levels of abstraction, providing a firm start point for the subsequent analysis. SRRs may or may not be considered formal reviews.

In addition to reviewing the requirements' flowdown, the SRRs may also review other information such as manufacturing plans, design schedules, and personnel requirements plans that will assist in fine-tuning the preliminary life-cycle concepts. The number of SRRs will depend on the size and complexity of the system of interest. Large and complex systems may require multiple SRRs during conceptual design, whereas simple systems may require very few. However, it is recommended that at least one SRR is held, even in simple projects, for each of the major activities described above, the development of BNR, SNR, and system requirements, so that business management and stakeholders have some formal involvement in the transition from the BNR (in particular, the BRS) to the SNR (in particular, the StRS), and finally to the SyRS. In some organizational development methodologies, there is only one SRR held before tendering.

Requirements reviewed at this stage are likely to contain margins of acceptable performance and may be categorized according to priority. A possible categorization may see requirements listed as mandatory, important, or desirable where mandatory requirements are absolutely essential, desirable requirements are nice to have, and important requirements are somewhere in between. SRR will review the margins and category of each requirement to ensure that each is being afforded appropriate weighting. Figure 3.16 [16] illustrates that, while there is a natural flow to the activities, iteration is required.

Although a significant number of requirements can be elicited and elaborated upon early on, the set must be analyzed and negotiated and then allocated to logical groupings so that the work to date can be validated. Having started at the inner point of the spiral, the first pass elicits informal requirements

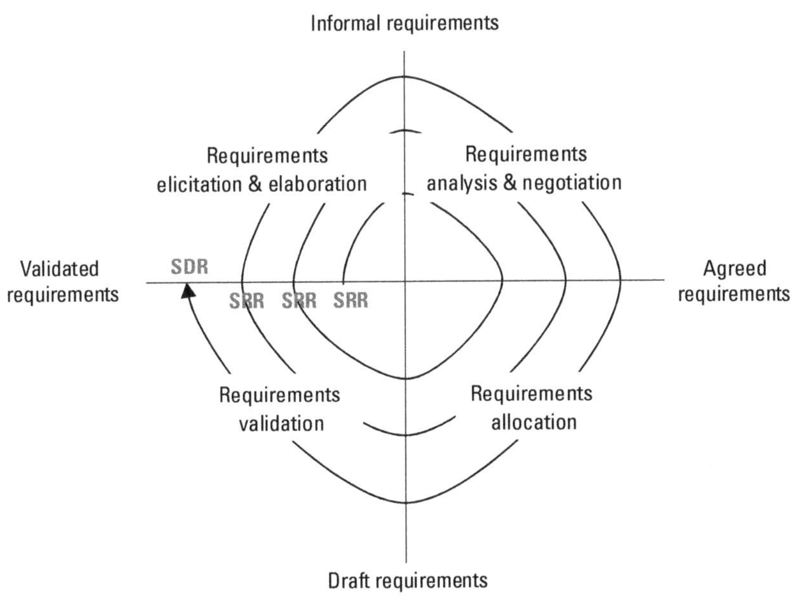

Figure 3.16 Iterative requirements engineering activities.

and negotiates agreed requirements that are then allocated and validated to form an early draft of the set of requirements available for stakeholder review. Requirements definition then continues through another pass of elicitation and elaboration, analysis and negotiation, allocation, and validation to a second review in Figure 3.16. The process continues for as many passes as is required. Note that at each pass we have a draft set of requirements that can be validated at an SRR. Once designers are satisfied that the set of requirements is sufficiently complete at the system level, the last review is the SDR.

3.4.6 Management Considerations

Although the system-level requirements are captured in the draft SyRS, there are additional project management and systems engineering considerations that need to be captured in terms of how the system development is to be conducted and managed. The following nonexhaustive list provides a small number of illustrations:

- Responsibilities for project management, systems engineering, logistics management, operation, and support should be determined and clearly articulated so that all stakeholders are aware of their responsibilities at all stages of the system life cycle.
- Project management considerations will include principal responsibilities, stakeholder engagement, schedule issues, budget issues, risk management, project management milestones, and payment schedules.
- Systems engineering management aspects will include expectations for test and evaluation, configuration management, artifact deliverables, technical reviews and audits, and adherence to standards; see Chapter 8 for more details.

These requirements need to be identified and planned because they will dictate how the design and development effort will be organized and managed. All the examples above (and many more) are necessary to reduce the technical risk of the development effort, but they require work and therefore must be accounted for by the project in the WBS.

Some management activities will be covered by work conducted in accordance with internal customer processes and procedures. When development of the system is to be outsourced, the management aspects must be contained in a contractual artifact such as an SOW. Broadly then, in a customer-contractor relationship, the central artifacts in a contract will be the SyRS, which articulates what is required of the system to be delivered by the contract, and the SOW, which describes the work to be conducted by the contractor during the contract.

3.5 C4. Conduct System-Level Synthesis

System design has now progressed to the stage where some of the system-level design decisions can be made based on the draft system requirements. Synthesis at the conceptual design level establishes a system configuration that is representative of the final system form at the subsystem level. At this point, the configuration is not assumed to be final as the design may be adjusted later in the design process as the design becomes physical.

In system-level synthesis, potential system solutions are developed (Figure 3.16, activity C4.1). Information is collected (Figure 3.17, activity C4.2) regarding how the potential solutions meet the draft SyRS. The potential solutions are then evaluated (Figure 3.17, activity C4.3) by the acquirer, which requires the development of suitable evaluation criteria as well as an evaluation

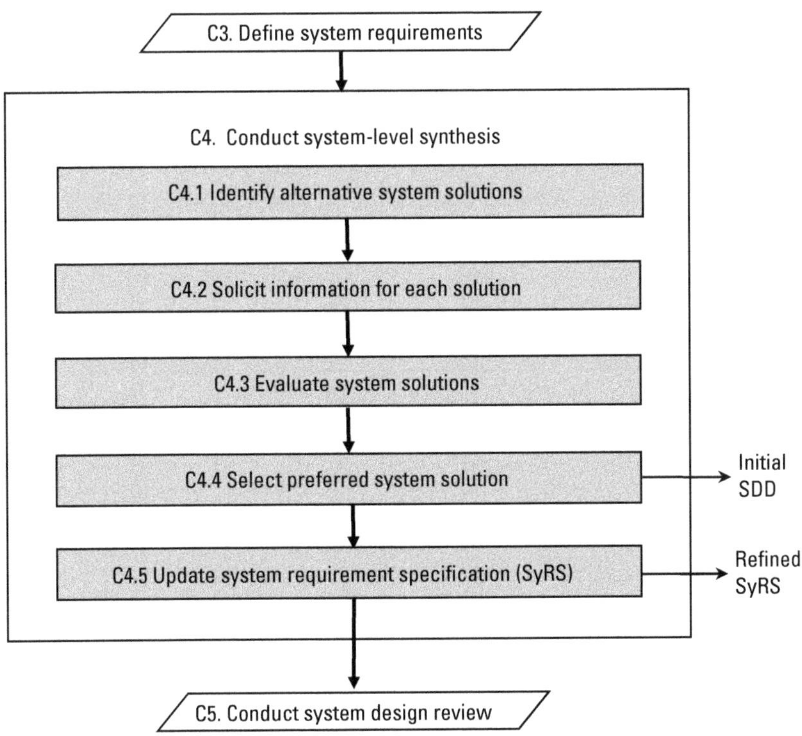

Figure 3.17 A summary of C4. Conduct system-level synthesis.

framework. Any discrepancies in the systems engineering products are noted and fed back to previous activities. The preferred solution is chosen (Figure 3.17, activity C4.4) and the system architecture is generally captured in a system design description (SDD) [17] or equivalent, which is used to show compliance against the requirements in the draft SyRS.

The system synthesis activities described in activities C4.1 to C4.4 in Figure 3.17 can also be illustrated by Figure 3.18 in which architectural options are developed, they are evaluated in the context of an evaluation framework using selected criteria with input from the stakeholders involved, and a preferred system architecture is developed.

Although system-level synthesis can be conducted internally by the customer, it often takes the form of a request for tender (RFT) in which tenderers are provided with a draft SyRS and an SOW. Each tender response shows how the solution offered provides a feasible system-level design that meets (complies with) the SyRS and how the work requested in the SOW will be conducted. The acquirer evaluates the tender responses before selecting the preferred system-level solution and inviting the preferred tenderer to negotiations. If negotiations are successful, a contract is signed.

The draft SyRS will contain margins of acceptable performance and priorities providing a sufficiently broad design space to allow a range of solutions to be tendered. As part of tender evaluation (synthesis), the draft SyRS is tested against a range of feasible system solutions and a preferred solution is selected. The draft SyRS requirements now need to be refined to remove the ranges of acceptable performance and to reflect more precisely what is to be

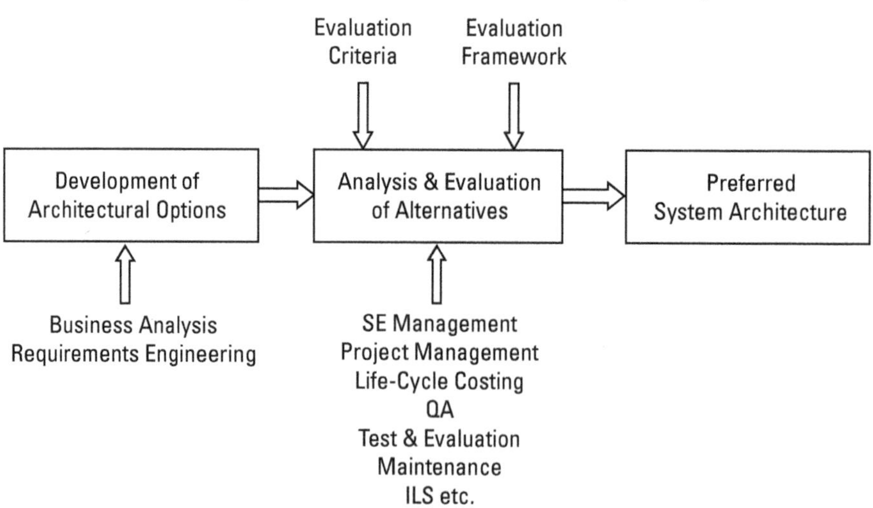

Figure 3.18 System synthesis activities.

achieved by the selected system and the minimum acceptable level of performance. New system-level requirements may need to be added to the SyRS and errors or conflicts in existing requirements need to be clarified. The priority of requirements is often removed so that each requirement statement is essential. As feasible alternatives are presented, the customer may also need to revisit the BNR/SNR and seek endorsement from the business or stakeholders for proposed scope changes to the draft SyRS if any of the endorsed requirements are not feasible within the approved schedule or cost constraints.

So the final activity in process (Figure 3.17, activity C4.5) refines the SyRS, updates BNR and SNR, and establishes the initial FBL. The main products of synthesis are the refined SyRS and the broad system solution described in the SDD.

The SyRS is perhaps the most important systems engineering artifact because it becomes the source of reference for all the subordinate specifications that are produced during the later stages of the design process. A solid SyRS forms an effective foundation for the remainder of the design and development effort. Errors or omissions in the SyRS flow into the remaining design effort. The later these errors are discovered, the more expensive and time-consuming the rectification will be. Additionally, the SOW is perhaps the most important project management artifact as the work described in it will be translated into the work packages in the developer's WBS; again, any errors or omissions will have consequences in terms of the project budget and schedule.

3.6 C5. Conduct System Design Review (SDR)

At the end of conceptual design, the SDR provides the following from a systems engineering perspective:

- Formal confirmation that the logical design meets the business and stakeholder requirements;
- Formal record of design decisions and acceptance;
- Formalized communication of the intended design approach, as articulated by the SDD or equivalent, to the major players in the design effort;
- Approval of plans for verification of the system;
- Approval of the Systems Engineering Management Plan (SEMP) and supporting plans.

In addition to review of the systems engineering effort, a number of review activities will most likely be performed at SDR to support project management:

- Confirm that the system to be procured aligns with the customer's organizational goals.
- The SOW is refined and the work required is better understood.
- The project management plan (PMP) is refined.
- Other early plans such as production plans are refined.
- Cost estimates are refined.
- The schedule is refined and is confirmed to be consistent with the cost and risk goals for the project.
- Confirmation that all required project resources are available.

Figure 3.19 summarizes the activities involved in the conduct of the SDR. As with all reviews, it is essential to establish the entry and exit criteria for the review; that is, it must be clear what must be in place before the review begins and what must have been achieved before the review can be considered

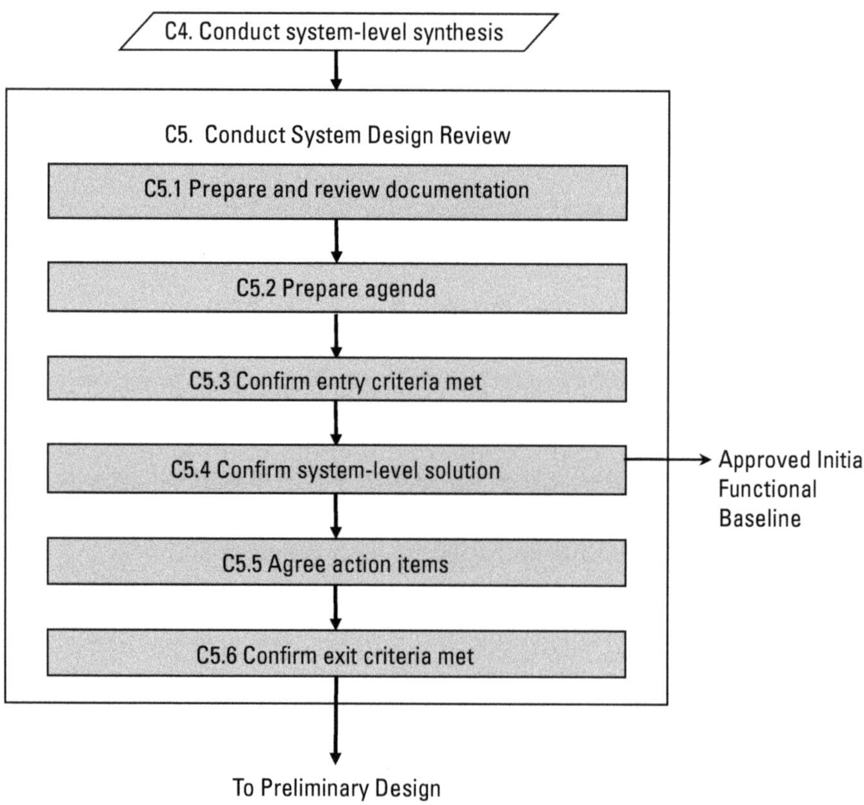

Figure 3.19 A summary of C5. Conduct System Design Review.

as complete. Consequently, the review cannot begin until the entry criteria have been met (Figure 3.19, activity C5.3). Part of that confirmation is to ensure that all of the necessary artifacts have been prepared and reviewed (Figure 3.19, activity C5.1) in advance of the review and that all parties have agreed to the agenda for the review (Figure 3.19, activity C5.2).

The principal purpose of SDR is to give confidence that the system solution described in the SDD addresses the requirements of the refined SyRS (Figure 3.19, activity C5.4). If all is satisfactory, the SyRS is approved and baselined as part of the initial FBL. It is rare that SDR ends with all outstanding issues resolved during the review since many will require some investigation and analysis. Rather than hold up subsequent design, it is common for a number of action items to be agreed (Figure 3.19, activity C5.5) with agreed timelines to account for any outstanding action (providing that resolution of those issues will not affect the remainder of the design). These outstanding actions are then completed in parallel with the early preliminary design activities and are reviewed for completeness in conjunction with a subsequent review or audit. Before the review can be considered to be complete, the exit criteria must be met (Figure 3.19, activity C5.6).

The boundary between conceptual design and preliminary design is often not clear. From a system design perspective, SDR aims to provide demarcation by approving the SyRS and establishing the initial FBL from which the remainder of the design will follow. For simplicity, we are focused on the SyRS, but we should keep in mind that the initial FBL includes considerably more than the SyRS in terms of systems engineering and project management artifacts. At the very least, there will also be the PMP, the SEMP, the major plans outlined in Chapter 8, as well as production plans and so on. Consequently, one of the other major aims of SDR is to mark the transition of responsibility from the customer, acquirer, or tasking activity to the contractor, developer, or performing activity.

3.7 Summary

In this first major process chapter, we have outlined the major elements of conceptual design, which is the most important of all activities in the system life cycle because it starts with a short statement of the mission statement and ends with an expansive functional description of the system (in the SyRS or a number of SyRS) that meets the need. Since the remaining activities in the acquisition phase are traceable back to the SyRS, it follows that it is the single most critical artifact in the system life cycle; any errors or omissions in the SyRS will invariably result in errors or omissions in the resultant system when it is fielded. The later these issues are discovered, the more expensive and time-consuming the rectification becomes.

The SyRS (and the remainder of the initial FBL) is also critical because it represents the transition from the customer, acquirer, or tasking activity to the contractor, developer, or performing activity, that is, the responsible parties change. It follows that the handover from one to the other, particularly in a formal contracting environment, must be unambiguous and correct if the subsequent work is to be correct.

From a project management perspective, the initial FBL represents the start point for the formalization of the project scope, which then allows the initial budget and schedule to be developed. However, before the scope, budget, and schedule can be finalized, the WBS must be developed, which requires a functional-to-physical transformation from the functions in the SyRS and SOW to the contracted WBS. That is the principal purpose of the next activity in the acquisition phase, preliminary design.

Review Questions

1. Briefly explain the five major processes conducted during conceptual design, including their major products.
2. Draw a diagram to illustrate the process of C1. Define Business Needs and Requirements. Briefly describe each of the major activities in the process.
3. Define a stakeholder and explain why a stakeholder is not simply someone who is affected by, or can affect, a system.
4. Briefly describe what is meant by business constraints, project constraints, external constraints, and design constraints.
5. Briefly list and describe the nine desirable properties of a good system mission statement.
6. Briefly list and describe the six guidelines for the format of a good system mission statement.
7. Briefly describe the step C1.2.2 Define Preliminary Life-Cycle Concepts including the role and likely content of the five major preliminary OpsCons.
8. Briefly describe the step C1.2.3 Define Preliminary Validation Criteria.
9. With the aid of diagrams, illustrate what is involved in creating a context diagram for a system.

10. Briefly discuss the issues associated with the definition of the system boundary.
11. Briefly describe the three steps associated with interface definition: interface description, interface impact analysis, and interface control analysis.
12. Briefly list and describe the four main steps in a feasibility analysis (in steps C1.4.1 to C1.4.4).
13. Briefly explain why the conduct of a feasibility analysis is an important activity in conceptual design.
14. Explain how an RBS assists with requirements analysis.
15. How do the statements in the StRS differ from those in the SyRS? Provide three examples of each type of statement for a motor vehicle system.
16. Explain the difference between functional, nonfunctional, performance, and verification requirements. Give an example of each for a motor vehicle system.
17. Explain the value associated with recording attributes against each requirement.
18. Briefly describe TPMs and how they are used to support the systems engineering effort.
19. Briefly describe the role and conduct of the SRR.
20. Briefly describe the step C4. Conduct System-Level Synthesis.
21. Briefly describe the step C5. Conduct System Design Review.

Endnotes

[1] Maier, M., and E. Rechtin, *The Art of Systems Architecting*, Boca Raton, FL: CRC Press, 2009.

[2] ISO/IEC/IEEE 29148, *Systems and Software Engineering—Lifecycle Processes—Requirements Engineering*, 2018.

[3] Bahill, A. T., and C. Briggs, "The Systems Engineering Started in the Middle Process: A Consensus of Systems Engineers and Project Managers," *Systems Engineering*, Vol. 4, No. 2, 2001, pp. 156–167.

[4] Grady, J. O., *System Requirements Analysis*, Burlington, MA: Academic Press, 2006.

[5] Miller, G. A., "The Magical Number Seven, Plus or Minus Two: Some Limits on Our Capacity for Processing Information," *The Psychological Review*, 1956, pp. 81–97.

[6] Ryan, M. J., *Requirements Practice in Conceptual Design: A Practical Guide*, Norwood, MA: Artech House, 2026.

[7] ANSI/AIAA G-043A-2012, *Guide for the Preparation of Operational Concept Descriptions*, American National Standards Institute, American Institute of Aeronautics and Astronautics (sponsor), 2012.

[8] ISO/IEC/IEEE 29148, *Systems and Software Engineering—Lifecycle Processes—Requirements Engineering*, 2018.

[9] See, for example: Rechtin, M. W. and E. Maier, *The Art of System Architecting*, Boca Raton: CRC Press, 2009.

[10] Wheatcraft, L. S., M. J. Ryan, and T. E. Katz, *Needs and Requirements Manual: Needs, Requirements, Verification, Validation Across the Lifecycle*, Hoboken, NJ: Wiley, 2024.

[11] International Council on Systems Engineering (INCOSE), INCOSE-TP-2010-004-04, *Guide to Writing Requirements*, 2024.

[12] Wheatcraft, L. S., M. J. Ryan, and T. E. Katz, *Needs and Requirements Manual: Needs, Requirements, Verification, Validation Across the Lifecycle*, Hoboken, NJ: Wiley, 2024.

[13] IEEE-STD-1220, *IEEE Standard for Application and Management of the Systems Engineering Process*, New York: IEEE Computer Society, 2005.

[14] Sources of information on the content of the SyRS can be found in:

Data Item Description (DID), *System/Subsystem Specification (SSS)*, DI-IPSC-81431, DD Form 1664, US Department of Defense, April 1989.

ECSS-E-10A, *Space Engineering—System Engineering*, Noordwijk, The Netherlands: European Cooperation for Space Standardization, 1996.

MIL-STD-961E, *Department of Defense Standard Practice for Defense and Program-Unique Specifications—Format and Content*, Washington, DC: US Department of Defense, August 1, 2003.

MIL-STD-490A, *Military StandardSpecification Practices*, Washington, DC: US Department of Defense, 1985.

[15] IEEE-STD-1220, *IEEE Standard for Application and Management of the Systems Engineering Process*, New York: IEEE Computer Society, 2005.

[16] Kotonya, G., and I. Sommerville, *Requirements Engineering*, New York: John Wiley & Sons, 1998.

[17] Information on the content of the SDD can be found in Data Item Description (DID), *System/Subsystem Design Description (SSDD), DI-IPSC-81432, DD Form 1664*, US Department of Defense, April 1989.

4

Preliminary Design

4.1 Introduction

Preliminary design starts with the initial FBL, as defined during conceptual design, and continues to translate system-level requirements into design requirements for the system elements that will combine to form the system. This translation requires continuation of the requirements analysis started in conceptual design in order to determine the specific requirements for the elements making up the system. Trade-off studies are conducted, and the result of the preliminary design effort is the establishment of an allocated baseline (ABL), in which requirements are "allocated" to specific physical system elements that combine to form the system.

Responsibility for performing preliminary design normally rests with the developer (the contractor), who develops the system to meet the requirements of the FBL (normally prepared by the customer/acquirer). The acquirer's role now increasingly becomes one of monitoring, reviewing, and supporting contractor progress. Although the acquirer normally avoids becoming actively involved in design decisions made during preliminary design (or any subsequent activity, for that matter), they remain very interested in the outcomes at each point and may be needed to make decisions with regard to the output of the trade studies. In some cases, the acquirer may opt to add an additional level of rigor to the process by engaging independent systems engineering consultants to provide independent reviews of the many and varied engineering artifacts produced by developers during a systems development, contributing to what is often called independent verification and validation (IV&V).

The processes conducted during the preliminary design effort include subsystem requirements analysis, requirements allocation, interface identification/design, subsystem-level synthesis, and the preliminary design review (PDR). These processes are illustrated in Figure 4.1 and are described in detail in the following sections.

4.2 P1. Conduct Subsystem Requirements Analysis

In the preceding chapters, we described requirements analysis as the process through which requirements are progressively elaborated (decomposed and derived) from the BNR to the system level. In preliminary design, we are now interested in the elaboration from the system level to the subsystem level and so on until all functions, parameters, and interfaces have been defined, and the necessary resources have been identified. The major activities in subsystem requirements analysis are described in Figure 4.2.

Because preliminary design will most probably conducted by a different team than that which conducted conceptual design (that is, most likely by

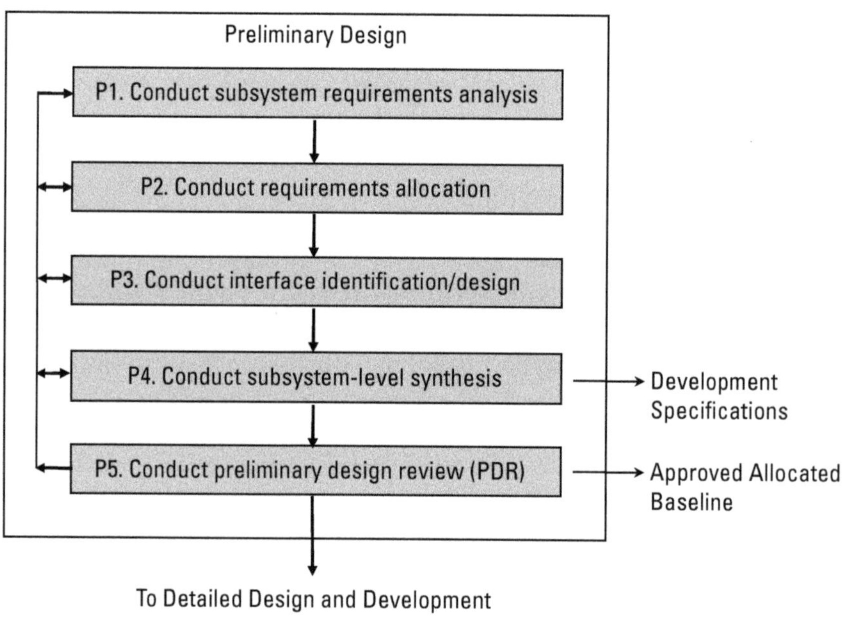

Figure 4.1 The five major processes performed during preliminary design.

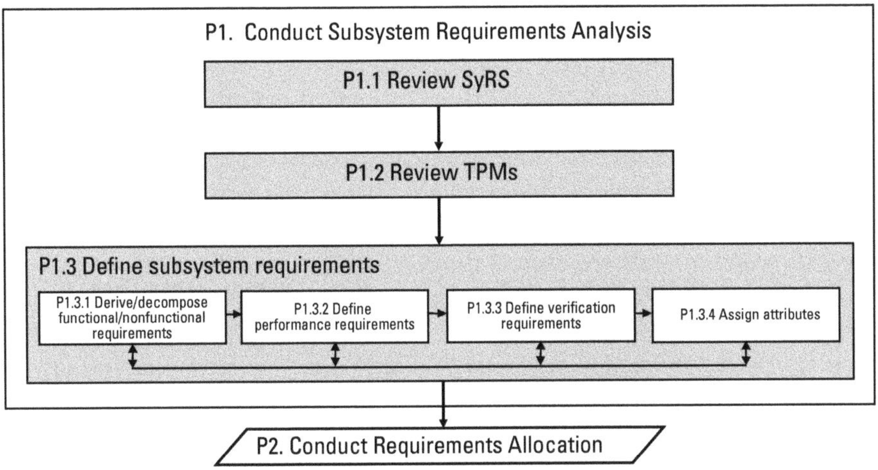

Figure 4.2 A summary of process P1. Conduct Subsystem Requirements Analysis.

the developer rather than the customer/acquirer), the subsystem requirements analysis activity begins with a review of the SyRS (Figure 4.2, activity P1.1) and a review of the relevant TPMs (Figure 4.2, activity P1.2). The next step is to begin defining requirements at the subsystem level (Figure 4.2, activity P1.3) by decomposing/deriving requirements from the SyRS (Figure 4.2, step P1.3.1) to provide more-detailed requirement statements that can be allocated to physical CIs. Performance (Figure 4.2, step P1.3.2) and verification (Figure 4.2, step P1.3.3) requirements are then assigned, as are attributes (Figure 4.2, step P1.3.4) for each requirement.

It is not possible to prescribe the level of detail that needs to be decomposed/derived during subsystem requirements analysis. The detail must be sufficient to define completely the functions required of the major subsystems comprising the system and must be able to be used either to procure, or to design and produce, the major subsystems and their assemblies and components during subsequent stages of the systems engineering process. Therefore, the level of detail that is required during preliminary design will be influenced by whether the design is making use of off-the-shelf (OTS) items or will require new or modified items to be created. This issue is closely related to the issues discussed in Section 4.6.

The RBS and the FFBD are relatively simple, intuitive concepts that were introduced in Chapter 3 as a means of performing an analysis of the functions required of the system. FFBDs also assist in defining the sequences and interrelationships between the blocks (e.g., series versus parallel) as well as the major interfaces and dependencies between functions. Each block of the FFBD is eventually considered individually in determining the necessary inputs, the expected outputs, the constraints placed on the function, and, finally, the mechanisms by which the function could be achieved. The diagrams that are generated during the preliminary design effort are more detailed than in conceptual design and are aimed at defining the subsystems (rather than the system). Note that this stage is therefore the beginning of the conversion of what the system needs to do into the how it is going to do it, which is more commonly referred to as the translation of logical design into physical design (or from the problem domain into the solution domain). Sometimes, as described in Section 4.3, several different requirements may be achieved by the same piece of equipment or piece of software.

Example 4.1: Detailed Aircraft System FFBD

This example illustrates how an FFBD may be used during preliminary design to define system functionality at varying levels of detail.

In Chapter 3, we suggested that an FFBD describing the actual turnaround operation of the aircraft may take the form shown in Figure 4.3 (a copy of the figure from Chapter 3). The StRS requirement to turn the aircraft around in 30 minutes resulted in several system-level functions, including 3.3 Conduct Refueling.

Let's investigate this function further as we would during preliminary design. By using the FFBD process to investigate the requirement further in Figure 4.4, we can see that refueling an aircraft involves several steps prior to and after the actual refueling process. We can also see that once we have taken the other activities into account, the actual refueling process, 3.3.6 Refuel Aircraft, must now take place within 17 minutes. In the worst case, therefore, we must be able to refuel the aircraft from a completely empty state to maximum fuel capacity in less than or equal to 17 minutes.

The requirement to refuel in less than or equal to 17 minutes is an example of allocation of the upper-level performance requirement of 30 minutes down to 17 minutes for the refueling process. We do not have space here, but in a real project we would continue the FFBD process further and investigate what is involved in the fuel-transfer function (Function 3.3.6), and so on.

Note that the numbering of blocks is hierarchical in nature, which allows lower-level functions to be related to system-level functions. For example, the numbering system makes it easy to associate Function 3.3.3 Provide Fire Safety Precautions with its parent 3.3 Conduct Refueling and its parent function 3.0 Turn Around in 30 Minutes.

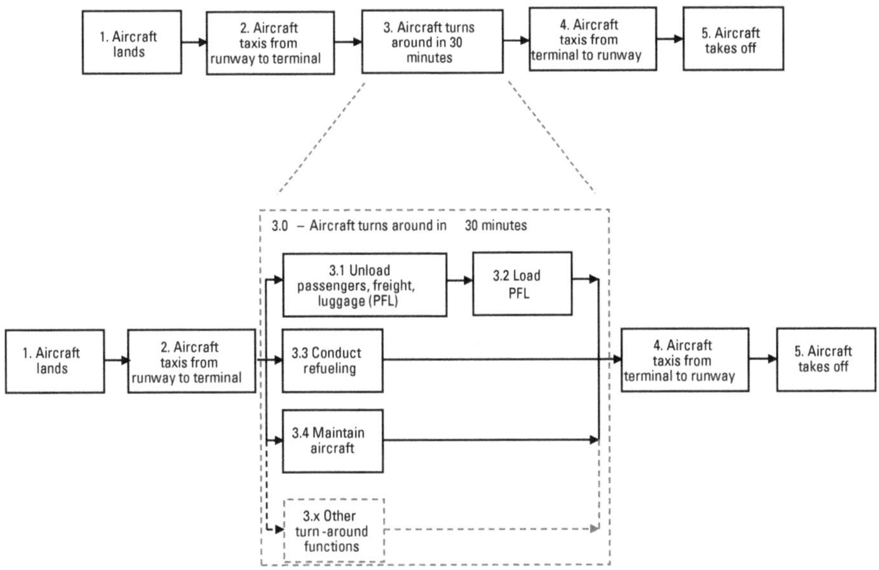

Figure 4.3 An FFBD describing the functions involved in aircraft turn-around.

This example has demonstrated how StRS and SyRS requirements could be investigated further during preliminary design using FFBDs. The requirement will be satisfied by whatever subsystem of the aircraft design is responsible for fuel-related function and performance. This may be called the fuel subsystem. Determining what physical subsystem will satisfy each requirement is the start of the translation of logical design into physical design, which is one of the key activities of preliminary design.

Some additional points to note:

- *If a refueling time of 17 minutes is impossible to achieve, some additional time may be gained by investigating the other functions involved in the 3.3*

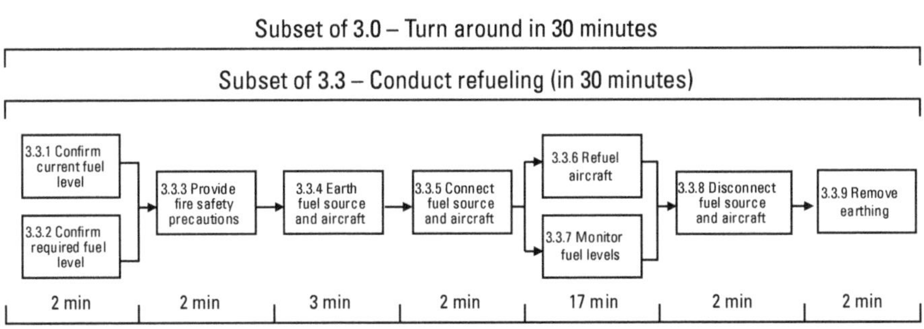

Figure 4.4 An FFBD describing aircraft refueling during turnaround.

Conduct Refueling function. It may be possible to conduct some functions in parallel, or it may be possible to speed up some functions. Either way, additional time may be provided to Function 3.3.6. If it is still impossible to perform the task in the allotted time, at least our traceability allows us to trace back up to the StRS requirement to turn the aircraft around in 30 minutes and revisit this imperative.

- *The FFBD process has also highlighted some interfaces that will need to be considered later in the process. Designers will investigate each of the functions in* 3.3 *Conduct Refueling and identify the interfaces. Remember that interfaces include human-machine interfaces (HMI), not just physical interfaces. An HMI will exist, for example, to allow the current fuel level in the aircraft to be determined (Function 3.3.1). A physical interface will exist to allow the aircraft to be correctly earthed (Function 3.3.4) and allow the fuel source to connect to the aircraft (Function 3.3.5).*

This concept is also a small snapshot of why we must also take a system-level view at all times. The balance and distribution of the timings in the lower-level functions supporting 3.3 Conduct Refueling are in the province of the system designer; that is, the 17-minute constraint on 3.3.6 Refuel Aircraft is the responsibility of the system designer, not the subsystem designer (the constraint has been chosen to optimize the system, not the subsystem).

It is vitally important that all subsystem requirements analysis is performed from the same FBL, that is, from the same SyRS. Although this sounds obvious, different functional baselines versions have been known to exist on projects, resulting in incompatible and conflicting preliminary and detailed designs. This reinforces the need to review the SyRS and the TPMs before embarking on preliminary design.

Once satisfied with the level of decomposition/derivation, each functional block can be considered independently, as was the case during system-level requirements analysis. This consideration should focus on the necessary inputs to the function, the expected outputs from it and any parametric requirements associated with the function. Some thought may also be given to how the function may be performed and what constraints exist in terms of the function's implementation.

4.3 P2. Conduct Requirements Allocation

As illustrated in Figure 4.5, requirements allocation refers to the process of grouping or combining similar requirements (identified during the requirements

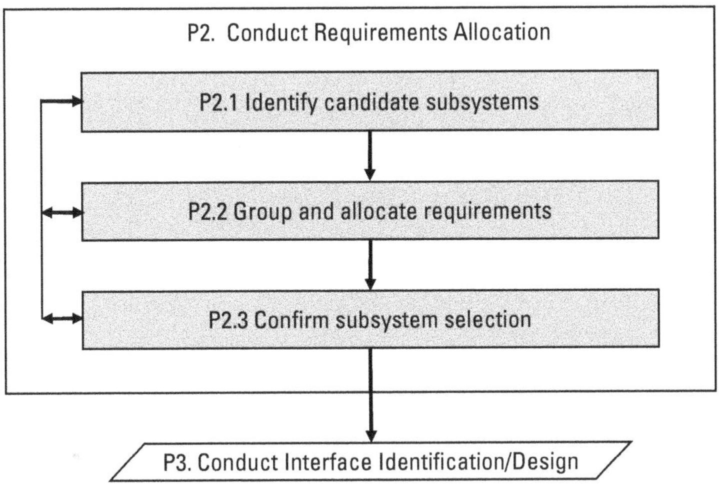

Figure 4.5 A summary of process P2. Conduct Requirements Allocation.

analysis) into logical subdivisions. We have done this before during conceptual design where the groups were based on the RBS headings. During preliminary design, the groups are based around a preliminary physical architecture formulated by the designers, rather than logical RBS headings. The groups of similar functions then assist the designer to determine the design of the major system elements (sometimes called subsystems) that are required to make up the system. This represents the translation from logical design to physical design. The subsystems consist of one or more configuration items (CIs) that are selected and/or designed to perform the group of requirements that are assigned to them. It is the assignment or allocation of requirements to subsystems and CIs that is the major focus of requirements allocation.

4.3.1 Identify Candidate Subsystems

The first step in requirements allocation during preliminary design is to identify the possible ways in which we could combine the requirements into logical sets or groups based on a preliminary physical architecture, that is, we need to identify the candidate subsystems. The logic used relies on experience and expertise in the relevant design domain. In our example, this domain is the aerospace design domain. That is, to perform requirements allocation during preliminary design in our example, we rely on knowledge of the likely physical composition of the aircraft system. Even though the detailed design is yet to be determined, experienced designers have in their minds the range of possible elements of a preliminary design.

Example 4.2: Subsystems for Our Aircraft System

Let's now look at the subsystems that comprise any aircraft (remember that our aircraft example has been pitched at a sufficiently general level to allow anyone familiar with aircraft to follow the generic process; if your expertise is in the aerospace domain, please be tolerant with the level of detail provided). First, it is worth looking at what the aircraft industry calls the airframe, which consists of a range of subsystems:

- *The Undercarriage subsystem includes the structures, tires, and brakes necessary to support the aircraft while on the ground.*
- *The Wings/Fuselage subsystem includes all the components necessary to accommodate the remainder of the subsystems and to form the aerodynamic portion of the aircraft system.*
- *The Fuel subsystem includes all the tanks, refueling/defueling receptacles, fuel pumps, and fuel lines necessary to support the engines.*
- *The Hydraulic subsystem includes all the hydraulic pumps, lines, and accessories needed to provide hydraulic power to all other aircraft systems such as flight controls and undercarriage.*
- *The Flight Control subsystem includes all the primary and secondary flight controls necessary for the control of the aircraft system in the air and on the ground.*
- *The Engine subsystem is the next major subsystem of any aircraft system, consisting of the engine itself including all the components that come together to form each individual engine such as turbines, gearboxes, and so on. The engine will also interface to an engine management subsystem including all the peripheral controls necessary to manage the engine during start-up, operation, and shutdown.*
- *The Avionics (aviation electronics) subsystem assists in the control of the aircraft, performing functions such as electrical generation and distribution, communications, and navigation. Most modern avionics subsystems comprise:*
    - *A computer (or collection of computers) that includes the hardware and software necessary to perform all aircraft operational functions;*
    - *Navigation subsystems that include the hardware and software necessary to perform the navigation functions for the aircraft;*
    - *Communications subsystems that include the hardware and software necessary to perform the communications functions for the aircraft;*
    - *Automatic flight-control subsystems that include the hardware and software necessary to perform the flight-control functions such as automatic pilot;*

- *An electrical generation and distribution subsystem responsible for the generation and distribution of all necessary electrical power to the aircraft systems;*

- *The Interior subsystem of the aircraft system will comprise:*
  - *The Cockpit, which includes all the elements to support the aircrew in their operation of the aircraft;*
  - *The Passenger space, which includes all of the elements necessary to support the passengers throughout their flights;*
  - *The Cargo space, which includes all the elements necessary to support the cargo throughout its flights;*
  - *The Galley space, which includes all the elements necessary to support the provision of food, drink, and refreshments to crew and passengers throughout their flights.*

Each of these major subsystems needs to be considered individually during preliminary design. Depending on how the designers intend to realize the subsystems, they may be broken down further into CIs, which comprise the hardware, software, or a combination of both designed to satisfy an allocated group of requirements. Sometimes a subsystem will be a single CI, but usually a subsystem will comprise several CIs. As the name suggests, the configuration of each CI is managed as a separate item for design, development, description, construction, review, audit, and test. The same CIs may also be used during the utilization phase for through-life support although, for reasons discussed in Chapter 8, the in-service CIs may be different to the development CIs.

The selection of CIs is a major part of the overall configuration management function called configuration identification (see more details in Chapter 8). CIs can vary in size and complexity from an aircraft, ship, or electronic system to a test meter or a round of ammunition [1]. In general, however, items may be identified as CIs because of:

- *Complexity.* One of the aims of the systems engineering process is to reduce the design problem to be manageable. Large, complex CIs defeat this purpose, so CIs should be small enough to be able to be designed and developed by an individual or a small team.
- *Interfaces.* The difficulty with creating small CIs is that there will be a lot of them in a complex system. For example, the Space Shuttle had 2.5 million parts; the Airbus A380 has 2.5 million part numbers produced by 1,500 companies from 30 countries; a complex military system such as a submarine will have some 4 million parts; and the Large Hadron Collider (LHC) accelerator complex has approximately 100 million

configurable components. The consequence of such a large number of CIs will generally be an increased number of internal interfaces. The selection of CIs is therefore a trade-off between the complexity of each CI and the number of interfaces required.

- *Use/function.* Each CI should perform a single function or a combination of related functions.
- *Existing items available.* Rarely will a system design require all components to be designed and developed totally. Many of the components required by a system will already exist in a certain configuration and the design should ensure that such components could be used rather than unnecessarily reinventing products. For example, if a reconnaissance aircraft has a functional requirement to record video, it makes sense for the functions to be allocated to a CI that looks something like a commercially available video recorder.
- *Commonality.* In addition to being chosen to match an existing item, a CI may be chosen by designers to be common with other CIs throughout the customer organization (or even a broader community of interest). It makes considerable sense in many cases to reuse CIs, or for the enterprise/business to mandate common CIs resulting in significant reduction in nonrecurring engineering (NRE), lower risk, shorter schedules, significant improvements in through-life support, and reduction in inventory.
- *Provided by a single supplier.* Similarly, unnecessary complexity is introduced into the design if a single CI is provided by more than one supplier.
- *Criticality.* A CI may be chosen to be smaller than would otherwise be required if the designers feel that the item is critical and therefore needs to be managed separately.
- *Maintenance and documentation needs.* Through-life support is an essential consideration in design and CIs should be chosen to facilitate maintenance and documentation.

Example 4.3: CI Selection

This example illustrates how a design team may determine the CIs within their specific area of design responsibility. We will use the example of the navigation subsystem to illustrate some possibilities.

It is possible that the navigation subsystem design team will use an OTS navigation system supplied by a third-party supplier to meet all requirements of the navigation subsystem. The design team will source and select a suitable OTS solution and integrate it with the aircraft. If our design team decides to do this, the design team does not have engineering control over the design of that navigation system (the third-party supplier does). In this case, the design team will consider the navigation subsystem to be satisfied by a single CI, supplied by the third party. There will be a one-to-one relationship between the subsystem and the CI.

Alternatively, it is possible that the design team will not be able to find any OTS navigation systems that meet the requirements allocated to the navigation subsystem. The design team may have access to in-house navigation software that is very capable and can do the job and may decide to integrate that software with a generic avionics computer from a third-party supplier. The designers might decide also to use a suite of sensors (for example, accelerometers, gyroscopes, GPS terminals, and air data sensors) from different suppliers to provide the necessary navigation inputs and data to the navigation software. In this case, their proposed navigation subsystem may consist of the following CIs (all from different suppliers), that is, there is a one-to-many relationship:

- *Navigation software CI;*
- *Navigation avionics computer hardware CI;*
- *Navigation sensor, accelerometer CI;*
- *Navigation sensor, gyroscope CI;*
- *Navigation sensor, GPS CI;*
- *Navigation sensor, air data CI.*

So there are at least two ways in which acquisition CIs can be selected for the navigation subsystem, depending on whether the subsystem is acquired as an OTS subsystem or developed as several CIs. In either case of acquisition CI choice, the in-service CI could be the single navigation subsystem CI or it could be several smaller CIs. Again, the choice is up to the designer or, in this case, the in-service support manager, who must consider the same selection criteria, albeit with a different perspective. For sustainment, a single CI may be preferred if the intention is for the CI to be maintained by one support contract. However, if there is to be a software support contract, a GPS support contract, and an avionics computer support contract, then it makes sense for the in-service CIs to be smaller and to match the desired support arrangements.

The determination and selection of CIs is a design decision made by the design team (developer) in their attempt to satisfy the system level requirements in the best possible way. The customer may influence the decision, especially if there are mandated solutions or constraints involving customer-furnished equipment that must be used in the design. The customer may also have additional "business rules" that they want applied to the CI selection decision. If so, these will be contained in the contractual documentation. For example, the customer may mandate that any single item that contains software of any form (even firmware) is to be split into a hardware CI and a software CI, which can then be managed separately. Even if the customer does not influence the decision, the customer needs to be aware of the CI selection process because it impacts on acquisition activities such as documentation and other artifacts, design reviews, configuration audits, and test and evaluation.

The selection of CIs also has an impact on operational use activities and system support activities such as maintenance and modification programs. In service, the CIs may be slightly different to those defined in acquisition, principally because the CIs in service are chosen to facilitate through-life support and will depend on the support philosophy adopted. Nonetheless, it is useful to minimize the differences between the acquisition and utilization CIs to reduce the impact of the transition to the in-service support agency as the system is fielded.

4.3.2 Group and Allocate Requirements

So, in the case of our aircraft example, we now know the requirements for the aircraft system as they are grouped logically in the populated RBS. We also have decided what the aircraft system is to consist of physically in terms of CIs (to simplify the example and to focus on the aspect of allocation, we assume in the following that each subsystem is a single CI). What is left to do is to connect the dots and relate the requirements from the logical architecture to the physical aircraft subsystem/CI architecture. This is the process of requirements allocation as performed in preliminary design and represents the translation from a logical architecture to a physical architecture. Requirements allocation tells us exactly what each subsystem/CI needs to do in terms of requirements, performance, and verification.

A proposed allocation for our aircraft system is shown in Figure 4.6. On the left side, we see the system-level requirements of the aircraft system (logical representation), and on the right side, we see the major subsystems, or CIs (physical representation). The two sides are related by the multitude of lines in the middle. Figure 4.6 illustrates one method of showing logical to physical relationships, which becomes unreadable very quickly. A more useful way of showing the same information is via a matrix, sometimes called an allocation matrix or traceability matrix. The information in Figure 4.6 is

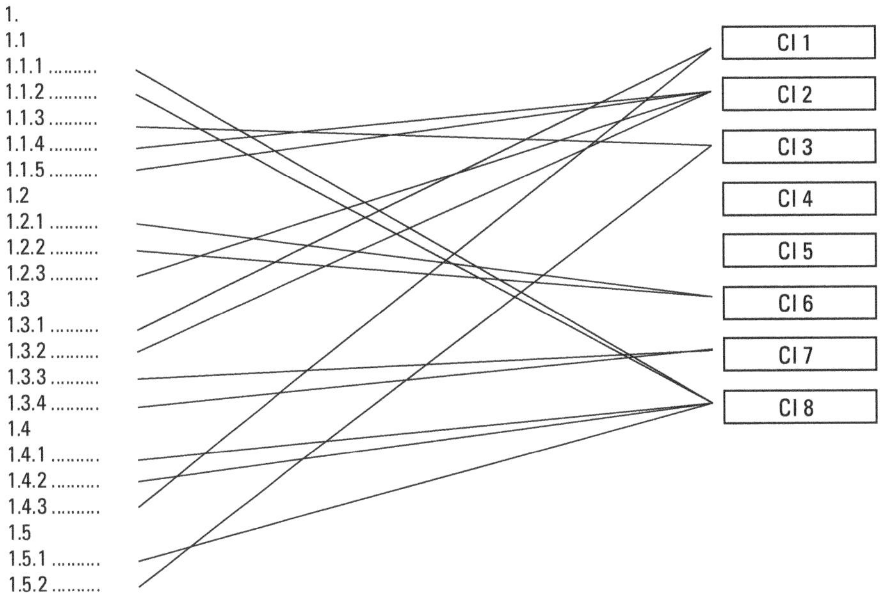

Figure 4.6 Example of grouping and allocation; the requirement statements in the RBS are allocated to one of the subsystems (and thence to one of the CIs).

shown in Figure 4.7 in the form of an allocation matrix for the CIs in our aircraft system example.

The RBS logical requirements (representing the logical design of the aircraft system) are shown on the left side of the matrix. The physical design is shown across the top of the matrix as a series of CIs as described previously. The designers of the individual CIs can clearly identify what requirements have been assigned to their CIs. The designers decompose/derive a host of additional requirements for their individual CIs to ensure that the CIs will be capable of meeting the assigned system-level and subsystem-level requirements. The higher-level requirements and the decomposed/derived requirements for each CI will form the basis of the requirement specification for that CI. The exact form of the CI requirement specification depends on the design alternative selected to produce the CI, but for developmental CIs, the specification is usually called a development specification.

The allocation matrix concept is also a very powerful tool when changes to either the logical or physical design are proposed. The impact of making changes to particular functions or CIs can be determined very quickly.

The test of the logical-to-physical allocation process is that there must be a one-to-one mapping between each RBS element and only one subsystem

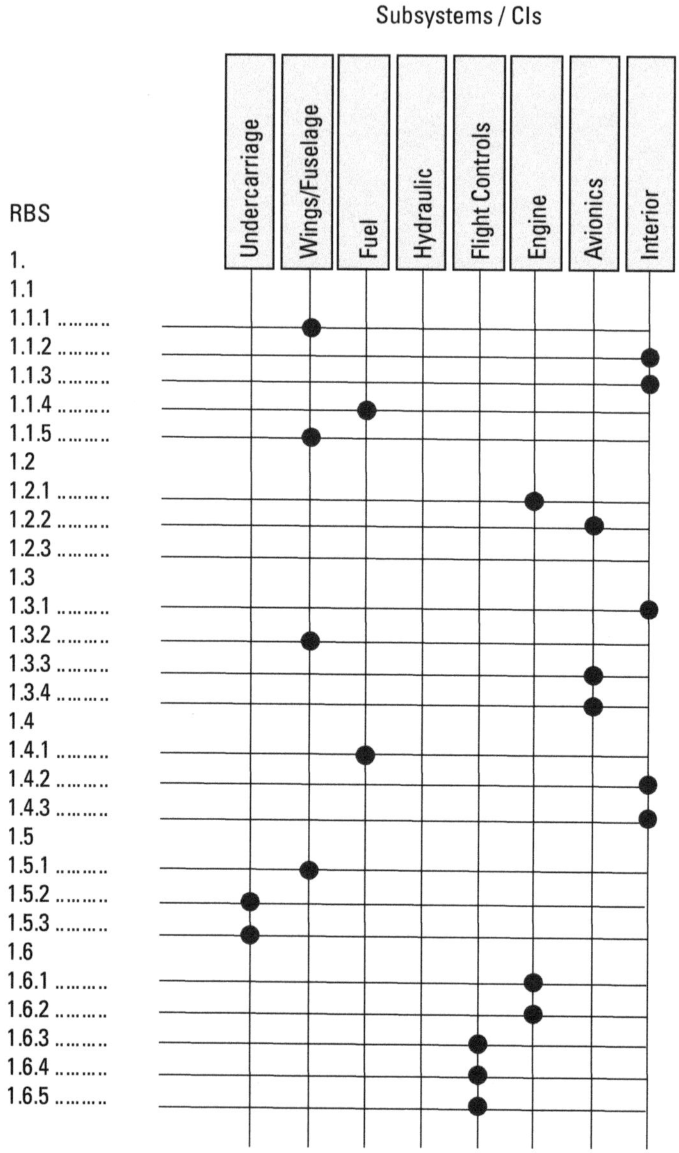

Figure 4.7 Simple example of an allocation matrix for the aircraft system showing allocation of RBS requirements (logically grouped) to physical subsystems (and subsequently to CIs within those subsystems).

(and only one CI within that subsystem); that is, the RBS function is performed by only one CI. In some cases, it is difficult to perform a one-to-one mapping, particularly for what are often called transverse requirements, such as safety, reliability, and availability, which must be allocated to multiple CIs within the discretion of the system designer.

Example 4.4: Engine CI

> *Mass is a good example of a system-level requirement that impacts on the design of all the CIs (apart from software CIs). In the aircraft example, we are limited in mass due to the requirement to operate from Class X airports. Since each CI contributes to the mass of the aircraft, a design decision is made to allocate to each CI a portion of that mass. For example, the engine CI may be allocated 15% of the maximum all-up aircraft mass and design of the engines must be achieved within that allocation to ensure that aircraft all-up mass is acceptable. To understand what else they must achieve with their engine design, the designers refer (vertically) to the allocation matrix and identify that their engine design must be capable of:*
>
> - *Propelling the aircraft sufficiently to take off on a runway of a certain length;*
> - *Achieving a certain fuel efficiency;*
> - *Achieving and maintaining a minimum cruising speed;*
> - *Enabling the aircraft to achieve minimum rates of climb and descent;*
> - *Being maintainable and reliable enough to allow operational-level maintenance to be performed in a very short amount of time.*
>
> *From these requirements that have been allocated to the engine CI, the designers derive requirements specific to the engine including the interfacing and integration requirements. For the engine CI, designers may arrive at the type of engine to be used, the required levels of thrust and other performance imperatives, air intake requirements, electrical requirements, mounting and interface requirements, and so on. These derived requirements and allocated system-level requirements form part of the development specification for the engine CI that, when complete, forms the ABL for that CI.*

The process of allocation has ensured that each RBS requirement (shown on the left of the allocation matrix in Figure 4.7) has been assigned to at least one subsystem/CI. By working horizontally in the matrix, we are effectively working from logical to physical design (top down). By working vertically in the matrix, we can appreciate the responsibility each CI has for achieving allocated requirements (bottom up). Bottom-up analysis shows how the requirements of the system drive the design of each of the subsystems/CIs (not the other way around).

As with all requirements, the TPMs determined during conceptual design also have to be analyzed and allocated to the various CIs in the design. The analysis helps to determine specific quantitative parameters that must be met by the CIs in order that the system-level TPM is met. The parameters are included in the specifications for the elements bearing in mind that these elements will either be procured (off-the-shelf), purpose-designed, or a combination thereof. Either way, the elements must meet the parameters assigned during preliminary design so that the integrated system is able to meet the TPMs.

4.3.3 Confirm Subsystem Selection

Before continuing, designers will be concerned to validate their choice of the set of subsystems/CIs by confirming as best as possible that the selected set meets the SNR. It is important to note the advantages of the top-down design approach as borne out by the selection of physical architecture to meet the decomposed/derived requirements and the TPMs assigned during preliminary design. The subsystems/CIs procured or designed for the system are done so with the system-level requirements and TPMs in mind. These subsystems/CIs are also procured or designed with a good understanding of the major interfacing and integration issues. By following the top-down approach, the subsystems/CIs stand a much better chance of being compatible with one another and combining to successfully meet the requirements of the system.

4.4 RBS Versus WBS

At this point, when considering the allocation of logical requirements to physical subsystems/CIs, it is useful to revisit the relationship of the well-known work breakdown structure (WBS) to the RBS. The previous examples have shown the process of establishing the logical design of the aircraft system and the eventual translation of that logical design to physical design. The logical design was articulated in the RBS and the physical design was illustrated graphically using an allocation matrix and the subsystem/CI list. The concept of a WBS is not far removed from the idea of the allocation matrix. The WBS identifies the products and the associated work packages necessary to produce the system. The work packages necessarily address the work required to produce all the subsystems/CIs listed in the allocation matrix. In addition, however, the WBS adds to the subsystems/CIs the work required to design, develop, integrate, and test these subsystems/CIs. The breakdown of the work and products in the WBS is based largely on project management imperatives and concepts such as scope, cost and schedule management, and resource allocation and management. To that end, there is a relationship between the WBS and the RBS that is not dissimilar to the relationship between the subsystem/CI list and the RBS shown in Figure 4.7.

Example 4.5: Example WBS for the Aircraft System

> *An example of an aircraft system WBS is shown in Figure 4.8 (based on [2]), including an illustration of the relative location of the subsystems/CIs (from the allocation matrix).*

Figure 4.9 provides a more complete view of the relationships among the elements of the project RBS (which includes system requirements in the SyRS and project requirements in the SOW) and the project WBS (which includes the CIs that make up the system in the system breakdown structure (SBS) and the enabling products produced to assist in the development of the system and its through-life management and support).

Note that although the subsystems/CIs do make up a significant portion of the WBS, there is a substantial amount of additional work recorded in the WBS; this work is identified in the contracted Statement of Work (SOW). Note also that the WBS is similar in structure to the RBS, although the WBS is organized physically, not logically.

4.5 P3. Conduct Interface Identification and Design

During the selection of the subsystems/CIs, the interfaces between the subsystems/CIs must be identified, as illustrated in Figure 4.10. The identification of interfaces is a critical part of preliminary design because interfaces not only determine successful operation of the system once integrated, but they also place additional limitations and requirements on the design of the individual subsystems/CIs.

4.5.1 Interface Control Document (ICD)

The task of recording the interfaces that must exist between system elements is managed primarily through an artifact, the system ICD, recognizing that the artifact could be something other than a textual document, which contains sufficient details to define completely the interfaces between the different CIs, or

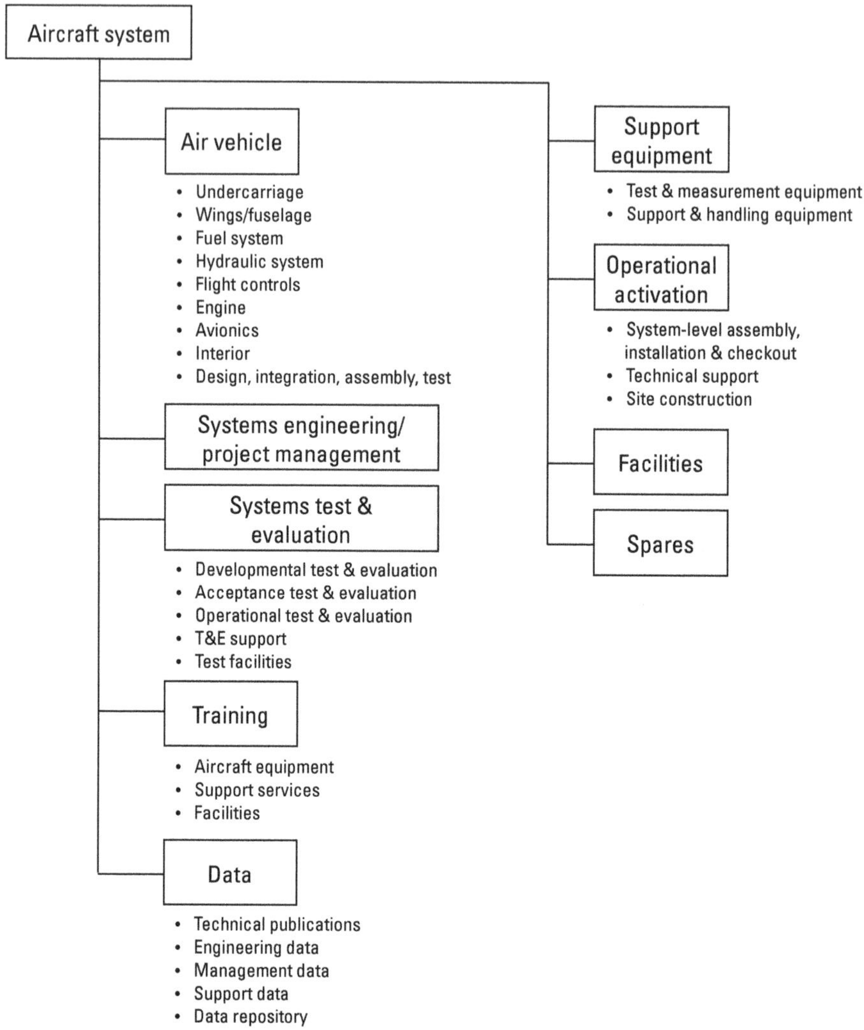

Figure 4.8 Aircraft system WBS.

at least points to where such information is recorded. A primary source of information for identifying interfaces is the context diagram produced as part of the requirements analysis effort in the conceptual and preliminary design phases.

Successful integration relies on accurate and complete interface definition during the preceding phases of the design. The ICD and individual interface requirement statements contained in development specifications and the product specifications all play an important role in the integration task. In addition to recording, cooperation between the various design groups (such as hardware

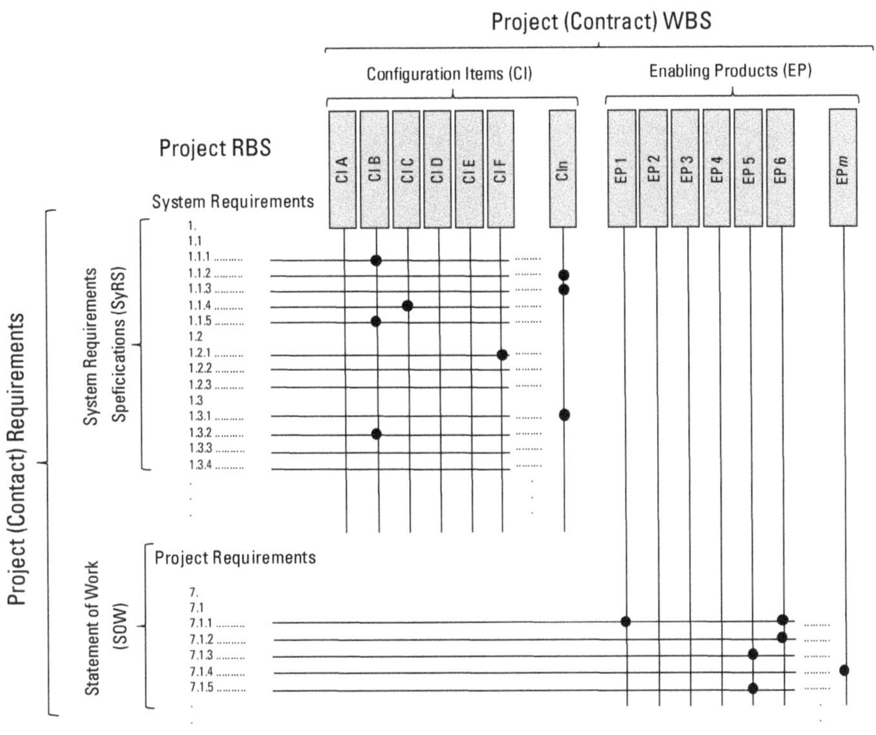

Figure 4.9 Relationship among the elements of a project RBS and WBS.

and software) is vitally important. Note that some system components may be completely designed and developed in geographically separate locations and only brought together during the integration stages.

The ICD is responsible for specifying the logical and physical interface requirements that exist between the different CI elements within the system. It should be remembered that interface requirements are defined in both the development and product specifications for the respective CIs. However, the ICD identifies the CIs between which there needs to be an interface and, through the definition of that interface, ensures that communication occurs between the different design teams. This is a critical process because interfaces often result in additional derived requirements being placed on the CIs. Additionally, CI-level interfaces become critical when considering engineering changes to a particular CI. The interfaces between that particular CI and other CIs in the design help to identify design dependencies and likely flow-on impacts of the change on the other CIs in the design. A formal interface control process is therefore a prerequisite for both successful change management and system integration.

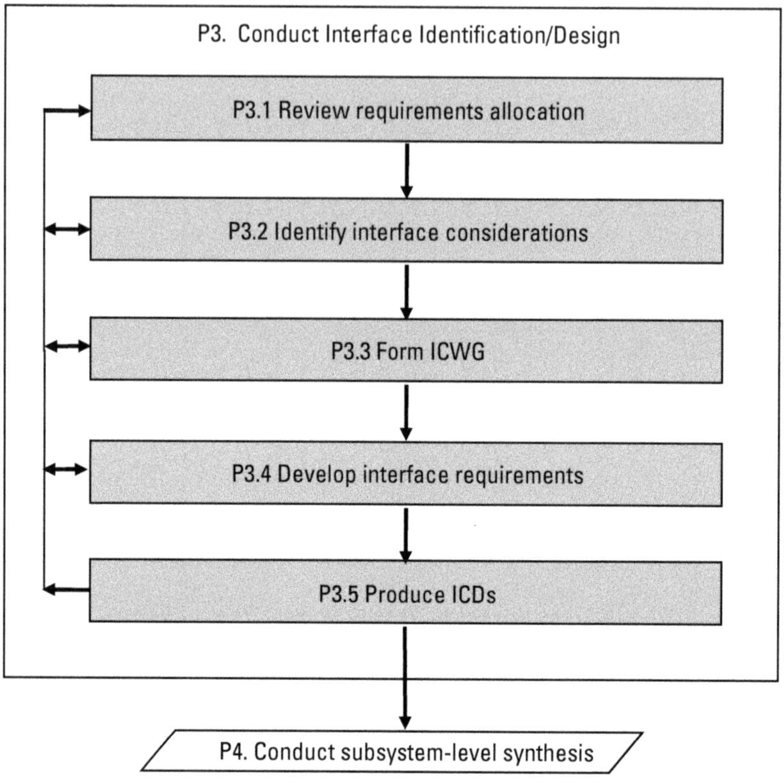

Figure 4.10 A summary of process P3. Conduct Interface Identification/Design.

Example 4.6: Importance of the ICD

> *Suppose a system consists of CI A and CI B. There would be separate development specifications for both A and B. The team responsible for A would not necessarily access the development specifications for B and vice versa. A single ICD would be produced (called the A-B ICD or similar) defining the common interface requirements. Both teams would need to refer to this ICD because it will refer to both products. Both teams would also be responsible for cooperatively producing and maintaining the A-B ICD (although, more usefully, one of the teams would be given principal responsibility for managing the interface). The ICD becomes the conduit through which the A and B teams communicate. In this way, the ICD assists in the systems integration effort.*

The interfaces are defined in gradually increasing levels of detail as the design progresses from conceptual design, through preliminary design, and onto detailed design and development. ICDs are also formally reviewed by the acquirer as part of the formal technical reviews and audits process outlined in Chapter 8.

4.5.2 Types of Interfaces

The ICD contains various types of information depending on the nature of the interface. Next we provide examples of some of the different types of interfaces that exist. Note that, to define completely some interfaces, an ICD may require information on more than one type of interface.

- *Physical interfaces.* Physical interfaces are normally the most obvious of interfaces because they need to exist in real terms. Examples include pipes through which a fluid or gas flows, fiber-optic cables passing digital information, or a structural interface such as a rope or a chain through which force is applied. In defining the physical interface, factors such as mechanical connection type and specification, physical size of the interface (such as diameter of pipe), and layout of the interface are defined. Physical interfaces often coexist with one of the other types of interfaces.
- *Electronic interfaces.* An electronic interface is the name given to the flow of electronic signals (analog or digital) between two points. These signals may flow over a physical interface such as a cable or fiber, in which case the physical cable or fiber interface is also defined. Alternatively, the interface may make use of short-range RF transmission such as Wi-Fi or Bluetooth or longer-range RF transmission such as microwave or satellite. The characteristics of the data flow between the two points need to be completely defined. The information requirements for this definition depend entirely on the nature of the interface but may include factors such as modulation, frequency, data format, and data rates.
- *Electrical interfaces.* Electrical interfaces are normally associated with a physical interface of some sort such as a power cable or data cable. The type of information required in this sort of interface will include voltage levels, voltage type (for example, 110V, 60 Hz), and any associated factors such as fault tolerance requirements. In the case of a data cable, pin assignments and descriptions may be required. If there are any electromagnetic compatibility issues relating to the interface such as shielding and earthing, these requirements will also be defined.

- *Hydraulic/pneumatic interfaces.* Hydraulic and pneumatic interfaces are the mechanical equivalent of electronic or electrical interfaces. A physical interface definition will coexist with this type of interface. Additional information such as flow rate, pressure, temperature, and fluid type will be required to define completely the hydraulic and pneumatic interfaces.
- *Software interfaces.* Software interfaces refer to passing data between different computer software CIs. This may occur within one piece of hardware using internal hardware buses or it may be between two different pieces of hardware. If the latter is the case, a physical interface definition is also required to define the physical link between the two pieces of hardware. Software interfaces require information such as expected inputs, generated outputs, format of the messages, and any protocols used during the transmission. Protocol definition includes data rate, error detection and correction, and start and stop codes.
- *Environmental interfaces.* Environmental interfaces are sometimes not considered to be interfaces, but they form the interface between the system and its external environment. In some cases, the environment provides the system with important information such as air data in the case of an aircraft. The environment may also provide the system with challenging interfaces such as structural (vibration, shock, and acoustics), thermal (ranges of temperatures), magnetic (flux densities), and radiation (flux densities and types). All environmental interfaces need to be considered and recorded to ensure that the most important integration of them all (system with its operational environment) is successful.

4.5.3 Interface Control Working Group (ICWG)

In large and complicated system acquisitions, a wide range of technical specialists (or experts) will be involved with producing a wide variety of subsystems. The specialists perform work in the traditional engineering fields such as electrical, civil, mechanical, and aeronautical engineering and in very specialized and detailed areas such as safety, human factors, electromagnetic interference/electromagnetic compatibility (EMI/EMC), and corrosion control. With all these specialized areas working on the same project, there is a significant requirement for a coordinated approach to the integration effort. The integration effort needs to be managed to ensure that compatibility exists between the different areas. The task of systems integration falls within the bounds of systems engineering management and can be managed effectively by ensuring the adequate attention is paid both to the logical and physical interfaces between the different areas.

When two or more CIs have an interface, the teams developing those CIs will no doubt have an interest in and specific requirements for that interface. To ensure that the teams developing the CIs have their needs met and to facilitate communication between the teams, an ICWG may be formed to bring together those involved in the interface definition and design.

ICWGs must meet regularly to generate the initial ICD and should then meet periodically or as the need arises to address any ongoing issues. These meetings and reviews may be informal, but the products of the meetings are formal interface design issues and decisions and updated ICDs.

Example 4.7: Interfaces and CI Design

A previous example examined the design of the engine CI using the allocated set of requirements to derive additional requirements. The allocated requirements are not the only requirements that will impact on the design of the engine.

The engine must interface with several other CIs, including the fuel CI. An interface between the fuel CI and the engine CI will be identified during preliminary design (see also integration management in Chapter 8). The identification of this interface includes details of the physical interface between the two CIs and also describes critical performance factors such as maximum flow rates, pressures and physical means of connection. Since both the engine CI designers and the fuel system CI designers have specific requirements from the interface, an ICWG may be formed to participate in the design of the interface.

Characteristics of the interface design such as flow rates, fuel pressure, and physical connections place additional constraints on the design of both the fuel and engine CIs. These constraints produce several derived requirements that must be included in the relevant specifications for the CIs.

4.5.4 Interface Definition Using the N^2 Diagram

An effective interface definition and design tool is known as the N^2 (or N2) diagram [3]. Although originally a software engineering tool, the diagrams are extremely useful in all types of interfaces by identifying the interfaces, identifying any conflicts, and highlighting any assumptions regarding dependencies. In simple terms, the diagram represents a matrix with system functions (or physical elements) running down the diagonal and the remainder of the cells in the matrix represent the interfaces. An example of an N^2 diagram is illustrated in Figure 4.11.

The interfaces are shown on either side of the diagonal. The interfaces are identified in both directions to take account of the fact that an interface may represent an input to one function and an output from another. If no interfaces exist between certain functions, the relevant box in the diagram is left blank.

The N^2 diagram can also use physical elements of the design in place of functions. During preliminary design, these physical items will be the CIs that have been identified, and the interfaces will represent interfaces between the CIs. As the design becomes more detailed during detailed design and development, these interfaces will be refined further. In this way, the N^2 diagram can support the top-down identification and design of interfaces throughout the development life cycle from very high-level logical interface identification during conceptual design to very specific and detailed interface design during detailed design and development.

Function 1 (F1)	F1 → F2	F1 → F3	F1 → F4	F1 → F5
F2 → F1	**Function 2 (F2)**	F2 → F3	F2 → F4	F2 → F5
F3 → F1	F3 → F2	**Function 3 (F3)**	F3 → F4	F3 → F5
F4 → F1	F4 → F2	F4 → F3	**Function 4 (F4)**	F4 → F5
F5 → F1	F5 → F2	F5 → F3	F5 → F4	**Function 5 (F5)**

Figure 4.11 An example N^2 diagram for five functions.

Example 4.8: Aircraft System N2 Diagram

An example of a partial CI-level N^2 diagram for the aircraft system is shown in Figure 4.12. Note the inclusion of "external systems" to allow the identification of interfaces with external systems and the environment.

Each of the interfaces identified in the N^2 diagram in Figure 4.12 is considered separately using the ICDs and interface requirements in the development specification. A detailed design of each interface is eventually recorded in the relevant product specification.

For example, interface requirements associated with the hydraulic pressure to the fuel pump would be contained in the interface design section of the hydraulic CI development specification and referred to in the fuel CI development specification.

Avionics	• Throttle controls • Electrical controls	Fuel pump control		HMI
Mechanical power to drive generators	**Engine**		Mechanical power	
Inputs to fuel gauges	Fuel pressure/ supply	**Fuel**		Current quantity indication
Inputs to hydraulic gauges		Pressure to fuel pumps	**Hydraulic**	Current quantity indication
Landing & navigation inputs		Refueling equipment	Refilling equipment	**External systems**

Figure 4.12 Example of a partial N^2 diagram for selected physical elements of our example aircraft system.

4.6 P4. Conduct Subsystem-Level Synthesis and Evaluation

Once the allocation of requirements has been performed, it is time to continue the familiar loop of analysis, synthesis, and evaluation and concentrate on synthesis. Subsystem-level synthesis results in a preliminary design that meets the subsystem-level requirements. Figure 4.13 illustrates the process.

4.6.1 P4.1 Review Requirements Allocation and Interfaces

The allocation process that has allocated requirements to individual CIs has completed the initial analysis. Designers responsible for each CI continued the analysis to an additional level of detail necessary to support the subsystem-level synthesis. Technical specifications and ICDs have been produced to detail the requirements, performance, and interface requirements for each CI. As a first step in the synthesis process, the specification and ICDs for each

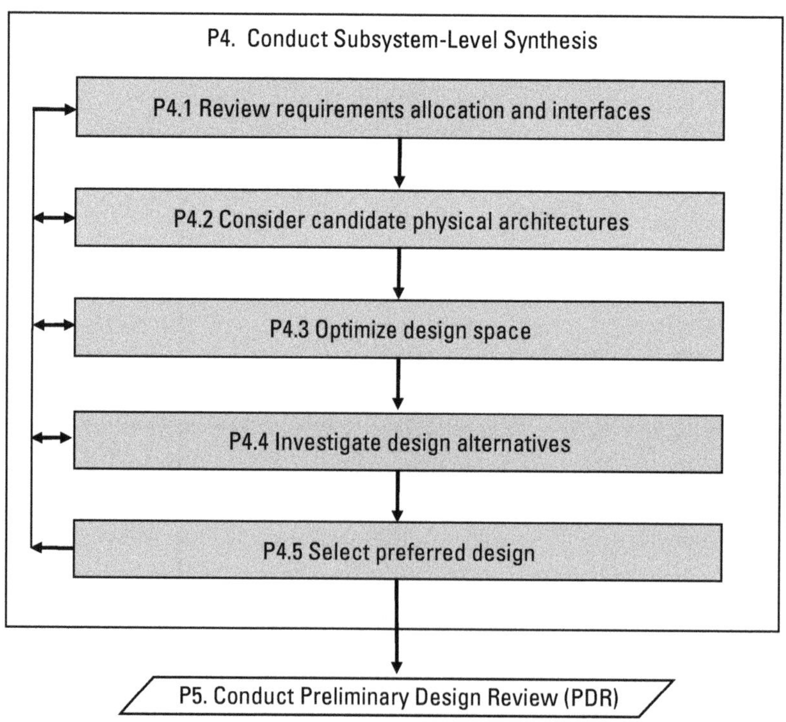

Figure 4.13 A summary of process P4. Conduct Subsystem-Level Synthesis.

CI should be reviewed, because it is these specifications that must be satisfied by subsystem synthesis.

4.6.2 P4.2 Consider Candidate Physical Architectures

The initial synthesis is conducted to determine a preliminary design that satisfies the specification using one of the three broad design options available. There are three broad design options available including OTS, modified OTS, and developmental items. Factors that impact on the designer's decision to use OTS, modified OTS, or specially developed items include the specific requirements to be performed by the item, the availability and stability of the current technology, the size of the market, supportability and cost, and any contractual directives. The following sections discuss each of these options.

4.6.2.1 OTS

Commercial OTS (COTS) equipment, as the name suggests, refers to pieces of equipment that are commercially available. Military OTS (MOTS) and government OTS (GOTS) similarly refer to equipment that is readily available and probably in service in the respective domains. Because there are several sources of such items, the term OTS is now commonly used. An OTS item can also be referred to as a nondevelopmental item (NDI) to distinguish it from one that is developed (and therefore subject of a development specification).

If the OTS equipment meets or exceeds the requirements detailed in the development specifications, there may be some considerable advantages to using OTS equipment.

- The most obvious advantage is that OTS equipment is that it is likely to be readily available with limited or no delay.
- Technical risk may be reduced since the OTS product is a known quantity whose performance can be readily verified.
- An OTS option is most likely cheaper than a developmental item (on the assumption that the cost of design and development has been amortized over a broader context).
- If the OTS item has a large user base, maintenance and support will probably be already in place for the item. If not, at least data may be available to facilitate the planning of maintenance and support (to facilitate the quantification of spares holdings, for example).
- The OTS item may already be validated in the operational environment, saving considerable costs in validation and verification.

The disadvantages often associated with OTS items include:

- The OTS product may not be perfectly suitable. There may be problems with form, fit, or function, including inappropriate size, shape, mass, or color or excessive or insufficient functionality.
- The OTS item may be based on outdated technology and may soon face technical obsolescence; if the item has been in the marketplace for several years, the technology incorporated will have been developed several years before that. The designers of the current system will therefore need to trade off the robustness of a proven item against the age of the incorporated technology.
- Alternatively, an OTS item may be very new and only recently released, meaning that it may be relatively unproven; designers should look for evidence of robustness. Most people will have direct experience with buying a new piece of technology or updating their personal devices with the latest operating system only to find problems that had not been detected or addressed prior to release.
- Use or support of the OTS product may be constrained by warranty, ownership, or intellectual property issues.
- OTS products may come with very limited documentation (functional or physical) that may limit understanding of their design and particularly their interfaces, making it difficult to integrate these products with other elements of the design.
- The OTS product may come with additional functionality that may be undesired or may be unknown and therefore represent a security risk. For example, modern concerns around cyber-security should prompt systems engineers to take a close look at OTS products before incorporating them into a design.
- Support of OTS will most likely need to be undertaken through the original supplier; any repair/modification by the acquirer may void warranty.
- Additionally, the provider of the OTS item may not be stable and may not be around to continue to support the item in the long term.
- The need to qualify new pieces of OTS equipment in the system's operational environment may also cause problems. Qualification is a time-consuming and expensive exercise and may reduce considerably the benefits associated with the use of OTS equipment. For example, before using an OTS radio in our aircraft example, the radio system would need to be proven or qualified using a set of standard

test procedures to ensure the radio could survive and function under the temperature, vibration, shock, and humidity extremes associated with aircraft operations. As noted earlier, however, these problems will dissipate if we choose an OTS item from a range of appropriately qualified candidates that have already been verified and validated for aircraft operations.

OTS items will ideally be described by a product specification delivered with the equipment, but the information available may well be substantially less detailed than a standard product specification. However, the description does need to be detailed enough to support tasks such as integration, verification, maintenance, and support.

It should also be noted that not all OTS will be provided by an outside supplier. Sometimes it is provided by the customer (as customer-furnished equipment) or it may be an in-house element from the contractor's existing inventory. In those cases, some of the disadvantages of OTS may be alleviated because the item is presumably better understood, original design description may be available, support may be well defined, and the element may be validated in the desired environment.

4.6.2.2 Modified OTS

OTS equipment that meets most requirements contained in the development specification for a CI may be able to be modified to better suit the needs of the designer. Similar advantages will exist with modified OTS as for OTS; however, there are the following additional disadvantages:

- Warranties and maintenance and support agreements may be voided if the OTS item is modified.
- The effort involved in modifying the OTS item can be easily underestimated, and the modification process can sometimes take much longer and cost much more than initially estimated.

Modified OTS equipment may be described in several ways. One way is to describe the original OTS equipment using the equipment's own documentation augmented by an OTS modification specification to describe the necessary modifications to the OTS item. An OTS modification specification may include a combination of process specifications to describe how to modify the OTS item and materials specifications detailing the materials needed to carry out the modification. These specifications are described in more detail in Chapter 9.

4.6.2.3 Developmental Items

If suitable OTS equipment is not available (even if modified), the designer may opt to design and develop the item from the ground up to meet the specific requirements and characteristics detailed in the relevant development specifications. Developmental items have the following advantages:

- An item developed from the ground up is likely to match the desired criteria precisely in terms of form, fit, and function.
- All aspects of the developed items will be understood (from a security perspective, for example, we will have much more confidence in the software written by our programmers).

The following disadvantages may result from a developmental approach:

- The effort involved in this development will not be insignificant.
- The desired advantages may never accrue due to the increased technical risk of developing a novel item.
- Maintenance and support issues will also need to be developed and deployed when items have been designed specifically for the system.

Hardware and software development specifications and hardware and software product specifications will describe developmental items in progressively more detail. These descriptions will contain information on the specific item and its interfaces. Hardware items may also require accompanying fabrication/production information including production process specifications, material specifications, and parts lists to complete the description.

4.6.2.4 Use of Architectures

We saw earlier that the logical architecture and the physical architecture are separate descriptions of the system and that we need to develop the logical architecture before choosing the physical architecture. However, designers are careful to be cognizant of candidate physical architectures when developing the logical architecture. Maier and Rechtin [4] observed that the two architectures should be as close as possible. Figure 4.14 shows (across the top) the process that we discussed earlier as the problem domain is gradually coalesced into the BRS, then the StRS, and then the SyRS, which describes the logical architecture of the system. This logical architecture in the FBL is then allocated to a physical architecture in the form of the CIs as they are described in the ABL.

As illustrated at the bottom right in Figure 4.14, the physical solution domain contains all the available technologies as well as the various types of

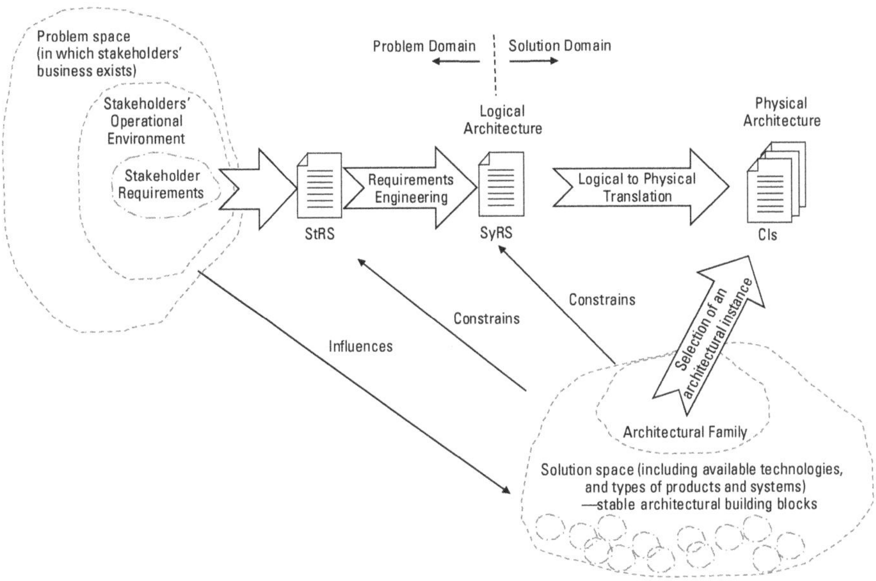

Figure 4.14 The use of architecture to support system development.

components, assemblies, products, subsystems and systems that are known to designers.

An architectural family is a way of combining elements of the solution domain to be useful (for example, in an ICT system, designers may choose between a client-server architecture or a distributed database architecture). It is from these families that designers choose an architectural instance to describe the way in which the system is to be described as an aggregation of CIs.

Across the bottom, Figure 4.14 shows that our desire to solve problems in the problem domain influences suppliers to undertake research and development to investigate technologies and create products that can be used in the solution domain to create new systems. Sometimes that can happen in the context of the development of the system; other times, it happens independently so that the technologies are already available when the need for the system becomes apparent.

So, as illustrated in Figure 4.14, the available physical design options are considered when designers allocate the functionality required to meet the SyRS to the various configuration items that will make up the ABL in the physical architecture. Designers draw on available architectures to make ready use of OTS CIs, they can modify OTS, or they can develop novel CIs for this system.

Largely, the detailed designers are engineers who draw on their engineering training which has given them experience in the design, development, and integration of standard (largely extant) CIs, which, for example, for a vehicle may be such items as differentials, drive trains, suspension systems, engines, and engine management systems.

When the system designers developed the SyRS, they were cognizant of the design tasks that they would be creating for lower-level designers so they took the available physical architectures in mind when they developed the logical architecture; there is little point in developing a logical architecture that is infeasible (that is, it cannot be met by any reasonable physical architecture). Similarly, when stakeholders were developing their requirements, requirement engineers provided advice on the feasibility and risk of achieving the desired business requirements from extant technologies and engineering methodologies and tools. New business requirements should be able to drive new technologies and new physical architectures, so the range of products available is influenced by the solutions required to meet the business problems faced.

The consideration of physical solutions for any system is best undertaken within the context set by trade-offs conducted at the logical level so that changes in technology can be accommodated appropriately. A suitable logical architecture changes slowly and remains valid so long as the fundamental business remains constant, and the functional trade-offs still apply. However, the same is not true for the physical architecture, which will commonly change several times within one generation of the logical architecture.

4.6.2.5 Technical Maturity

One of the most significant risks to a system development is the maturity of the technology being utilized. There are several ways to assess technical maturity but, in any case, it is common to rely on the notion of technical readiness levels (TRL). The US GAO [5] recommends the following nine levels:

- *TRL 1:* Basic principles observed and reported. Scientific research can be transformed into applied research and development and a material or process can be shown to have promise for application in system development.
- *TRL 2:* Technology concept and/or application formulated. Practical application can be demonstrated, although there may not be detailed analysis or validation in an integrated system.
- *TRL 3:* Analytical and experimental critical function and/or characteristic proof of concept. Formal research and development have begun in the form of rigorous analytical or experimental evaluations to validate the technology, normally in isolation (that is, without integration).

- *TRL 4:* Component and/or breadboard validation in a laboratory environment. Basic technology components are integrated in an ad hoc manner to investigate interfaces.
- *TRL 5:* Component and/or breadboard validation in relevant environment. Basic technology components are integrated in a more realistic environment with reasonably realistic support elements.
- *TRL 6:* System/subsystem model or prototype demonstration in a relevant environment. System/process prototype or representative model demonstrated in a relevant environment.
- *TRL 7:* System prototype demonstration in an operational environment. Near-final system/process prototype demonstration in an operational environment.
- *TRL 8:* Actual system completed and qualified through test and demonstration. The system has matured to its final form and is proven to work through DT&E and AT&E.
- *TRL 9:* Actual system proven through successful mission operations. Actual system in its final form proven through successful operations (OT&E) under mission conditions in its operating environment.

While maturity of technology is assessed continuously throughout a system life cycle, there are commonly several formal technical maturity assessment activities associated with major milestones, particularly acquisition. For example, there would be an initial general assessment to inform the risk management plan at the outset of the project to inform the choice of the functional architecture, which is confirmed at SDR, assessment of CIs at PDR to inform acceptance of the ABL, assessment at FQR, OT&E to inform operational capability, and at key OT&E points throughout the life cycle.

Although not meaning to be completely quantitative, TRLs are a very useful way of communicating technical maturity. For example, a customer may communicate to tenderers that they would expect that all elements of a tendered solution are assessed at TRL 8 or higher.

Assignment of a TRL can be to the entire system or to elements of the system. When the assessment is undertaken at the system element level, the TRL assigned is the lowest TRL of the constituent subsystems; the TRL assigned to the subsystem is the lowest TRL of the constituent components.

4.6.3 P4.3/P4.4 Optimize Design Space and Investigate Design Alternatives

During the selection of the different CIs making up the system, it is important to remember that it is the system performance that is of vital importance, not the performance of the individual CIs and components. That is, to achieve optimal system performance, the performance of the CIs may need to be constrained and balanced (that is, have suboptimal performance). In many cases, none of the CIs may perform as well as they might be able to so that their combined performance (the performance of the system) is optimized. A simple example illustrates this concept.

Example 4.9: System/Subsystem Optimality

> *A traffic management system (Figure 4.15) is used to control the traffic flow through a central business district. The CBD has North/South traffic flow along the roads 11, 22, and 33, and East/West traffic flow along the roads AA, BB, and CC. Each of these roads could be considered a subsystem.*
>
> *For each road subsystem to operate optimally with maximum traffic flow, all traffic lights on that subsystem must remain green. While that may be a reasonable design decision in the context of any single subsystem, this situation clearly results in suboptimal performance of the system as all of the other traffic lights must be red. Most design space problems are multidimensional, representing nonlinear problems that require multidisciplinary teams to solve.*

To cater for the concept of optimal system performance, trade-offs need to be considered during the selection of the preferred design approach. These trade-offs consider the performance requirements of the different subsystems and that of the system and consider any "design space" available for trade-off purposes. This helps to ensure that the most important system-level performance measures (as ranked during conceptual design) are met first followed by the less important parameters.

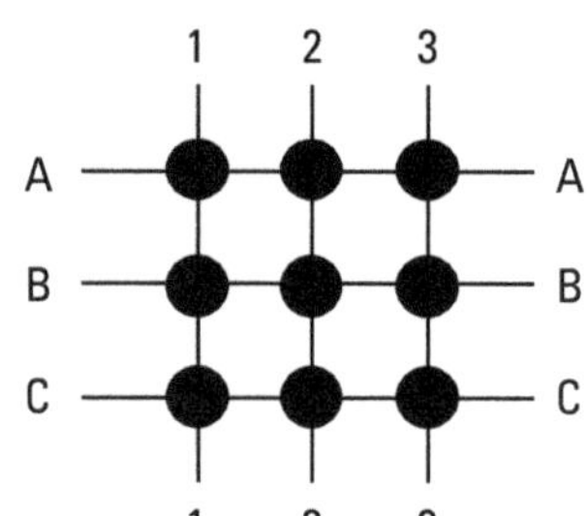

Figure 4.15 A road system comprising several subsystems.

Example 4.10: Aircraft System Design Space

The aircraft system development has reached a stage where the design of the CIs can commence. This example investigates the design of the interior CI and the actual passenger seat. The example demonstrates the concept of design space and its use during preliminary design (and subsequent design activities).

We know that seat design is of critical importance to the class-leading comfort required for the aircraft system and we can trace the seat dimensions back to the StRS. Let's say for this example that the minimum acceptable seat size is 576 in^2. Seat dimensions have been allocated to the interior CI. When speaking to the designers responsible for the seat design, they explain that the number of passengers on board critically limits the possible seat dimensions. Naturally, the more passengers, the smaller each seat becomes for a given overall length allocated to the interior CI. We also have the requirement for the number of passengers that the aircraft system must be capable of transporting. For the purposes of this example, the minimum acceptable passenger load is 150 people.

The relationship between seat dimensions and passenger load (as determined by the designers of the interior CI) is shown simplistically in Figure 4.16.

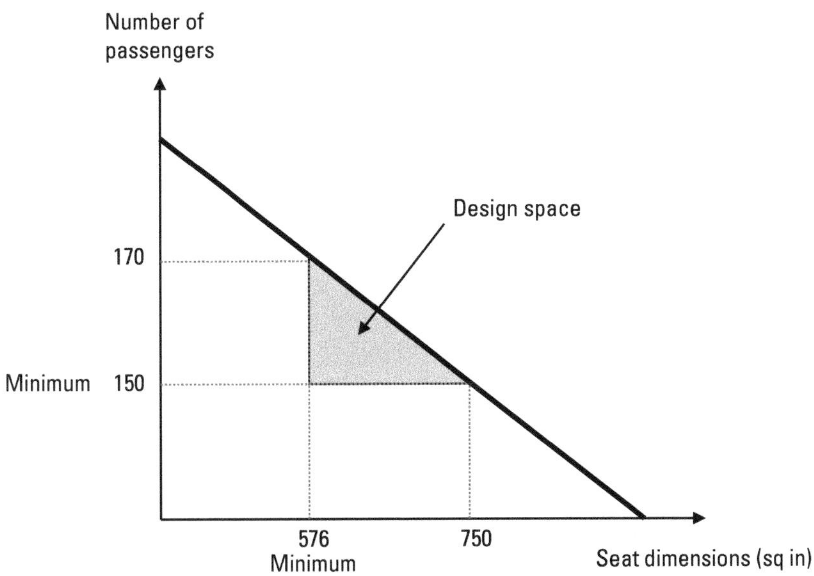

Figure 4.16 The design space represented by the relationship between the number of passengers and the seat dimensions.

Based on Figure 4.16 alone, we feel confident that we can meet both requirements. We can either increase the number of passengers to 170 (exceeding the minimum requirement for 150) and still meet the seat dimension requirement, or we could meet the passenger requirement and provide each passenger with a much bigger seat, or we could choose somewhere in between. In short, we have some "design space." The commercial viability of the aircraft (from which the passenger capacity requirements were derived) may be deemed more important than exceeding passenger comfort levels. If this is the case, then the decision may be to maximize the number of passengers on the aircraft and opt for 170 passengers with 576 in^2 seats.

Provided that we keep within the design space available, we will meet or exceed all requirements. The aim in using design space is not to exceed some requirements at the expense of others, but rather to strike a balance and use the design space to at least satisfy all requirements. To use design space appropriately, however, the stakeholders must be consulted. Designers do not necessarily own the design space; the stakeholders do. In this example, the commercial stakeholders were given priority over the stakeholders vying for additional seat size.

We also know that, to meet the StRS requirement to operate from all Class X airfields in the world, a maximum aircraft mass constraint was determined. For this example, we assume that the aircraft mass is not to exceed 130,000 lbs. Weight was allocated to all CIs, including the interior CI. Figure 4.17 illustrates how passenger numbers impact on aircraft all-up mass.

The design space relating to passenger numbers has now shrunk somewhat from that described in Figure 4.16. Based only on seat dimensions, the aircraft could carry 170 passengers. However, when all-up mass is considered, only 160 passengers can be carried. The designers of the interior CI will be happy, because they will now be able to exceed the minimum seat dimensions set for them (see Figure 4.16). Note also that we have a minimum achievable all-up mass of 120,000 lbs if we only carry the minimum acceptable number of passengers.

Often, however, this type of analysis reveals unavoidable conflicts between requirements and the design of individual CIs. For example, the designers of the engine CI may identify an OTS engine that meets or exceeds all allocated and derived requirements, except for the minimum thrust requirement. The engine does not produce sufficient levels of thrust to support the allocated requirements of minimum runway length, cruising speed, and rates of climb/descent for a 130,000-lb aircraft. The designers are attracted to the OTS engine because it offers all the classic advantages of OTS design:

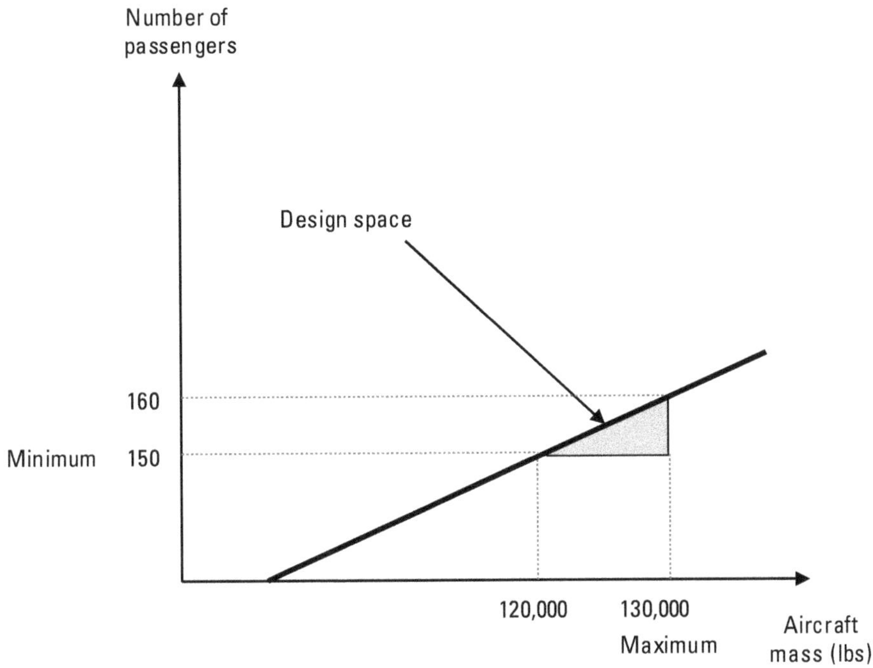

Figure 4.17 The design space represented by the relationship between the number of passengers and the aircraft mass.

reduced schedule, cost, and technical risk and well-established logistics support. Further to this information, the designers indicate that it is not technically feasible to provide the necessary levels of thrust from an aircraft engine using current technology for a 130,000-lb aircraft. The engine designers illustrate their case, with Figure 4.18 showing thrust versus mass necessary to achieve the allocated requirements of runway length, cruising speed, and rates of climb/descent.

Figure 4.18 clearly shows that the engine will be incapable of supporting the performance requirements for an aircraft that exceeds 115,000 lbs. In isolation, this is not a problem because 130,000 lbs is, after all, a maximum and a 115,000-lb aircraft will obviously meet that requirement. However, when passenger load is revisited, it is seen from Figure 4.17 that the 115,000-lb upper limit is not sufficient to meet the minimum passenger load of 150. This is an important conflict in the requirements.

Traceability now becomes critical as the systems engineer traces the requirement for minimum passenger load back to the requirements in FBL

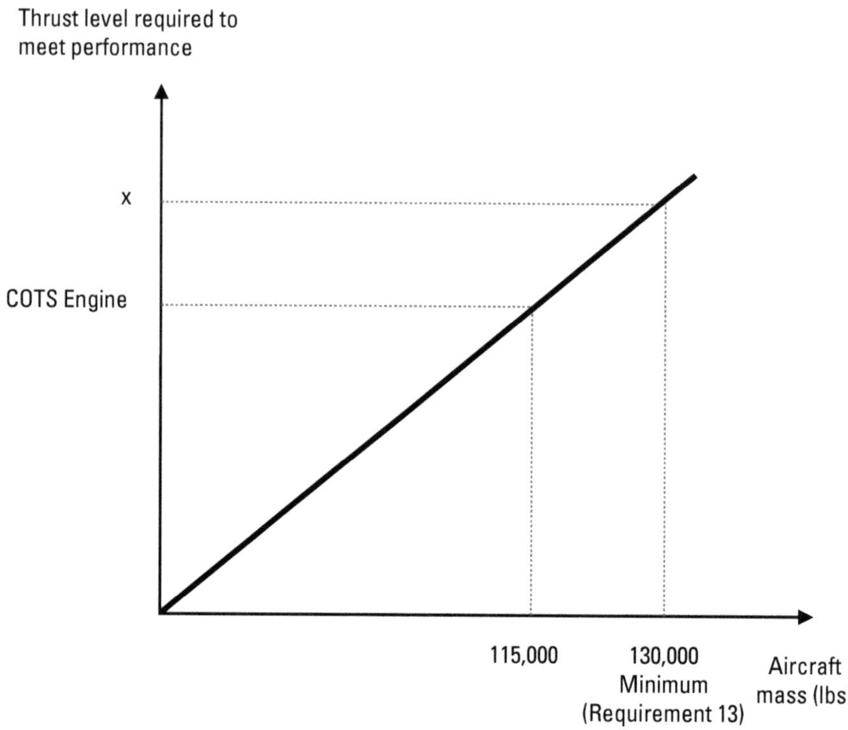

Figure 4.18 Revised power versus capacity function.

(at the system level) from which that requirement was derived. The performance requirements came, in part, from regulatory requirements that are not negotiable. In this simple example, it is easy to remember that passenger load was related to commercial viability. The systems engineer must alert the owners (the relevant stakeholders) of this requirement as soon as the conflict becomes apparent, seeking guidance, clarification, and prioritization. Based on the feedback, design decisions can be made while keeping the owners informed. The system-level designers may be able to save some mass in other areas of design (by referring to the design space) to allow the minimum passenger load to be achieved while keeping the aircraft all-up mass within the 115,000-lb limit set by the engine design.

This exercise has demonstrated the concept of design space and trade-offs to meet the original requirements of the system. The example has also demonstrated the concept of suboptimality where subsystem design must not take priority over achievement of system-level performance requirements.

Trade-offs within design space must be cognizant of the subsystems making up the system (as defined during the subsystem requirements analysis and requirements allocation effort). Each subsystem affects different parts of the functionality and has different levels of integration and interdependency. To that end, each unit has its own set of trade-offs and design spaces within which it must operate.

4.6.4 P4.5 Select Preferred Design

The purpose of writing development specifications is to specify the requirements for each of the subsystems/CIs making up the system to allow design engineers to select the subsystems/CIs that will be compatible with, and contribute to satisfying, the system-level requirements. This process is iterative involving the familiar activities of analysis, synthesis, and evaluation. It may not be possible to meet all the allocated requirements exactly on the first run through the design. The design (or synthesis) is evaluated to determine the level of success in meeting the requirements. Shortfalls are identified and the process must be repeated with different variables being traded off until as many of the allocated requirements as possible are met.

Once the preliminary design process is mature, the result is a preliminary architecture made up of various subsystems/CIs designed and organized in such a way to meet a series of requirements that have been assigned to each element of the preliminary design. Each subsystem/CI has an individual description of its intended purpose and the requirements that it must be capable of meeting to fit into the system. These individual specifications must be detailed enough for a hardware engineer to design and build a piece of equipment (if that is deemed to be an appropriate approach) or for a software engineer to write a piece of software to perform the functionality, or for a logistics person to source a piece of OTS hardware of software equipment to satisfy the requirements. Clearly, the level of definition for these individual specifications will vary. This allocation of requirements to system elements, leading to the production of the ABL, is the major output of preliminary design.

4.7 P5. Conduct PDR

The PDR ensures the adequacy of the preliminary design effort prior to focusing on detailed design. PDR is designed to assess the technical adequacy of the proposed solution in terms of technical risk and the likely satisfaction of the FBL. PDR also investigates the identification of CI interfaces and the compatibility of each of the CIs. The major PDR activities are illustrated in Figure 4.19.

As with all reviews, PDR must be conducted only when the acquirer is satisfied that the system design has progressed to such an extent as to justify holding the review. Prior to commencing PDR, the acquirer must be confident that the system architecture is in place, with all CIs identified and described to an appropriate level of detail in the development specifications. Interface identification and design should also be confirmed by investigating the state of the ICDs and the content of the development specifications prior to the review. The acquirer also needs to be satisfied that other entry criteria such as financial and project management criteria have been satisfied.

PDR investigates the results of the CI requirements analysis, the requirements allocation process, and the various trade-off studies conducted during the synthesis stage. Again, the results of these activities should be seen to address all requirements. Any deviations from the requirements are noted and corrective action determined and initiated.

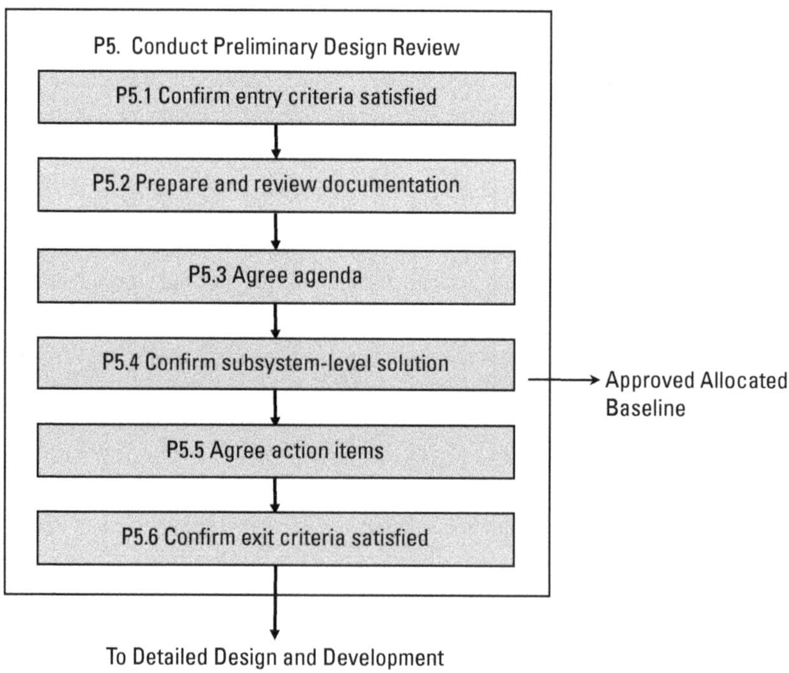

Figure 4.19 A summary of process P5. Conduct Preliminary Design Review.

At the conclusion of a successful PDR, the participants should be confident that all the FBL requirements are being addressed adequately by the CI design, and that the logical and physical interfaces that need to be in place between the CIs and between the CIs and the external environment have been identified and described as ready for detailed design. Participants should also be satisfied that the ABL is in a fit state to be used during detailed design and development and that the design approach for each CI has been determined. Any additional exit criteria listed against PDR should also have been adequately demonstrated. The successful conclusion of PDR results in the establishment of the ABL against which detailed design is subsequently authorized. Sometimes, an issue is raised during PDR that needs resolution prior to commencing detailed design and development. Either the PDR can be stopped until the issue is resolved or a decision is made to continue with the PDR and deal with the issue in a separate forum.

Three broad approaches can be taken with respect to PDR. A separate PDR can be conducted for each subsystem/CI of the system, separate PDRs can be conducted for logical groups of subsystems/CIs, or a single PDR can be conducted covering all subsystems/CIs. The approach taken should depend on the size and complexity of the system under investigation. In some cases, design dependencies between subsystems/CIs dictate the order in which PDRs can be conducted. The principal artifacts reviewed and approved at PDR are the development specifications relating to each of the subsystems/CIs and the ICDs describing internal and external interfaces.

The aims of the PDR include:

- *Evaluation of the preliminary design.* PDR investigates preliminary design activities to ensure that they have been conducted thoroughly and that the resultant design is appropriate and adequate margins exist regarding TPMs.
- *Approval of development specifications and ICDs.* The subsystems/CIs defined during preliminary design are recorded in the development specifications and ICDs. PDR evaluates these artifacts and approves them prior to detailed design and development.
- *Assessment of traceability.* PDR will ensure that all the requirements in the FBL have been adequately captured and addressed in the ABL; that is, that the requirements flowdown is appropriate and backwards and forwards traceability is provided.
- *Establishment of the ABL.* The development specifications and top-level design developed during preliminary design are the central elements

of the ABL. This baseline is formally established at the conclusion of PDR and paves the way for the detailed design effort.

Other artifacts supporting the top-level design are also reviewed during PDR if available. Examples of additional artifacts include test plans, production plans, and support plans. Particular attention should be paid to high-risk areas, interface requirements, and system-level trade-off analyses. If appropriate, TPMs identified earlier in the system design effort should be discussed and progress should be traced during PDR.

4.8 Tools

There are several tools that may be used during preliminary design (in some cases, these tools will have already been used in conceptual design and will also be used in later design activities). This section aims to introduce the more common tools that may be used during preliminary design and beyond and briefly explain their characteristics.

4.8.1 Schematic Block Diagrams

Schematic block diagrams (or simply schematics) are perhaps the most basic and popular synthesis tool. Schematics allow designers to visually describe a solution to a given logical need as defined during the analysis stage of the process. Schematics show all the elements of the solution and show their interrelationships and interfaces. These interfaces will include internal interfaces (between schematic elements or CIs) and the external interfaces (with external systems beyond the system of interest). Schematics can be drawn at different levels of detail during systems engineering processes to serve different purposes. For example, during conceptual design, the context diagram shows the system of interest as a "black box" and concentrates on boundary definition and external interfaces but does not show any detail of the system architecture. Later in conceptual design and throughout preliminary design, a system-level schematic block diagram can be developed to provide a "white box" view showing the subsystems and CIs and their internal interfaces. Each of those subsystems and CIs may be then visualized in increasing levels of detail during detailed design and development using more detailed schematic block diagrams.

A major advantage of schematic block diagrams is their ability to communicate clearly the proposed solution to technical and nontechnical persons alike: *a picture is worth a thousand words.*

An example of an avionics subsystem schematic block diagram is shown in Figure 4.20, showing the major CIs, their interrelationships, internal interfaces, and external interfaces. Notice how readily the design and operation of this avionics CI can be understood.

4.8.2 Physical Modeling

Physical modeling can take any one of several forms. Perhaps the best-known physical models are engineering models, such as the wind-tunnel models used during vehicle design to determine aerodynamic stability and characteristics. These models can include scale models of aircraft and land vehicles. Scale models of ocean-going vehicles such as ships and submarines can also be tested in wave pools to confirm buoyancy and stability characteristics. In a third, possibly less well-known example, scale models of aircraft, vehicles, and ships can be used in anechoic chambers to help to measure and estimate radar signatures. All of these models are built to exacting standards to ensure that the results obtained during the trials will be representative of the full-scale variant.

Full-scale models or mock-ups of our systems may be built to allow human factors engineers to experiment with design decisions such as cable routing, access points, and operator equipment.

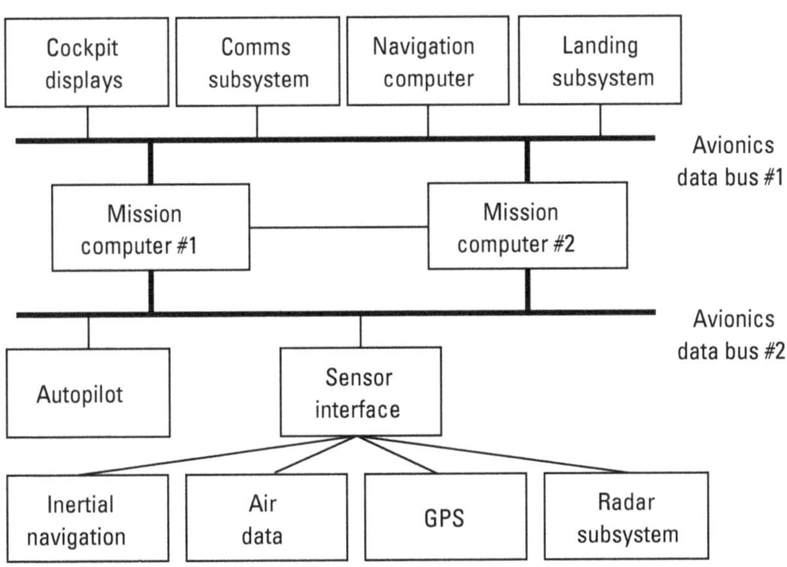

Figure 4.20 An example schematic block diagram.

Electrical and electronic design can also benefit from different forms of physical modeling. Examples include breadboards of electrical and electronic circuits and brass boards used during the design of radio frequency (RF) elements. These mock-ups can be used to help prove electronic designs options by using real components and real input/output signals.

It is fair to say that the use of physical models in today's design and development environment is gradually reducing as the power and capability of computer-based tools increases. For example, the use of physical scale models in wind tunnels, wave pools, and anechoic chambers is very expensive and time-consuming. Computer-based tools are now used to at least supplement this form of physical modeling to help to reduce the time and cost associated with the design and development effort. Electrical and electronic design software is continuing to improve and is replacing the requirement to develop circuits using mock-ups like breadboards.

However sophisticated computer tools become, some form of physical modeling will still play a role in the synthesis of certain design solutions prior to moving to the construction and production phase. The production of a system prototypes (or pre-production systems) for the purposes of test and evaluation may be an example of the continued use of physical models.

4.8.3 Mathematical Modeling and Simulation

Mathematical modeling and simulation were discussed in Chapter 2 as being useful tools for requirements exploration; they are also used during the physical design stages of system developments. Mathematical modeling and simulation are widely used and are becoming increasingly popular as computers and software become more capable. As the name suggests, mathematical models rely on a mathematical representation of a physical system or design. The models can then be used to determine both static and dynamic characteristics of the system. An example of a static model would be one that represents a bridge's structural characteristics. Loads and stresses can be changed in magnitude and position to ensure that the structure will withstand the loads associated with its intended purpose. An example of a dynamic system is an aircraft's autopilot system. This is a classic example of a control system. Characteristics such as stability and response times can be determined using this sort of model.

Finite element analysis (FEA) [6] is a very well-known mathematical modeling technique and serves as a good example of how mathematical modeling can assist and enhance synthesis throughout the systems engineering process. FEA is used to study the static and dynamic behavior of physical structures such as aircraft wings when they are placed under varying conditions. FEA is used to understand this behavior without the need for expensive and time-consuming prototypes and tests. The structure under consideration is assumed to consist of a finite arrangement of individual elements. Each element is modeled mathematically using relatively straightforward physical and engineering laws. The individual models are then combined in such a way as to represent and model the entire structure. The model of the entire structure is analyzed to assist the iterative design (synthesis) of the structure.

FEA assists by ensuring that the design of the structure adequately meets the requirements assigned to it during the systems engineering process. It also ensures that the structure is not unnecessarily overengineered, saving valuable time and money during design.

Models and simulations are excellent tools for performing what-if analysis and determining the design limits of the design. Variables can be changed and conditions altered easily to allow designers to investigate the likely behavior of the design in varying conditions. Simulators are widely used for a variety of different synthesis tasks. Environmental design and HMI design are two examples of synthesis tasks that can benefit from simulation. The fidelity of simulators must be defined and confirmed prior to their use as synthesis tools.

To build and use a mathematical model, designers need to define the problem and formulate the model to replicate the problem. This model can be used with pen and paper if desired, but the more common approach is to convert the model into a software product for ease of use and speed. Once this is done, the step that is perhaps the most crucial must be performed. The performance of the model must be verified to ensure that the model is an accurate representation of the problem. It is pointless collecting data from a model if the model is incorrect (not representative). Once the model has been built and verified, designers can experiment with it and collect important design data.

4.8.4 Trade-Off Analysis

Throughout this text, references are made to the evaluation of alternative solutions against multiple criteria and the selection of the preferred approach. The process of analyzing alternatives and selecting the preferred solution is often generically called trade-off analysis (or more generically, multicriteria decision analysis or decision-making). The analysis process must consider the different alternatives and balance the many (potentially conflicting) requirements prior to selecting the preferred solution. It must be emphasized that trade-off analysis is a decision-making aid only and should be used by the decision-makers as a tool in making decisions. Trade-off analysis should not be the decision-maker.

4.8.4.1 Trade-Off Analysis Utility

Trade-off analysis is used to assist decision-making throughout all phases of the system life cycle. Table 4.1 summarizes typical activities conducted during the different life-cycle phases that may benefit from the use of trade-off analysis [7].

4.8.4.2 Trade-Off Analysis Process

The process employed when conducting a trade-off analysis will vary with the individual needs of the decision-making process. The following steps provide a guide to the general process that may be employed.

Step 1: Definition of Objective and Scope

The definition of the scope of the problem to be investigated and the objective and of the trade study must be identified in precise terms. Systems engineering artifacts, including specifications and design aids such as FFBD, will provide a good source of this information. Sources will vary as the development moves through the system life cycle. This step also defines the processes and methodologies to be used for the trade study including data and information collection requirements as well as a schedule of activities and a budget.

Step 2: Identification of Alternative Solutions

The trade-off analysis must identify the widest range of alternative solutions possible to maximize the chance of selecting the optimal solution to the

Table 4.1
Trade-Off Analysis Examples

Acquisition Phase	Potential Applications
Conceptual Design	Investigation of candidate new and emerging technology
	Selection of the preferred system design concept
	Identification of potential system configurations
Preliminary Design	Selection of the technology to be employed
	OTS versus developmental options analysis
	Selection of the preferred system configuration
Detailed Design	Consideration/selection of subsystem and component design
	Consideration/selection of testing methodology
Construction/Production	Consideration of design changes
	Optimization use of design space

problem. There are many sources of alternative solutions, including systems engineering synthesis activities, OTS equipment, and other existing equipment. Once the range of solutions has been identified, the possible solutions should be passed through a screening process to eliminate infeasible or unsuitable solutions from further consideration. This will avoid investing valuable time and effort in solutions that ultimately cannot be implemented for one reason or another. The remaining (feasible) solutions should be described in sufficient detail to allow further detailed consideration at a later stage in the process.

Step 3: Nomination of Selection Criteria

Criteria must be formulated to provide measures of the effectiveness of the alternatives to meet the requirements of the problem. If possible, these measures should be quantitative in nature to avoid subjective assessment of the alternatives as much as possible. The different criteria should be independent from one another to avoid assessing the same characteristics several times, and the criteria should be directly related to the defined problem.

Step 4: Determination of Criteria Weighting

Once the criteria have been identified, they should be weighted according to their relative importance. Some criteria may be considered equally important as another and may be assigned the same weighting; however, in most cases, there will be a clear difference in the level of importance among the criteria. This should be reflected in weighting. If considered necessary, the weightings may be withheld from the individuals performing the evaluation to help to ensure the objectivity of the assessment. Numerical assessments with weightings will assist in the comparative assessment of the alternatives under consideration. By combining an assessment of how well an alternative meets a certain criterion with a measure of the importance of that criterion, the evaluation will provide an excellent indication of the alternative's performance.

Step 5: Definition of Scoring Functions

It may be worthwhile to assign scoring functions to each criterion to ensure that the performance of each of the alternatives against the criteria is scored consistently. The scoring functions will be used by the scorers during the assessment of each alternative; scoring functions can be linear or nonlinear, depending on the nature of the criterion. A range of scores should be selected (for example, from 0 to 10) and this range applied to all criteria. By applying different scoring ranges to different criteria, inadvertent weighting will be introduced. Weighting has already been assigned during the previous step.

Example 4.11: Aircraft System Weighting

Figure 4.21 illustrates how the criteria weights may be allocated for our aircraft example.

Example 4.12: Aircraft System Scoring Function

Using the aircraft example, the scoring function in Figure 4.22 may be applicable to the cruising speed criterion.

In both examples, candidate aircraft that have a cruising speed of greater or equal to 500 knots will score 10 points. Speeds of less than 300 knots will be given a score of 0. In the linear example on the left in Figure 4.22, a score is awarded in accordance with a linear interpolation for speeds between 300 and 500 knots, which scores 10; speeds greater than 500 knots score no greater than 10. In the step-function example on the right, a higher score of 5 is available for speeds between 300 and 400 knots, a score of 10 for speeds greater than 500 knots, but no score greater than 10 for speeds greater than 500 knots.

Step 6: Evaluation of Alternatives

Once all the weights and scoring functions have been produced, the alternatives can be assessed. Normally, the scores and weightings will be recorded in a tabulated form and the total score for each alternative will be calculated using the following (or similar) expression:

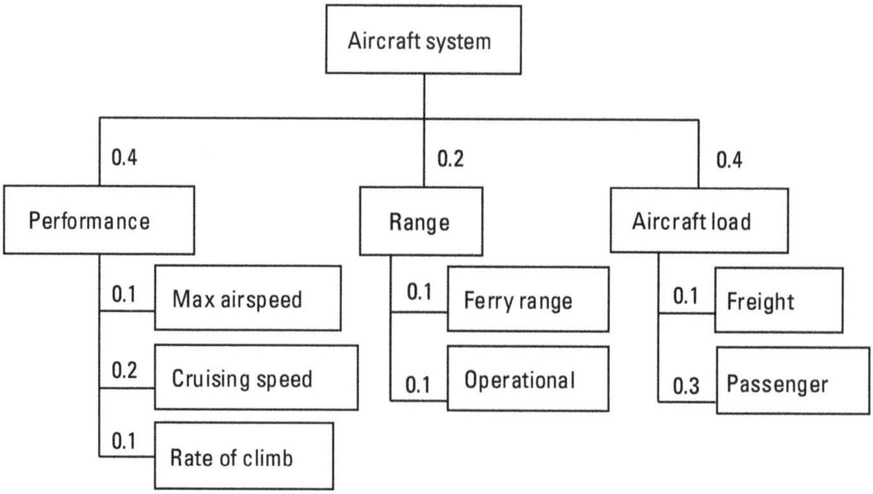

Figure 4.21 An illustration of criteria weights for our aircraft system.

$$\text{Weighted score} = \sum_{i=1}^{N} w_i s_i$$

where

N = number of criteria

w_i = weighting of ith criteria

s_i = score of ith criteria

Provided the criteria, weightings, and scoring functions have been correctly compiled, the weighted scores will then provide an indication of the relative order of merit of each of the alternatives.

Step 7: Sensitivity Study

A sensitivity study, or sensitivity analysis, should be performed to confirm the validity of the results of the overall trade-off analysis. The sensitivity check will concentrate on the allocation of weights to each criterion and the range of possible scores that could have been achieved by each alternative against each criterion (given that the score assessment may have been subjective to an extent). The allocated weights and the range of scores will then translate into a range of weighted scores for each alternative. Sensitivity studies attempt to avoid the situation where the incorrect alternative is selected due to poorly informed assessment by investigating the situation where even a small change in a single score can have significant effect on the overall weighted score. For example, consider again the scoring function in Figure 4.22(b). If we were using this scoring function to compare two alternative solutions (one that delivers a cruising speed of 300 knots and another that delivers a cruising speed of 299 knots), the 300-knot option would score 5 (out of 10), but the 299-knot option would score 0. If cruising speed was a heavily weighted criteria, this small difference in claimed cruising speed would create a large difference in weighted scores for the two options.

A sensitivity study should highlight these kinds of situations to guard against misleading outputs caused by the mathematical process. If the score ranges still clearly identify the preferred alternative, the sensitivity study will have reinforced the validity of that alternative. If, however, the score ranges overlap, the decision-makers need to consider the following courses of action:

- Obtain additional information for each alternative to further differentiate between them.
- Review the criteria, weightings, and scoring functions to ensure that they are correct.

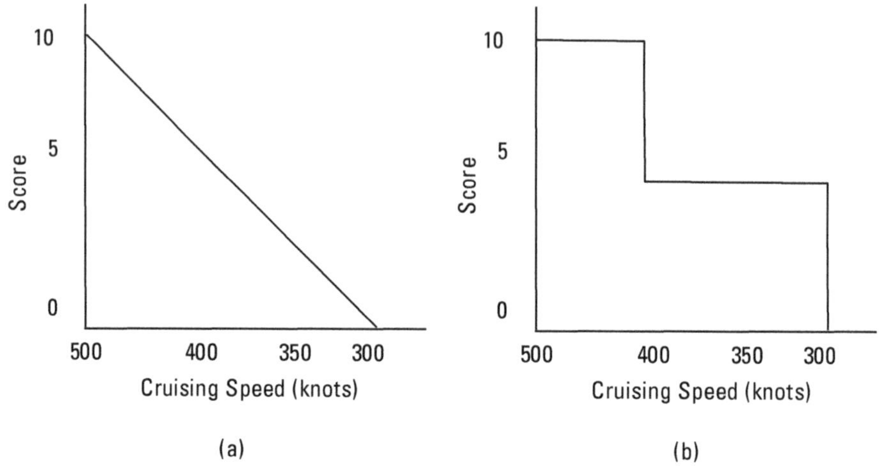

Figure 4.22 Illustration of a scoring function for the cruising speed criterion for our aircraft system: (a) linear, and (b) step function.

- If the results are still inconclusive, proceed with caution until evidence becomes available to allow a clear differentiation among alternatives.

4.8.5 Model-Based Systems Engineering (MBSE)

In Section 4.8.3, we discussed the use of modeling and simulation and defined what a model is. When modeling becomes the central part of a systems engineering process, the process becomes the MBSE [8]. INCOSE defines MBSE as "the formalized application of modelling to support system requirements, design, analysis, verification and validation activities beginning in the conceptual design phase and continuing throughout development and later lifecycle phases" [9].

MBSE is often contrasted with what some call "document-based" systems engineering. In document-based systems engineering, critical systems engineering information is captured in artifacts like the ones described in this text. For example, system-level requirements are captured in the SyRS, interface information is captured in ICDs, system-level design information is captured in the SDD, verification procedures are captured in planning documents, and so on. In an MBSE process, this information is contained in a model (or models). Stakeholders view the information in which they are interested by using an array of views (or diagrams) extracted from that model. The model therefore becomes the single source of truth because the same model is being viewed regardless of whether the stakeholder is interested in system requirements, interface requirements, system design, or verification. If a stakeholder detects an error in their view, they can correct the model and that correction becomes available immediately to the other stakeholders who are using the same model for their own purposes.

When used successfully, MBSE offers the potential to improve stakeholder decision-making and understanding, enhance communications, support trade-off analyses, and capture system behavior, structure, and requirements across all levels of the system hierarchy. MBSE also promises to allow systems engineers to measure progressively the completeness and correctness of their work by detecting errors and omissions in the systems engineering effort. MBSE may also enhance the ability of systems engineers to deal with change management efficiently and effectively as part of a wider configuration management process.

Adopting an MBSE approach requires stakeholders to agree on three fundamental things: the underlying systems engineering process to be used, the MBSE toolset to be used, and the formal MBSE language to be used in the modeling effort. Standard systems engineering processes such as the one described in this book can be adopted to MBSE, but commercially available MBSE process models also exist. These commercially available processes are generally associated with a particular MBSE toolset that usually must be purchased before it can be used. For example, IBM published their Harmony methodology that is designed to be used in conjunction with their MBSE tool called Rhapsody. Dassault Systèmes, Siemens, and Thales are other examples of vendors that offer both an MBSE methodology and accompanying toolsets. The formal language that is used by a majority of the MBSE toolsets is called System Modeling Language (SysML) [10] and could be considered as the default standard MBSE language. The formal language is critical because it determines views and diagrams, the notation and syntax to be applied, and rules associated with both static and dynamic models of the system. Using a standardized language also allows for sharing of models across different organizations, different organizational levels, and potentially across MBSE toolsets from different vendors.

As in more traditional systems engineering processes, the model at the heart of the MBSE process can be developed progressively: from an operational to a component level, from functional to physical views, and across various levels of a system hierarchy via a standard systems engineering life cycle. Alternatively, organizations may start the systems engineering process for a new or improved system by modeling the current system that needs to be replaced or improved.

As organizations around the world continue to develop their systems engineering processes and adopt concepts such as MBSE, they are experiencing and overcoming common challenges. For example, to interact with and understand the model(s), expertise is required in both the modeling language (such as SysML) and in the selected MBSE methodology and tool. While this may not be a challenge for an MBSE practitioner, it creates a communication and comprehension challenge for a host of other technical and nontechnical stakeholders for whom the modeling may have been performed. Effort is required to become sufficiently proficient in both the language and the toolset to understand the different views and usages of the model. This effort is not just time-related because commercially available tools require additional financial investment in terms of both licensing and training costs. An integrated MBSE approach to system development can also be difficult in a classic customer/contractor procurement model, especially if there are significant differences in how the customer and contractor are performing and using MBSE. The same challenge is replicated within a developer's organization in the handover of systems engineering work where there are differences between how the systems engineers are working and how the more traditional detailed design and development disciplines are working.

Excellent coverage of MBSE and its application can be found in a number of texts [11].

4.9 Summary

Preliminary design starts with the initial FBL, as defined during conceptual design, and continues to translate system-level requirements into design requirements for the system elements that will combine to form the system. This translation requires continuation of the requirements analysis started in conceptual design. Specific requirements for the elements making up the system need to be determined. Trade-off studies are conducted, and the result of the preliminary design effort is the establishment of an ABL, in which requirements are "allocated" to specific physical system elements that combine to form the system. The processes conducted during the preliminary design effort include subsystem requirements analysis, requirements allocation, interface identification/design, subsystem-level synthesis, and PDR. Once PDR is completed, we are ready to proceed to detailed design and development.

Review Questions

1. Briefly describe the major activities and artifacts of preliminary design.
2. Briefly describe the P1. Conduct Subsystem Requirements Analysis process within preliminary design.
3. Briefly describe the P2. Conduct Requirements Allocation process within preliminary design.
4. Briefly describe the eight factors that guide the identification of CIs.
5. Explain the relationship between an RBS and a WBS.
6. An allocation matrix shows the relationship between logical and physical design. Explain why this relationship is important to: (a) the customer, and (b) the contractor responsible for design.
7. Describe the three broad design options (including their advantages and disadvantages) for CIs: OTS, modified OTS, and developmental items.
8. Briefly describe the role of the ICD.
9. Briefly explain the role of an ICWG in the management of interfaces.
10. List and briefly describe the types of interfaces between subsystems.
11. Briefly explain the use of architectures in system development.
12. Briefly explain what is meant by the notion of design space.
13. Briefly explain the use of TRLs.
14. Briefly explain why a systems designer may have to force a subsystem designer to accept a suboptimal design space for their subsystem.
15. Briefly explain the role and conduct of the PDR.
16. Briefly describe the seven steps of a trade-off analysis.
17. Briefly describe MBSE and its utility.

Endnotes

[1] MIL-HDBK-61A(SE), *Military Handbook—Configuration Management Guidance*, Washington, DC: US Department of Defense, 2001.

[2] This example is derived from that given in MIL-HDBK-881A, *Work Breakdown Structures for Defense Materiel Items*, Washington, DC: US Department of Defense, 2005, which provides an excellent reference to WBS.

[3] INCOSE, *Systems Engineering Handbook: A Guide for System Lifecycle Processes and Activities*, Walden, D. D., et al., (eds.), INCOSE-TP-2003-002-05 2023, New York: Wiley, 2023.

[4] Maier, M., and E. Rechtin, *The Art of Systems Architecting*, Boca Raton, FL: CRC Press, 2009.

[5] US Governance Accountability Office (GAO), *Technology Readiness Assessment Guide: Best Practices for Evaluating the Readiness of Technology for Use in Acquisition Programs and Projects*, GAO-20-48G, January 2020.

[6] For more detail, see, for example: Cook, R., D. Malkus, and M. Plesha, *Concepts and Applications of Finite Element Analysis*, New York: John Wiley & Sons, 1989.

[7] Defense Systems Management College (DSMC), *Systems Engineering Management Guide*, Washington, DC: US Government Printing Office, 1990.

[8] Vitech Corporation, *A Primer for Model-Based Systems Engineering*, 2nd Edition, 2011.

[9] INCOSE, *Systems Engineering Vision 2020*, (INCOSE-TP-2004-004-02), International Council on Systems Engineering, 2007.

[10] Object Management Group, *OMG Systems Modeling Language (SysML®) Version 2.0*, Needham, MA: Object Management Group, 2025.

[11] Excellent coverage of MBSE and its application can be found in a number of texts:

Buede, D. M., and W. D. Miller, *The Engineering Design of Systems: Models and Methods*, New York: Wiley, 2006.

Delligatti, L., *SysML Distilled: A Brief Guide to the Systems Modeling Language*, Boston, MA: Pearson Education, 2010.

Dori, D., *Model-Based Systems Engineering with OPM and SysML*, New York: Springer, 2016.

Friedenhal, S., A. Moore, and P. Stein, *A Practical Guide to SysML: The Systems Modelling Language*, Boston, MA: The MK/OMG Press, 2015.

Jenney, J., et al., *Modern Methods of Systems Engineering: With an Introduction to Pattern and Model Based Methods*, Createspace Independent Publishing Platform, 2011.

Hetherington, D., *Simple SySML for Beginners: Using Sparx Enterprise Architect*, Austin, TX: Asatte Press, 2019.

Holt, J., *Systems Engineering Demystified: A Practitioner's Handbook for Developing Complex Systems Using a Model-Based Approach*, Birmingham, UK: Packt, 2021.

Holt, J., and S. Perry, *SysML for Systems Engineering: A Model-Based Approach*, Institution of Engineering and Technology, 2013.

Micouin, P., *Model-Based Systems Engineering: Fundamentals and Methods*, New York: Wiley, 2014.

Vitech Corporation, *A Primer for Model-Based Systems Engineering*, 2nd Edition, 2011.

Weilkiens, T., *Systems Engineering with SysML/UML: Modeling Analysis, Design*, Boston, MA: MK/OMG Press, 2008.

5

Detailed Design and Development

5.1 Introduction

The detailed design and development activity continues the development effort and makes use of the FBL and ABL developed during conceptual design and preliminary design. The detailed design effort takes these definitions of the overall system (as contained in the FBL) and of the subsystems and CIs (contained in the ABL) and finalizes the design of specific components that make up the developmental CIs (and subsystems). The realization and description of individual components used to support production are referred to as the product baseline (PBL).

Several tasks are undertaken during detailed design and development. The major technical activities include:

- Describing the lower-level assemblies and components making up the developmental CIs (and their interrelationships);
- Defining the characteristics of the above items through specifications and design data;
- Finalizing the design of all interfaces necessary to support system integration;
- Either procuring the above items off-the-shelf, or designing them if they are unique to the system under development;

- Developing prototypes or engineering models of the CIs (and eventually the final system) by integrating the above items (for the purpose of design verification, and system-level test and evaluation);
- Redesigning and retesting (as required);
- Conducting the critical design review (CDR) to confirm that the design is ready for construction and production.

5.2 Detailed Design Requirements

System-level requirements and constraints were derived and recorded in the SyRS during conceptual design. A system-level solution was selected during conceptual design as the preferred way of delivering the requirements and recorded in an SDD. The solution was developed further during preliminary design where details of the various CIs and interfaces were refined eventually. The requirements associated with each of the CIs, including the internal and external interfaces, were recorded in development specifications.

The design approach to realizing the CIs was also considered during preliminary design, including what elements of the design would be satisfied by hardware-dominated CIs and what elements of the design would be allocated to software CIs. Regardless of whether the requirements were allocated to hardware, software, or a combination of the two, there were three broad options available to realize the design: off-the-shelf (OTS), modified OTS, and developmental items. Based on the considerations discussed in Chapter 4 (Preliminary Design), a design decision was made as to how each CI would be realized. The CI design is therefore likely to consist of a mixture of the three broad options, which are iterated and reviewed to achieve system-level optimization and balance.

The detailed design and development effort concentrates on the design of the modifications to OTS CIs, and/or the design of the developmental CIs. Detailed design and development also concentrates on the integration issues associated with pulling all of the CIs together to create an effective system-level solution. Apart from integration effort, there is no need for OTS CIs to undergo detailed design and development because they already exist and have been selected based on their ability to meet the requirements in the relevant development specification.

More detailed specifications will be written during detailed design and development to address modifications, developmental CIs, and integration issues. The detailed design specifications contain allocated requirements and constraints, derived requirements that arise from the design process or production constraints, and associated characteristics that must be incorporated into the design of specific components. The specifications also necessarily contain requirements and design descriptions of the interfaces relevant to the products, components, and assemblies. In addition to detailed requirements and design, there will be a requirement to specify the fabrication processes and materials necessary to produce the hardware elements of the detailed design.

The specification process is an excellent example of top-down design process, which is responsible for establishing specific requirements at each level in the system's hierarchical design structure. The process evolves through the now familiar iterations of analysis, synthesis, and evaluation until the definition of all system components (recorded in the appropriate specifications) is complete and recorded in the PBL.

The PBL forms the basis of the next stage in the life cycle, which is construction and/or production. To have sufficient confidence in the PBL before going to production and construction, the detailed design phase is likely to involve activities such as prototyping, simulation, modeling, and integration to prove the design. These activities assist in identifying problems with the design, prior to releasing it for production. Problems with the design are rectified, and the baseline is updated until the results indicate that the baseline is sufficiently mature to support production and construction.

The ramifications of getting the detailed design work wrong are sometimes highlighted in the media, with events such as safety-critical recalls on cars being particularly newsworthy. Such a recall of products due to safety concerns (during or after production) are costly to the organization responsible and may be damaging to reputation and consumer confidence. Detailed design activities and reviews can go a long way to preventing these sorts of errors.

The next stage in our development life cycle is production and construction. It is likely that aspects of the production process will need to be proven during detailed design and development (and prior to entering full-scale production). In some cases, this may occur via a discrete pre-production phase or a low-rate initial production (LRIP) process where a small number of items are produced to prove both the design and production readiness. It may involve design reviews such as production readiness reviews (PRRs).

There are likely to be critical production-related issues that need to be resolved during detailed design to make sure that the design is producible using the intended production process. In some cases, pre-production may highlight design areas that need to be revisited. For example, the design process may assume a certain dimension (space claim) for a weld seam that is not achievable using existing production processes. In another example, the design may require use of a certain material that is not available within the timeframes associated with production or has been recently superseded by a superior material. These sorts of practical production issues may drive the need for designs to be altered prior to production commencing. Additionally issues such as manufacturing processes and tolerances, use of specialized materials and fabrication processes, and production finishes and quality may need to be proven prior to production.

Throughout this book, we have emphasized the concept of forward and backward traceability. Traceability connects and explains the transition of requirements from system-level to CI-level and onto component level. It should be noted that, while the requirements contained in lower-level specifications such as product specifications need to be derived from higher-level specifications, the lower-level specifications will also contain requirements that come from design and production requirements/constraints. These requirements will not be directly traceable to higher-level requirements but should be recorded with the same discipline; that is, they should have an explanation or rationale explaining why the requirement (or group of requirements) exists and an agreed-upon verification approach.

5.3 Detailed Design of Hardware

It is not possible to cover all categories of hardware associated with complex system developments. It may be useful to consider what the term may mean to engineers from different engineering disciplines as follows:

- *Electrical/electronic engineer.* Hardware may be an electrical or electronic system of some description that relies on a combination of electronic components arranged in such a way as to manage, condition, or distribute electrical power or signals.
- *Civil engineer.* Hardware may be the foundations and structural components of a bridge or other piece of physical infrastructure.
- *Mechanical engineer.* Hardware may be a physical system such as an engine, chassis, or drive train.
- *Aeronautical engineer.* Hardware may be the structures and components associated with the wings and fuselage of an aircraft.
- *Software engineer.* Hardware may be considered as anything that is not software including computer processors, network infrastructure, and human-machine interfaces.

Rather than attempting to describe a detailed design process for every engineering discipline, this section presents a generic design process that may be tailored and applied by different disciplines. This section then discusses the engineering design artifacts resulting from the detailed design and development of hardware. The design artifacts will support the production and construction of the hardware aspects of the system in subsequent phases of the life cycle.

5.3.1 Generic Hardware Detailed Design Process

A typical detailed design process is very iterative in nature and is dominated by reviews and feedback at each stage. Every organization that is involved with detailed design and development should have a defined detailed design process. If the organization has an accredited quality management system in place, that detailed design process will be subject to periodic audit to confirm that all design staff are complying with its requirements (noting, however, that the presence of a quality management system does not by itself guarantee that the detailed design process is effective, merely that it is being followed). A generic detailed design process is shown in Figure 5.1.

5.3.2 Preparing for Detailed Hardware Design

The detailed design process is initiated at Figure 5.1, activity D1, by the completion of the previous phases in the acquisition process (conceptual and preliminary design). The SyRS and the set of Development Specifications are provided by preliminary design and provide the designers with a clear understanding of what each CI needs to be able to do to meet the system-level requirements. Due to the flowdown of requirements from system to subsystem, the constraints associated with the design of the subsystem will also be available to the designers. The designers use this information in Figure 5.1, activity D2, and Figure 5.1, activity D3, to determine the detailed design and definition of the CIs (via the familiar analysis, synthesis, and evaluation process). If they are dealing with a modified OTS strategy, detailed design will be limited to determining the design of the modification, and the method of incorporating that modification during production and construction. If the designers are dealing with a developmental subsystem, the entire CI will need to be addressed by the process.

Figure 5.1, activity D4, discusses specifications. The detailed design requirements for each hardware subsystem will be recorded in the respective draft product specification, which will include the breakdown of the CI into its lower-level elements. Internal interfaces (within the CI will need to be determined and recorded here) as well as the CI-to-CI interfaces and any external interfaces have been allocated to the CI. These definitions are then reviewed

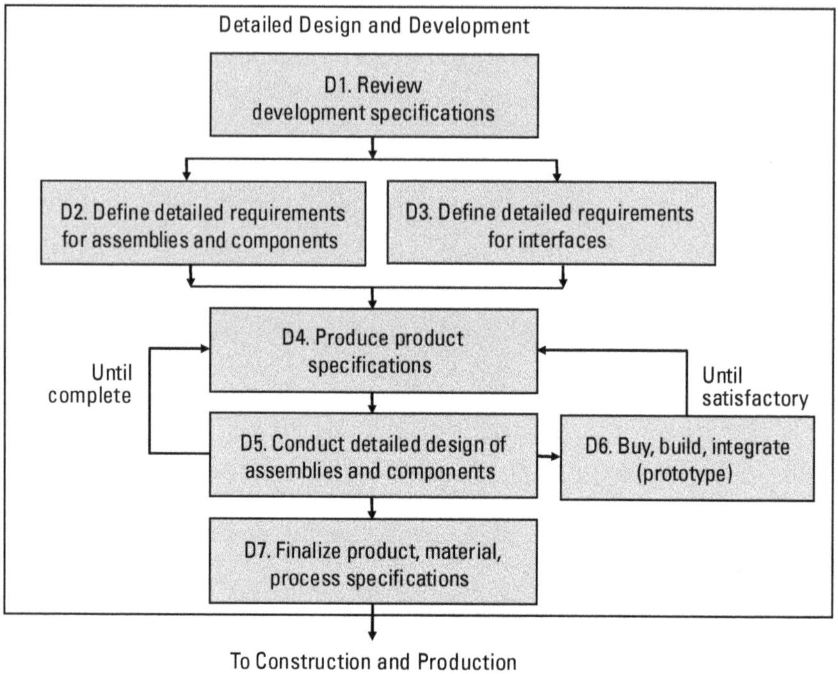

Figure 5.1 Detailed design and development process.

to ensure that the definitions are complete and meet the overall system-level requirements. If the review is negative, the process is repeated and the draft specifications updated until considered complete.

5.3.3 Performing Detailed Hardware Design

Activity D5 (Figure 5.1) is where the actual detailed design of the hardware occurs. Once the breakdown of the CI or its modification is determined, engineers and technicians can work towards a realizable design that can be produced in accordance with the draft product specification.

Detailed design is a challenging task and, fortunately, as technology advances, sophisticated design aids exist to help the system designer. These aids allow the designer to realize rapidly and inexpensively different design alternatives for the purpose of comparison and selection. The different alternatives can be traded off against one another to ensure that the most suitable design alternative is selected. This section provides some general examples of design aids used for hardware engineering.

All fields of hardware engineering design have benefited from the explosion in computing power available to the end user. Computer-aided hardware design tools allow engineers to perform analysis, synthesis, and evaluation at each stage of the detailed design process without having to resort to expensive and time-consuming physical engineering models and mock-ups [1]. Modern software packages are usually associated with mechanical and structural engineering designs and allow engineers to:

- Rapidly develop three-dimensional models of proposed hardware;
- Experiment with different layouts and designs;
- Conduct simulations and analyses using techniques such as finite element analysis (FEA) explained further below;
- Allow designs to be rotated and viewed from any direction or angle;
- Calculate and predict attributes such as strength;
- Support record requirements by producing drawings to support instructions in operator and maintenance manuals;
- Interface with computer-aided manufacture (CAM) systems to develop parts lists, bills of materials (BOMs), and so on for the final configuration of the design;
- Interface with manufacturing systems to support production (for example, such as CNC systems and additive manufacturing systems).

Electrical and electronic systems design has also benefited greatly from an explosion of computer software tools over the recent decades. Design packages exist to support electrical and electronic engineers to:

- Design electrical and electronic circuits of various types using libraries of basic component building blocks;
- Prepare draft layouts of circuits to support the production of printed circuit boards;
- Perform simulation of the designs, including input/output behaviors and stability under a range of simulated conditions to verify the correct operation of the design;
- Conduct analysis and prediction of the electrical power requirements and consumption of the design;
- Use tools such as design rule checking that help to prevent the development of designs that will be difficult to produce;
- Run specialist simulation tools such as FEA to further verify that the design will operate correctly when produced;
- Create the design descriptions and parts lists required to support production and through life support of the design.

It is common for engineering disciplines to make use of FEA during detailed design and development. An internet search reveals an enormous number of software packages available to support such analysis. The packages tend to offer support (or options for support) for specialized FEA such as electromagnetic, thermal, and structural designs. The computer packages tend to allow a visualization of aspects of the design such as bending, stretching, and twisting in structures, temperature gradients in thermal designs, and electromagnetic field strengths and densities in electromagnetic environments.

As with CAD, computerized FEA allows the design to be refined without the need to build (potentially expensive) prototypes and physical models. This can speed up the design process and reduce some of the costs associated with detailed design by reducing the need to manufacture prototypes. It also allows different design options to be considered and compared in a short period of time, supporting the desire to arrive at an optimal design solution.

5.3.4 Proving the Detailed Hardware Design

Computer models and simulations do have limitations and, in some cases, elements of the design may need to be realized (prototyped) to allow designers to verify that the detailed design is complete. Activity D6 in Figure 5.1 describes the need to realize the design in some form in order to prove that it is adequate by building, buying, or prototyping a working model of some description. This important step recognizes that it is unlikely that a design will ever go from a computer-aided environment straight into production without first being proven.

The use of models, mock-ups, and prototypes is diverse but may include determining human engineering requirements (ergonomics), physical layouts, cable routes, and access points, and so on. Models and mock-ups can also assist related disciplines such as logisticians to verify reliability, availability, and maintainability (RAM) estimates, and production engineers to confirm manufacturing processes [2]. The use of prototyping as a way of exploring requirements during the early stages of a life cycle was introduced in Chapter 2 and is discussed further in Chapter 11, but prototypes also play an important role during detailed design and development. We discuss the role of prototypes in proving detailed design later in this chapter.

Each engineering discipline will have variations on what different forms a physical model or mockup may take. In the motor vehicle industry, clay modeling of either scale models or full-size replicas of the vehicle is still used today to gain an appreciation of the overall aesthetic appeal of the vehicle. The same industry conducts extensive aerodynamic testing of vehicle performance using models in wind tunnels. This is becoming increasingly important as the industry strives to achieve increased levels of fuel efficiency from their vehicle designs. In the aircraft industry, models and mock-ups may be used to determine aerodynamic performance as in the car industry, but models may also be used to verify more complex issues such as electromagnetic requirements like radar cross-section and antenna performance. The commercial aircraft industry also uses full-scale models and simulation to confirm subjective design requirements such as ride quality and comfort for passengers. This is achieved in some cases by constructing a full-scale model of a section of the aircraft interior and placing that model onto a mechanical motion platform. A simulation of the aircraft's flight control system then controls the motion platform. The flight control algorithms can be tuned until the desired level of passenger comfort is achieved. In the electronics industry, circuit designs will be taken from the computer-aided environment and physically built in some form of prototyping environment to confirm the requirements of the circuit under different conditions. From the prototyping environment, the circuit may be transferred onto a pre-production circuit board to confirm constraints such as physical size and robustness.

It is left as an exercise for students and engineers within specific industries to discover what detailed design tools and processes are used in their organization and industry, including the role of physical modeling or mock-ups.

We leave this section acknowledging that the detailed design needs to be proven before finalizing the relevant design artifacts in readiness for production. Discovering design errors after moving into production can be very costly (and damaging to the organization's reputation).

5.3.5 Recording the Detailed Hardware Design

Detailed design needs to be recorded to support production and construction of the hardware and the through-life support of that hardware. As with many other aspects of hardware design, arriving at universally applicable list of data required to describe hardware design is not possible due to the infinite range of industries, technologies, and disciplines that could be categorized as hardware disciplines. Clearly, the data required to produce and support an aircraft is different from the data required to construct and support a high-rise office building.

IEEE-STD-1220 [3] provides the basis of such a list for hardware systems, as part of what it calls an integrated data package. The IEEE-STD-1220 integrated data package extends to include much more than just design data, but notes that the list is indicative, meaning that it needs to be tailored to specific projects. Where this text talks about a product baseline, IEEE-STD-1220 talks about the need to establish the "Build to Baseline," which is perhaps more descriptive of the level of detail expected at this point in the life cycle. The design-related, hardware data that IEEE-STD-1220 expects includes:

- Arrangement drawings showing the relationship between major CIs or components of the system;
- Assembly drawings showing the relationship between the lower-level items in a subsystem;
- Connection and interface drawings showing the connection to interfaces such as electrical power, water, communication, hydraulic, pneumatic, or other services;
- Construction drawings describe the design of a building or structure, including plans, elevations, and sections as required;
- Product specifications and drawings that describe the configuration, requirements, and constraints associated with a product that needs to be developed or procured to meet the stated requirements;
- Engineering drawings, documents, and artifacts that describe the detailed design of an item or product;
- Installation and integration drawings containing instructions on how an item or product is installed into its supporting structure and integrated with associated items or products;
- Logic diagrams showing the sequence and function of logic circuits;
- Numerical control drawings and other production specifications to facilitate production of an item or product;
- Parts lists detailing the parts and subassemblies required to produce the item or product;
- Schematic diagrams and circuit layouts showing the design and layout of electrical and electronic circuits;
- Wiring/cable harness drawings showing how groups of wires can be laced together to simplify installation;
- Models, simulations, or design databases showing how the previously mentioned data items had been determined and proven.

Once design has been completed, proven, and recorded, the hardware elements of the project are ready for the next stages. In some projects, this will involve a series of steps involving integration and verification leading to a pre-production system-level prototype (to verify that system-level requirements are achieved by the integrated system prior to committing to production).

Before discussing this, we will look at the detailed design of software that will have been occurring roughly in parallel with hardware design in systems involving software development.

5.4 Detailed Design of Software

5.4.1 Software Design Process

In parallel with the effort associated with determining, proving, and recording the hardware, software engineers will be completing their own detailed design process for the software elements of the design. Software development process standards exist to guide software developers in their quest to design successful software. Software development methodologies are an area of rapid change, significant research and publication, and heated debate. There appears to be supporters and detractors associated with just about every software development methodology or strategy. Software development, such as systems engineering, does not have a one-size-fits-all solution. The characteristics of the system within which the software is to be developed should help software engineers to choose the most appropriate methodology for the situation. Chapter 11 explores development methodologies from a system perspective, and the discussions in Chapter 11 can be applied equally to decisions about software developments.

This section does not aim to contribute further to the debate of the strengths and weaknesses of different software development methodologies. It is important to note that development methodologies each have their own strengths and weaknesses making them suitable in some circumstances and unsuitable in others. Systems engineers must select the most suitable approach based on circumstances. Therefore, it is dangerous to dismiss a methodology simply because it is old or it has failed when applied in the wrong circumstances.

Historically, DOD-STD-2167A [4] is a well-known software development standard that enforces a strict waterfall approach to software development (see Chapter 11 for a description of the waterfall approach). Many critics of the waterfall model exist, especially with respect to its use as the basis of a software-intensive development. Critics argue that this approach (dating back to the earliest days of software engineering) is not sufficiently flexible to account for the unique nature of many modern software developments, especially in fluid situations subject to changing requirements.

DOD-STD-2167A was replaced by MIL-STD-498 [5]; a standard that does not prescribe (nor discourage) use of the waterfall approach but rather allows developers to choose the most appropriate development methodology.

Some options available under MIL-STD-498 include waterfall development, incremental development, evolutionary development, or iterative development. MIL-STD-498 was canceled in 1998 and replaced by ISO/IEC/IEEE 12207, but it remains a valuable reference source for many engineers. An example of the valuable reference material in MIL-STD-498 is its comprehensive set of data item descriptions (DIDs) or templates for a wide array of specifications and manuals. Some of the DIDs, including the operational concept document (OCD), system/subsystem specification (SSS), and system/subsystem design descriptions (SSDD), are written to cater for systems not just software and therefore find application as templates for artifacts described in this book such as the StRS, SyRS, and SDD, respectively.

ISO/IEC/IEEE 12207 [6] is the latest revision of an international standard on software life cycle processes that defines a set of common software development processes that apply to the acquisition, maintenance and disposal of software products. The standard does not specify methods or procedures, as these come from organizations involved in the software acquisition/supply, and support. ISO/IEC/IEEE 12207 confirms the very strong nexus between systems engineering and software engineering but acknowledges that software must sit within the context of a system, even if that system simply consists of software running on a processor of some sort. ISO/IEC/IEEE 12207 also acknowledges that software engineering is based on the software requirements that have been extracted from system-level requirements and allocated to software for satisfaction. ISO/IEC/IEEE 12207 makes it clear that requirements management remains an important part of software development even if the waterfall approach to software development is not used. The standard does not prescribe a development strategy (it mentions waterfall, incremental, evolutionary, and spiral), but requires that a suitable development strategy be selected and used. The standard makes it clear that simply selecting the name of the model is insufficient. Developers must select a strategy, determine a sequence of stages in that strategy, define the purpose of each stage, and describe the software outcomes that will be achieved during that stage.

While ISO/IEC/IEEE 12207 covers the entire software life cycle and relates this to the overall system life cycle, this section concentrates on what it says about the detailed design of software products.

5.4.2 Preparing for Software Design

As described in IEEE-STD-12207, requirements for software development have come from the preliminary design phase where, in this book, software requirements were determined and recorded in subsystem-level requirements artifacts called development specifications. The software development specifications contain requirements that the software needs to be able to support, the software interfaces that must exist to support CI and system interfaces, and any derived requirements and constraints associated with the software design.

Therefore, the software development specifications allow the software developers to prepare for software design by starting with a clear understanding of what their software eventually needs to be able to do.

5.4.3 Performing Detailed Software Design

In terms of detailed software development, a wide (and ever-increasing) range of development environments and methodologies exist today that allow software engineers to rapidly develop or mock up user interfaces and functional software for the purposes of software design evaluation. Current software development techniques include rapid software development methods such as Agile Modeling [7], Extreme Programming (XP) [8], and Rapid Application Development (RAD) [9] where the line between software design and software production is becoming blurred. These techniques encourage software developers to address software requirements by designing, building, and delivering working software in small, incremental steps, rather than using the grand design approach described by the waterfall approach.

Development environments are helping software engineers by incorporating development tools, debuggers, editors, and compilers and documentation generators. Embedded software (used in micro-controllers, and so on) can be developed, debugged, and compiled in a single environment, and the same environment usually incorporates a simulator function to simulate the behavior of the hardware within which the embedded software will operate. Computer-aided software engineering (CASE) tools (the software equivalent of CAD discussed earlier) assist the developer in their efforts to design, test, and record software components of systems. The emergence of artificial intelligence (AI) tools promises to change the landscape of engineering in general and software engineering in particular. For example, it is now possible for someone with no software development expertise to use AI tools to generate code very quickly based on a description of what functionality is required from the software. The disciplines of systems engineering and software engineering may be useful in providing the function and performance descriptions that the AI-based software development tool uses to convert those descriptions into working software products.

This section does not go into detail about the design technologies and methodologies available to modern designers as this area of technology moves so quickly. It is sufficient to understand here that technology is helping to make the software design process quicker and more likely to arrive at the correct answers.

5.4.4 Recording the Detailed Software Design

Independently of how the detailed software design has been determined, ISO/IEC/IEEE 12207 confirms that some record of the detailed design is required. This record needs to be sufficiently detailed to support subsequent phases in the software life cycle. In some cases, the next phase involves software production, but in the case of rapid development methodologies, the next step may involve the customers using operational software. In the latter case, software support (including future increments of software development) will need some design information upon which to base the ongoing development effort. Rapid software development methodologies should not be viewed as an excuse not to record the software design in some way. Poorly recorded software becomes a major problem during the utilization phase of the life cycle, especially when the software needs to be modified or maintained in some way.

Software design records include information on any design decisions or constraints in relation to the inputs, outputs, and functionality of the software. The design philosophy and details relating to the user interfaces (if applicable) are also required. Characteristics of the software including response times and other performance characteristics should be described. If the software is to be divided into lower-level components, then this architecture should be recorded, showing how the components relate to one another and the development status of the components. If there are any assumptions made, or design approaches and conventions followed, these should be listed. For example, an assumption may include details of the minimum computer hardware resources assumed to be available to the operational software.

The software design records should include details of the design of any software interfaces provided to or from the software in question. This will include the type and priority of the interface, the nature of the information flowing over the interface, any special data requirements (units of measure for example), formatting requirements, and protocol selections. The design needs to address the entire interface requirements assigned to that part of the software by the relevant software requirements artifacts developed in the preceding phase.

The design requirements of any databases or underlying data repositories used by the software needs to be recorded. The design requirements extend to the design of any software that is required to access or manipulate the data within these databases also. The purpose of the database should also be explained by the design artifacts. In the case of commercial software products, the design artifacts may simply be a reference to where more detailed information about the database can be located.

5.5 Integrating System Elements

Detailed design will eventually reach a stage where all CIs (hardware and software) have been designed completely, as have the interfaces between the CIs and between the CIs and external systems. In the case of both software and hardware, the design teams have proven their design by releasing elements of the design and testing their design against the requirements. This process may have resulted in a series of redesigns and tests, but the result is a series of designs that have been completed. In many cases, subsystem and system-level design will need to be confirmed by a series of integration and test stages. Low-level items of the design (although individually tested and confirmed) may need to be integrated together to form the next higher level of assembly in the system hierarchy and tested against the relevant requirements artifacts.

While developmental elements of the design can be proven in this manner, the OTS elements of the design will also need to be involved in the integration process. The OTS items in the design can be procured and checked against the claimed specifications of the supplier. Modified OTS items can be procured and modified and then checked against the relevant requirements artifacts.

Eventually, all of the items in the design can be integrated together in accordance with the overriding system architecture to form a system model or prototype of the final system. This prototype can be used to confirm that the system-level design requirements have been met. The process of progressive integration and test to confirm the completion of detailed design is illustrated in Figure 5.2.

Prototypes can vary from very complete and almost final versions of the system in terms of form, fit, and function through to models that only represent the selected functional characteristics of the final system.

Example 5.1: Aircraft System Prototypes

A prototype of our avionics system may be no more than a series of networked personal computers running a mature version of the avionics software suite,

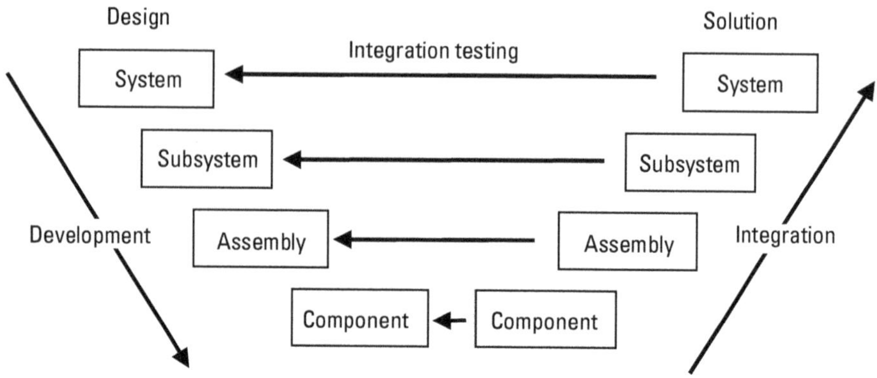

Figure 5.2 Top-down development and bottom-up integration process.

sometimes called an operational flight program (OFP). This network of computers may be operating in an office environment in the center of a major city.

Conversely, the same avionics system development may require a prototype that consists of the final aircraft hardware mounted in a mock-up of the aircraft equipment bay and wired together using the actual wiring looms to be used in the aircraft installation. Actual aircraft instrumentation may be used and simulated parameters such as basic air data may be used to stimulate the model.

The type of prototype used will depend very much on the nature of the project and system. In the case of safety-critical or mission-critical systems, the system prototype must offer stakeholders high levels of confidence in the robustness of design. For example, aircraft avionics impacts on the airworthiness of aircraft, so very sophisticated prototypes are required to verify the correct performance of aircraft avionics software under a range of conditions prior to testing the avionics subsystem in an actual aircraft. However, a prototype designed to confirm the layout of the aircraft galley is likely to be quite a simple arrangement.

In many industries, such as the car industry, several mature prototypes may be produced (including production prototypes to confirm the correct operation of the production system). Producing system-level prototypes may seem an expensive and time-consuming way of confirming the completion of detailed design, but these costs pale when compared to the costs associated with releasing unproven systems designs to production and then producing systems with undetected design flaws requiring subsequent recall and rectification.

5.6 Detailed Design Reviews

As with each of the preceding design phases, a design review (or a series of reviews) is normally conducted at the conclusion of detailed design and development to ensure that the system design is complete prior to construction and/or production.

5.6.1 Equipment/Software Design Reviews

Several low-level design reviews may be conducted throughout a detailed design. These design reviews, sometimes called equipment or software design reviews, are focused on items of software and hardware to ensure that the specific design approach meets the requirements. These reviews investigate the drawings, computer programs, material lists, and the necessary specifications to thoroughly review the item under investigation. In the case of software development, these reviews may provide important evaluation of increments of software under development to pave the way for additional increments to be designed. As such, these reviews may require input from key stakeholders within the customer's organization in the spirit of collaboration during development.

Associated artifacts such as trade-off reports that justify and explain the design approach will also normally be investigated during these reviews. If component-level prototypes or models have been produced and tested, the results of the testing and any functional discrepancy rectifications will also be tabled.

All equipment/software design reviews will be completed prior to the major formal review of the phase, which is called the critical design review (CDR).

5.6.2 Critical Design Review (CDR)

CDR is the final design review resulting in the official acceptance of the design and the subsequent commencement of construction and/or production activities. The result of the successful completion of CDR is the establishment of the PBL, which is the effective freezing of all design activity. Only discrepancies in the system design identified during testing result in further design activity following CDR. As with the PDR, CDR needs to be conducted on all developmental CIs of the system (hardware and software). Whether a separate CDR is conducted on each CI, on selected groups of CIs, or a single system-level CDR is conducted (reviewing all CIs at once) depends on the size and complexity of the system under investigation and whether there is parallel production of all CIs or some staged production process.

CDR aims to demonstrate that the CIs under review can satisfy the requirements allocated to them in the development and product specifications. CDR also confirms the compatibility of the CIs with themselves and with other parts of the system including other equipment, facilities, and personnel. Currently, test and evaluation activities are reaching a critical stage and CDR confirms that strategies, procedures, and support are in place to ensure a comprehensive test and evaluation process. CDR is also the forum in which plans for construction and production of the system are finalized and approved.

CDR will also include a review of the TPMs that were identified during conceptual design and whose progress has been tracked and monitored ever since. All TPMs should be either exceeding or meeting the necessary levels by CDR. By this mature stage in the design, there should be no surprises as far as TPMs are concerned. Departures from the accepted results should be noted and corrective action attempted, although at this stage in the project any departures are normally difficult to rectify.

As with all design reviews, prior to commencing the review, a clear demonstration of readiness should be provided through the satisfaction of all entry criteria. All necessary artifacts must be received and reviewed in advance of the design review. Examples of artifacts that would be expected to be reviewed prior to CDR includes revised interface descriptions such as ICDs, specifications such as development specifications and draft product specifications, test and evaluation procedures, relevant technical data including assembly diagrams and drawings, installation drawings and schematics, software source code listings, and comments. Reviewers should be convinced that all necessary equipment or software reviews have been conducted prior to CDR. There is little point in conducting a system CDR prior to the completion of all CI-level CDRs. The system architecture and detailed design should be mature prior to commencing CDR as CDR marks the effective freezing of design activities. If design effort is to continue, then the design is not ready for the CDR. This judgment will be possible following receipt and review of the artifacts already described. Once satisfied that the system design is ready for CDR, the review can commence.

The PBL is established by successful completion of CDR, and the system is then ready to move into construction and/or production in accordance with approved construction and/or production planning. It is likely that the result of CDR will prompt updates to the ABL and perhaps to the FBL.

The timing of CDR within detailed design and development is critical. If the review is planned too early in the process, the design may be immature and may result in several post-review changes. Post-review changes impact on the production planning and may introduce additional risks in the later stages of the acquisition phase. Holding CDR too late in the project may result in errors being detected too late to be incorporated into the design, which can adversely impact on the production schedule and may result in testing and delivery delays.

It is likely that some integration and prototyping will have occurred since the last review (PDR) and CDR investigates and assesses the success of this effort as a key indicator of design maturity.

The aims of the CDR include:

- *Evaluate the detailed design.* Detailed design is described in various forms including the mature product specifications and related engineering drawings. CDR should evaluate these artifacts and the detailed design to ensure that it adequately addresses the original requirements. Discrepancies raised during the PDR should be revisited to ensure that the rectifications are adequate. The review of TPM progress should be reviewed during CDR and all TPMs should be well within expected bounds. Particular attention should be paid to any TPMs that were highlighted during PDR as being potential problems.
- *Determine the readiness for production/construction.* In addition to the product specifications, ICDs, and drawings, CDR should review the process specifications, the material specifications, and the production plan to ensure that the detailed design for the hardware items can progress to construction and/or production. The system prototype (if produced) provides an excellent tool for assessing readiness for production and/or construction.
- *Determine the maturity of software.* In a classic development model, software will enter the coding stage following CDR (this can be considered the software equivalent of hardware fabrication). The software product specifications (including interface requirements and design) should be thoroughly reviewed prior to approving the software aspects of CDR. As discussed, it should be noted that current software development processes might allow software coding to have begun long before system-level construction and/or production. If this is the case, CDR may be a convenient forum within which to review the progress of software development against the planned progress, and to investigate any software-related test and evaluation results.

- *Determine design compatibility.* CDR should investigate and confirm design compatibility of CIs and other aspects of the system such as facilities. This requires a detailed investigation of all external and internal interfaces. The system prototype will provide an indication of design compatibility.
- *Establish the PBL.* The complete set of product specifications, once approved, will form the initial PBL for the system against which the system is built. Functional and physical configuration audits will be conducted on the production system prior to the formal approval of the system baselines. These audits are discussed in Chapter 6 as part of construction and/or production and Chapter 8 as part of configuration management.

5.7 Summary

Detailed design and development continues the development effort of the functional and allocated baselines to design of the developmental subsystem and components. The realization and recording of individual components are captured in the PBL, which is confirmed at CDR and the system is ready to enter construction and/or production stage.

Review Questions

1. Explain the role of system prototyping (including examples of different types) used during detailed design and development.
2. Describe the major review conducted at the conclusion of detailed design and development.
3. Explain why integration is the ultimate test of interface identification and design.
4. How detailed does the detailed design and development effort need to be?
5. Explain the importance of some form of preproduction process (including a production readiness review) prior to finalizing the design and commencing production.

Endnotes

[1] Further reading on aspects of CAD and CAM can be obtained from the following sources:

Amirouche, F., *Principles of Computer-Aided Design and Manufacturing*, Upper Saddle River, NJ: Pearson/Prentice Hall, 2004.

Ashley, K. L., *Analog Electronics with LabVIEW*, Upper Saddle River, NJ: Prentice Hall, 2002.

Brown, S., and Z. Vranesic, *Fundamentals of Digital Logic with Verilog Design*, New York: McGraw-Hill, 2007.

Chang, T. C., R. A. Wysk, and H. P. Wang, *Computer-Aided Manufacturing*, Upper Saddle River, NJ: Prentice Hall, 2005.

Fusco, V. F., *Microwave Circuits: Analysis and Computer-Aided Design*, Englewood Cliffs, NJ, Prentice Hall, 1987.

Kalay, Y. E., *Architecture's New Media: Principles, Theories, and Methods of Computer-Aided Design*, Cambridge, MA: MIT Press, 2004.

Krishnamoorthy, C. S., S. Rajeev, and A. Rajaraman, *Computer Aided Design: Software and Analytical Tools*, Harrow, UK: Alpha Science International, 2005.

Leondes, C., (ed.), *Computer Aided Design, Engineering, and Manufacturing: Systems Techniques and Applications*, Boca Raton, FL: CRC Press, 2001.

Saxena, A., and B. Sahay, *Computer Aided Engineering Design*, New York: Springer, 2005.

Schroeder, C., *Printed Circuit Board Design Using AutoCAD*, Boston, MA: Newnes, 1998.

Szalapaj, P., *The Digital Art of Architectural and Engineering Design*, Oxford, UK: Blackwell Publishing, 2009.

Tront, J. G., *PSpice for Basic Circuit Analysis*, New York: McGraw-Hill Higher Education, 2004.

[2] Further detail can be obtained in, for example: Blanchard, B., *Logistics Engineering and Management*, Englewood Cliffs, NJ: Prentice Hall, 1992.

[3] IEEE-STD-1220, *IEEE Standard for Application and Management of the Systems Engineering Process*, New York: IEEE Computer Society, 2005.

[4] DOD-STD-2167A, *Defense System Software Development*, Washington, DC: US Department of Defense, 1988.

[5] MIL-STD-498, *Software Development and Documentation*, Washington, DC: US Department of Defense, 1994.

[6] ISO/IEC/IEEE 12207, *Systems and Software Engineering—Software Lifecycle Processes*, New York: IEEE Computer Society, 2017.

[7] Kendall, K., and J. E. Kendall, *Systems Analysis and Design*, Upper Saddle River, NJ: Pearson/Prentice Hall, 2008.

[8] Bittner, K., and I. Spence, *Managing Iterative Software Development Projects*, Reading, MA: Addison-Wesley, 2007.

[9] Sangwan, R., et al., *Global Software Development Handbook*, Boca Raton, FL: Auerbach Publications, 2007.

6

Construction and/or Production

6.1 Introduction

At the end of detailed design and development, the PBL has been established, and the production process is in place and has been proven (most likely with the trial production of selected system elements). The system can now move into construction and/or production.

Some system development projects are aimed at producing only one copy of the system. There are many examples of one-off system developments including high-rise buildings and large ships such as cruise liners. In these projects, the construction process may blend with the detailed design and integration process. For example, when building a single ship, a prototype may be developed to support detailed design, development, and integration. This prototype may pass through stages of test and evaluation during which the design is refined. Ultimately, however, the design and development will be finished by which time the "prototype" may have matured into the final system that is offered to the customer for acceptance.

With other system developments, many copies of the system may be produced once prototypes have been proven. An aircraft system is an example of a system where many copies (aircraft) are likely to be produced after an initial aircraft (or even a small batch of aircraft) has been produced as prototypes to support design verification and user validation before the production is allowed to proceed.

6.2 Production Requirements

Production requirements must be considered early in the acquisition phase to ensure that production risks are identified and addressed as early as possible. This is particularly true when there are one or more of the system elements that make use of novel technologies (with which existing production processes may be unfamiliar). As with all other technical requirements, the earlier that production issues are identified in the acquisition, the easier they can be addressed. To that end, although construction and/or production is being considered following detailed design, production engineers should be working with the rest of the design team from the earliest possible stages of the system development to ensure that production and construction issues are appropriately addressed. In short, the systems engineering effort must make sure that the design that flows out of the preceding stages is able to be produced.

Production risks are mitigated by involving production specialists in the design team and as part of the design review process. These specialists would also be expected to be involved in the relevant design reviews to confirm that production considerations have been accounted for, and the production processes are in place for the particular system being developed. Production specialists would also need to be involved with the change management process (including participation in the configuration control process) to ensure that proposed changes to the design do not impact adversely on the producibility of the design.

Typical construction and production issues that need to be addressed and monitored throughout the entire systems engineering process include:

- Material availability (lead times), ordering, and handling;
- Availability of skill sets (including any training);
- Availability of production tools and equipment;
- Fabrication requirements including production requirements, assembly drawings, and instructions;
- Processing and process control;
- Assembly, inspection, and test;
- Packaging, storage, and handling.

Strictly speaking, construction and/or production could be considered a specialty program that supports the systems engineering processes. The systems engineering processes can be used before and during construction and/or production to support the construction and production planning activities.

The analysis-synthesis-evaluation process is drawn upon again to support the determination of the construction and production plan as shown in Figure 6.1. The production plan helps to ensure that all the necessary enablers are in place in readiness for production. The primary input to the production planning process is the design of the system. The production requirements are analyzed based on the system design, and a production process is determined (synthesized) and evaluated.

The production planning process also shows that production may limit or constrain the system design. This is shown as an input to the synthesis trade-offs that are made throughout the systems engineering process. This ensures that the system designers are mindful of the production capabilities of our organization, which helps to arrive at a producible design. These processes include producibility analyses (ensuring that designs are producible), consideration of producibility issues during trade-off studies and the selection of preferred design options, and life-cycle cost considerations that take the cost of production and construction into account.

The output of this process is a production plan that should be drawn up as early as possible in the systems engineering life cycle. The plan should mature with the design because the production system needs to be readied and proven prior to design freeze. An approved production plan is certainly

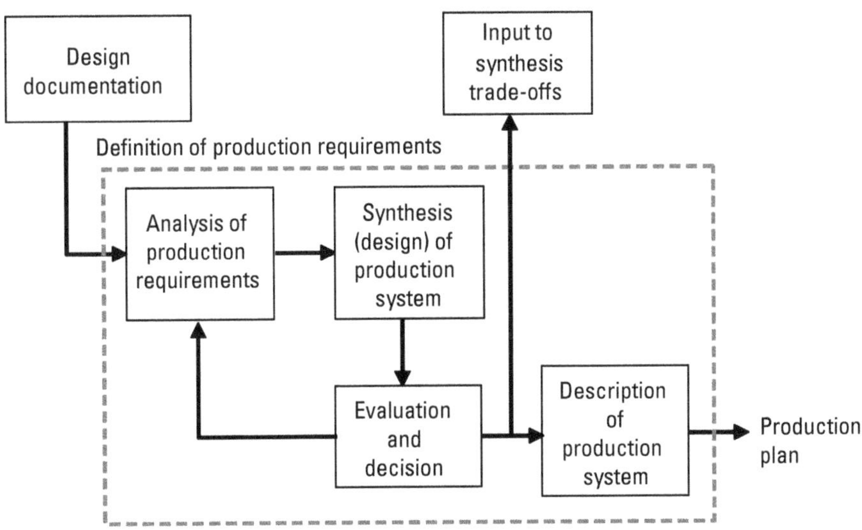

Figure 6.1 Systems engineering applied to planning for construction and/or production.

required prior to entering construction and/or production. The content of the production plan varies widely with the specific characteristics of the system, but should include the following considerations:

- *Resources.* The plan needs to consider the type and size of any plant required to produce and construct the system, details of the skills required by personnel and any training programs required.
- *Production engineering considerations.* Production and construction schedules are required and manufacturing methods and processes detailed. Special tooling, rigs, and test equipment requirements must be specified. Any special facility-related requirements should be detailed such as storage and layout requirements. If any automated processes or manufacturing information systems are being used, these should be specified and described. Additionally, any data requirements should be detailed, including assembly instructions and drawings.
- *Materials and purchased parts.* A complete bill of materials should be compiled and listed to be available following the freezing of the detailed design description. Any OTS or modified OTS equipment should also be included in the plan so that all necessary items are covered. Any items involving long procurement lead times should be identified and ordered in a timely manner. An inventory control system is probably necessary to track all the parts, and this system should be specified.
- *Management.* The standard management functions should also be outlined in the production plan, which should include details of process control, production testing procedures, subcontractor control, and problem prevention.
- *Logistics.* The plan should also detail any special logistics requirements such as packaging, handling, and storage requirements.

6.3 Engineering Management Issues

During the production and construction stage, the system will be subjected to several important systems engineering management activities. These include test and evaluation (T&E) activities, final reviews, and configuration audits. Technical reviews and T&E are discussed in Chapter 8, but the production-related activities are summarized here.

Before the system completes construction and/or production, it should be subjected to a range of acceptance tests aimed at confirming the ability of the system to satisfy the original user requirements that started the whole systems engineering process.

Acceptance testing, alone, is insufficient. Audits should be conducted during production and/or construction to ensure that the system has been built in accordance with the approved specifications and project documentation and artifacts; these are generically called *configuration audits*. The construction and/or production stage is often the only opportunity to conduct these audits.

6.3.1 Major Audits

There are two main types of configuration audit—the functional configuration audit (FCA) and the physical configuration audit (PCA) [1].

6.3.1.1 FCA

FCA is used to verify and certify that the performance of the CI meets the specified requirements [2]. That is, it is a check to ensure that the final as-built CI functionality as confirmed by the test and evaluation effort is the same as the specified functionality for that CI in the relevant development specifications and product specifications. The FCA may not be a complete review of CI functionality; it may be conducted incrementally throughout the development of the CI (although a final FCA is necessary for the CI). Some CIs depend upon the successful operation of other CIs and are not testable until after system integration. In these cases, the FCA may not be completed until after the integrated system testing is complete. FCAs for all CIs should be complete to support confirmation of the FBL before release to production.

6.3.1.2 PCA

The objective of the PCA is to provide confidence that the as-built CIs match the low-level specifications such as the product specifications, assembly specifications, drawings, and technical data [3]. PCAs are conducted on the final versions of the CIs (that is, the version of the CIs representative of the as-built items). Once physical correlation has been confirmed for all CIs, the PBL can be established.

The PCA is performed on the first production version of each of the CIs. In the case of a system where multiple copies are being produced, a partial PCA may be conducted on each subsequent production item to ensure that the production process remains constant. Normally, FCA for a given CI will precede PCA. This prevents the situation where the CI fails FCA and requires some engineering changes to be made, in which case, if PCA has already been conducted, it may be invalidated by the changes to the CI. Once a CI has passed FCA, PCA can be conducted with more certainty.

Example 6.1: Aircraft System Configuration Audits

When constructing our aircraft, we will have the perfect opportunity to confirm CI-level compliance with relevant PBL documentation. For example, we will be able to confirm that the wiring looms, fuel lines, and flight control connections are all installed in accordance with the layouts contained in the relevant engineering drawings. It is essential that we do this now because, once each aircraft has been completely assembled, it will not be possible to confirm that the looms, lines and connections are installed in accordance with the drawings since they will be obscured from sight by the aircraft structure and skin.

Some may argue that it does not matter where the looms, lines, and connections run, so long as the aircraft functions correctly. This is a shortsighted argument that fails to account for through-life activities such as modification programs and preventative maintenance. We will rely on accurate technical data such as engineering drawings throughout the life of the aircraft, and it is during construction and production that we are able to audit the system and its documentation to confirm its consistency and accuracy.

Example 6.2: FCA and PCA for a Domestic Dwelling

Understanding the roles of FCAs and PCAs may be assisted by a simple example relating to a domestic dwelling.

The construction of a family house provides a good illustration of the importance of conducting both functional and physical configuration audits on our systems. While a typical house consists of many CIs grouped logically in subsystems, this example uses only the following subsystems: plumbing and drainage, footings and structure, and electrical power.

When the house project commences, the designer, architect, and property owner will work together to determine the most suitable layout and design of the rooms and fittings within the house. They will also discuss additional issues such as compliance with building codes and regulations, as well as details such as the location and quantity of power outlets and options for lighting, hot water, and equipment layout. These discussions and decisions will be recorded in some form of top-level design or drawing set that may be used subsequently to communicate at a more technical level such as with structural engineers, builders, plumbers, and electricians.

The size of the property, the nature of the land upon which the house is to be built, and compliance with codes and regulations will inform the detailed design of the CIs that make up the various subsystems of the house. The detailed design needs to be sufficiently detailed to allow relevant tradesmen to provide a quotation for the work and then, if selected, to carry out the construction of the house. The detailed design requirements will be recorded in drawing sets, associated schedules of materials, and so on.

For the plumbing subsystem, the detailed design will include the locations of taps, toilets, and rain-water tanks; downpipes; layouts of stormwater drains and sewerage; and the location of the connection to the city stormwater and sewer systems. The electrical power subsystem will include details of connection to the city's mains power, the numbers and layout of different circuits, safety equipment requirements such as earth-leakage devices and circuit breakers, and details of any special requirements such as three-phase power and dedicated high-current circuits for stoves and machinery. The footings and structural subsystem will include details of the size, shape, and location of the footings, the nature of the reinforcing to be used in the footings, and the type of concrete to be used.

As house construction proceeds, it is important to confirm through a series of inspections that the as-built configuration of the building matches the detailed design artifacts. The inspections need to verify that the various subsystems function properly (in accordance with their allocated requirements) but they also need to verify that the physical as-built subsystems are reflected accurately and are relevant in the detailed design descriptions (such as drawings and schedules of materials).

The need to confirm functionality of the building subsystems may be self-evident. We need to confirm that the plumbing and drainage subsystems are able to manage simultaneous use of all constituent CIs (toilets, showers, and laundry facilities, for example). We need to confirm that the electrical subsystem can deliver without fault the total electrical load on each circuit that it was designed to provide. We also need to confirm that the footings provide the building with the structural integrity required. If there are any problems encountered during these checks, some rework is likely to be required. As we saw earlier, confirmation of the functionality of the CIs associated with the various subsystems against the allocated set of requirements is achieved through the conduct of FCAs.

In addition to conducting FCAs on all the CIs, we must also conduct audits that are focused on the physical construction of the subsystems. A great deal of this confirmation can only occur during the construction process (and is often not possible afterwards). For example, when the building's footings have been dug in accordance with the engineering drawings, an inspection should occur to confirm that the shape and dimensions of the footings comply with the drawings (it is not easy to confirm the depth of the footings once the concrete has been poured). Similarly, the reinforcing required in the footings (as specified in the detailed design drawings) must be inspected and confirmed prior to the pouring of concrete. Similar progressive inspections are conducted as the plumbing and drainage are laid (before the pipe work is buried in the earth or concealed behind walls), and for the electrical subsystem before the wall sheets are hung. In systems engineering parlance, these progressive inspections are the PCAs because they confirm the correlation between the physical nature of the CIs associated with the as-built subsystems and their detailed design information, in this case, probably a detailed drawing set.

Theoretically, there should be no difference between the detailed design descriptions and the as-built configuration of the house. However, building work does not always go to plan and the construction workers may be unable to follow exactly the detailed design documentation, artifacts, or drawings. For example, when digging footings, it is not uncommon to hit large areas of rock under the ground. In these cases, the builders might reengage the structural engineer, who might modify the design of the footings to account for the expanses of rock. When laying plumbing and drainage, it may be that the tie point to the city stormwater system or sewerage system is not exactly in the location specified on the drawings (again, perhaps, due to the location of large rocks). In these cases, the builder may depart from the detailed design documentation, artifacts, or drawings to complete the construction while taking account of the actual situations they encounter. In these cases, the drawings must be updated to reflect how the building was constructed. Otherwise, it may be possible to end up with a house that is perfectly functional (from a footings, sewerage, or stormwater perspective) but one that is not in constructed in accordance with the approved drawings, which may frustrate future plans for expansion or modification.

It is a common perception that it is not important if the physical construction is slightly different from the drawings, so long as the house functions correctly. However, accurate design descriptions become critical during periods of maintenance and/or modification. For example, when planning landscaping and gardening after construction, owners will refer to the drawings to ensure that large trees that are to be planted will not interfere with the sewerage or stormwater pipes. If the drawings are incorrect (due to variations that occurred during construction), the owners may plant trees over the top of the pipes or may damage the pipes during the planting process. When modifications are being planned, the modification must take account of (and be influenced by) an accurate understanding of the existing system. For example, the original house construction may have included a reinforced concrete slab to accommodate a carport. In subsequent years, the owners may decide to add a garage and workshop on top of that slab. When designing the garage, the owners need to be able to rely on accurate design descriptions relating to the structure of the slab in order to determine whether the slab is able to support the garage and workshop structure, or whether additional support is needed.

6.3.2 Major Reviews

6.3.2.1 Test Readiness Review (TRR)

Testing is an expensive and time-consuming exercise involving highly trained personnel, expensive facilities, and specialized test equipment. TRRs are sometimes contractually required by the acquirer to demonstrate that the system CIs are ready to enter formal test and evaluation. To that end, CI-level TRRs occur during construction and/or production and are designed to avoid the unnecessary expense involved with committing T&E resources to a CI or system that is not sufficiently mature to enter testing. A system-level TRR may be conducted before commencing major verification activities such as AT&E.

To assess the readiness of the CI or system to enter formal testing, the TRR normally reviews a range of artifacts including:

- *Test and evaluation master plan (TEMP).* The TEMP should be reviewed at each of the preceding design reviews but should also be rechecked at the TRR. Particular attention should be paid to the TEMP appendices, to ensure that the test procedures are correct and complete. The TEMP is covered in more detail in Chapter 8.
- *T&E plans.* The TRR should ensure that there are appropriate test plans and test procedures in place for the system element under test.
- *Formal and informal test results.* A major part of the TRR should be dedicated to reviewing the results of the formal and informal T&E activities. These results give an excellent insight into the current status and performance of the CI. The test results should be recorded in either the format specified in the TEMP or in some other approved way (for example, as recorded in the developer's quality assurance artifacts). More credibility can be given to formal testing witnessed by either the acquirer or the relevant quality assurance organization.
- *Supporting artifacts.* User and operator manuals are examples of artifacts that should be reviewed as part of the TRR. Design descriptions

such as development specifications and production specifications are also referred to during TRR. Acquirer test personnel most probably require these supporting artifacts in the performance of the testing.

- *Support, test equipment, and facilities.* The test procedures list the support, test equipment, and facilities required to conduct testing. TRR should ensure that the equipment is available, in place, and calibrated prior to commencing test and evaluation. Typically, a check should be made of the test equipment (oscilloscopes, meters, wiring looms, telemetry equipment, test interfaces such as cables and connectors); test software; recording devices and storage; associated platforms to support the test; test facilities; and availability of operators as well as specialist test personnel.

6.3.2.2 Formal Qualification Review (FQR)

The FQR may be required to verify that the performance of each of the CIs meets all the functional requirements when integrated together into the system. The FQR will demonstrate that the integrated system complies with all the specifications generated as part of the project including SyRS, hardware and software development specifications, and interface requirement specifications. From this point of view, the FQR could be considered a "system-level" configuration audit. For example, a FQR may be required to verify that the software performs as designed when integrated with the actual hardware to be employed by the final system. With this in mind, FQR needs to be sequenced at a logical point in the schedule around individual CI configuration audits and T&E effort.

6.3.3 Configuration Management

During construction and production, it may become clear that items of the system are not going to meet all of the minimum requirements specified in the baselines. To cover this possibility, the configuration management process uses a system of requests for variance and engineering changes. Engineering changes are required where the departure from the specification is going to result in either a change in design or a change in requirement. However, if the failure to meet the minimum requirements is temporary in nature, limited to only a small percentage of the supplies or a very minor departure, a request for variation may be approved by the acquirer. Configuration management is a critical component of systems engineering management and is covered in detail in Chapter 8.

6.4 Summary

At the end of detailed design and development, the PBL has been established, and the production process is in place and has been proven so that the system can move into construction and/or production producing one or a number of copies of the system. Once complete, the system can move into utilization.

Review Questions

1. Explain why production specialists should be involved with the detailed design process, even though production generally follows design.
2. Give an example of a situation where detailed design may force an update to an existing production process.
3. Give an example of a situation where detailed design might be constrained by an existing production process.
4. Provide an outline of a production plan describing the content of each section.
5. What is a TRR? Why and where would it be conducted?
6. List and explain three things that should be covered in a TRR.

Endnotes

[1] EIA-649-C, *Configuration Management Standard*, Warrendale, PA: SAE International, 2019.

[2] EIA-649-C, *Configuration Management Standard*, Warrendale, PA: SAE International, 2019.

[3] EIA-649-C, *Configuration Management Standard*, Warrendale, PA: SAE International, 2019.

7

Operational Use and System Support

7.1 Introduction

Once the system has passed the necessary testing and audits, it is ready to enter operational service or use: the utilization phase. The major activities during the utilization phase include operational use, system support, and modifications. The influence of systems engineering over these activities is relatively minor and is normally confined to modifications, which should be made to the system in accordance with sound configuration management procedures.

Configuration management continues to play a role during the utilization phase to ensure that the configuration of the system is managed and updated, especially as modifications occur. Differences between system descriptions and the physical system tend to result when systems are operated and maintained in an environment void of configuration management practices. These differences make maintenance and operation difficult and potentially dangerous. Accurate records support the need to understand the design and capability of the system, especially if there are slight differences in configuration across a fleet of systems (for example, an aircraft fleet). Modifications made in accordance with configuration management practices enhance supportability and operations, but unmanaged modifications adversely impact training, support, safety, and operations in the long run.

7.2 Modifications

Modifications may be required to ensure that the system continues to meet operational and support requirements. Modifications may be made for several reasons:

- Rectification may be required to address discrepancies with the performance of the system that were not identified during the acquisition phase. These discrepancies are often discovered during the OT&E effort when the system is placed in its operational environment and exercised in its intended purpose.
- Failures identified as part of the FRACAS process described in the next section may result in engineering changes to the system via the modification process.
- System-level requirements may change as a result of a range of factors including a changing operational or support environment.
- Opportunities may arise to increase the efficiency of the system, or perhaps reduce maintenance costs, through the replacement of system elements with improved designs or through the insertion of new technologies.

Regardless of their origins, modifications have the potential to significantly impact system performance and functionality; the degree of impact depends on the size and significance of the modification. As illustrated in Figure 7.1, significant modifications can be considered systems in their own right.

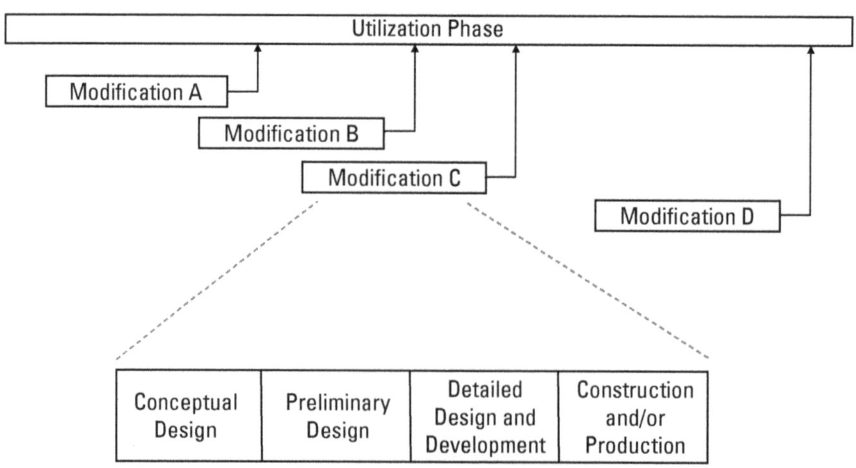

Figure 7.1 A number of system modifications may be required during the utilization phase; the systems engineering effort involved is highlighted for Modification C.

Example 7.1: New Engine for our Aircraft System

Large aircraft systems such as military aircraft or airline aircraft are often designed to have a useful life of 30 to 50 years. During this time, changes in technology areas such as engines and avionics will continue at a rapid pace. It is common for the operators of these aircraft systems to conduct major modifications to their fleets. Modifications may include replacing engines with more modern equivalents and replacing aging avionics with more sophisticated solutions.

Such modifications have a great impact on the aircraft system and may result in performance enhancements in the system, may allow additional tasks to be conducted using the aircraft, and may allow the aircraft to operate in a changed environment.

Although significant functionality enhancements often result from aircraft upgrade projects, the projects also help to ensure that the aircraft system remains maintainable through to the aircraft's design life of type.

Adding a new engine to our aircraft system would seem to be a simple process of changing a few details and revising the manufacturing process to produce the new variant. A simple modification, when viewed with respect to its impact on the original WBS can take on more significant proportions. Example 4.7 illustrated the concept of design space, and the balancing act required to ensure that all system-level requirements were met. Changing a major part of the system can significantly upset the function and performance of the overall system. In Example 4.3, our consideration of the engine CI illustrated how interface requirements are born. With these two examples in mind, a change in the engine has a large number of flow-on effects to other CIs and ultimately to system performance.

Consider the Aircraft System WBS in Figure 4.8 and the impact that a simple engine modification may have on all the elements in that WBS. The air vehicle element will be changed significantly because the engine is a major component of the aircraft. It interfaces with a range of other CIs so changing an engine will also require changes to other CIs. Test and evaluation on the engine itself and any system-level functions affected by the engine will need to be rewritten and performed. Training will be impacted as maintainers and operators will need to become familiar with the change. A new suite of technical and engineering data will be required to record the changes. Support equipment, spares, and facilities may also be affected.

Since systems engineering plays an important role in the utilization phase as modifications become more prevalent, Figure 1.16 could be revised to reflect the importance of systems engineering in utilization (as shown in Figure 7.2).

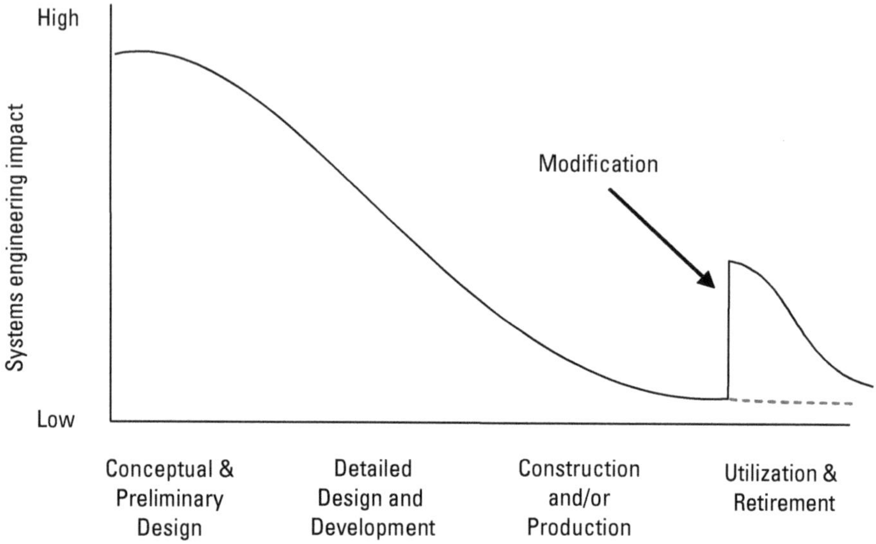

Figure 7.2 Effect of a modification on systems engineering impact.

System support also commences during the utilization phase to ensure that the system is maintained in operational order. The system support will be conducted in accordance with the approved support concept. As identified earlier, modifications to the system may be necessary to ensure that the system continues to be supportable.

7.3 Obsolescence Management

Obsolescence occurs when our systems have longer operational life cycles than the life cycle of the underlying technology components from which a system is constructed. Components become obsolete when they are no longer produced or supported by the organization from which they were originally procured. This can cause supportability challenges during the utilization phase as spare parts and components become increasingly unavailable. Complex systems are constructed from a wide range of technology groups (such as mechanical, electrical, electronic, and software components), and each of these technology groups tends to have different rates of obsolescence. For example, embedded electronic components, computers, and operational software tend to have shorter life cycles and become obsolete more rapidly than mechanical components.

Obsolescence management may be considered part of the integrated logistics support (ILS) program and is the act of accommodating the different rates of obsolescence within a complex system to continue to be able to maintain and support that complex system for as long as is required. Obsolescence management strategies can range from "brute force" approaches such as preemptively purchasing a lifetime supply of spare parts for our systems prior to them becoming unavailable, through to more sophisticated proactive approaches that predict approaching obsolescence issues and propose technology replacement/insertions at appropriate points in the life cycle. These technology insertions are typically in the form of modification programs that aim to replace elements of the system that are at risk of obsolescence with more modern elements prior to those elements becoming obsolete.

Systems engineering can certainly assist with obsolescence-focused modifications and upgrades during the utilization phase (as discussed in Section 7.2), but systems engineering can also design the system during the acquisition phase to more readily accommodate modifications and upgrades. For example, accounting for technology maturity and obsolescence risk when selecting candidate technology during conceptual and preliminary design can help to increase time before obsolescence becomes an issue. Also, using well-planned architectures with suitable modularity and partitioning can help isolate obsolescence problems to individual subsystems or components. Finally, ensuring that internal interfaces between subsystems and components are well defined and documented and making use of industry standards were practical can increase available options to address future obsolescence.

7.4 Failure Reporting, Analysis, and Corrective Action System (FRACAS)

A FRACAS is a closed-loop system designed to continually maintain visibility into system operation and support. The FRACAS is put in place to record, analyze, and rectify the cause of system failures, especially recurring or related failures. The important difference between a maintenance system and a FRACAS is that the maintenance system rectifies the failure, and the FRACAS attempts to rectify the cause. A successful FRACAS may highlight reliability and maintainability issues as a system proceeds through its life cycle; as it ages, a system normally becomes less reliable and more difficult to support and maintain. The FRACAS process is in place to help to counter this natural decline in reliability and maintainability.

The recurring theme throughout this text is that problems discovered early in the systems engineering process are generally the easiest, cheapest, and quickest problems to rectify. Therefore, the FRACAS should be established as soon as possible and becomes critical as the components, subsystems, and the system are produced, integrated, and tested. For OTS items, this could commence during preliminary design, and for developmental items, it could commence during detailed design and development.

There are several industry standards that call for a process similar to a FRACAS [1]. MIL-HDBK-2155 [2] is a freely available reference that describes a FRACAS as having six steps as follows:

- Failure reporting;
- Failure analysis;
- Failure verification;
- Corrective action;
- Failure report and close-out;
- Identification and control of failed items.

An adaptation of this process is shown in Figure 7.3.

The system under investigation is operated or tested in accordance with the standard procedures until a failure is noted. At this stage, the failure is logged and data collected to assist with the FRACAS process. The logging and data collection process needs to be recorded, and all personnel involved with the test or operation of the system need to be educated in the process to ensure that it is applied rigorously and consistently.

Testing personnel are normally well versed in the FRACAS process but sometimes failures go unreported during operations because operational personnel are not aware of the formal process in place. The success of the FRACAS relies on all failures being recorded so it is critical that operational failures are included in the FRACAS; after all, the system spends a majority of its life in the utilization phase of the life cycle.

The data collected regarding a failure event needs to be sufficiently detailed to support the FRACAS process. The data needs to describe the conditions under which the failure occurred, symptoms of the failure, and quantitative descriptions of the result of the failure. The data supports the analysis process and allows investigators to replicate the failure if necessary and practical. To that end, it makes sense to involve the investigators in the design of the form that will be used to collect the data. A failure report is generated to record the failure and its associated data. Modern FRACAS processes support

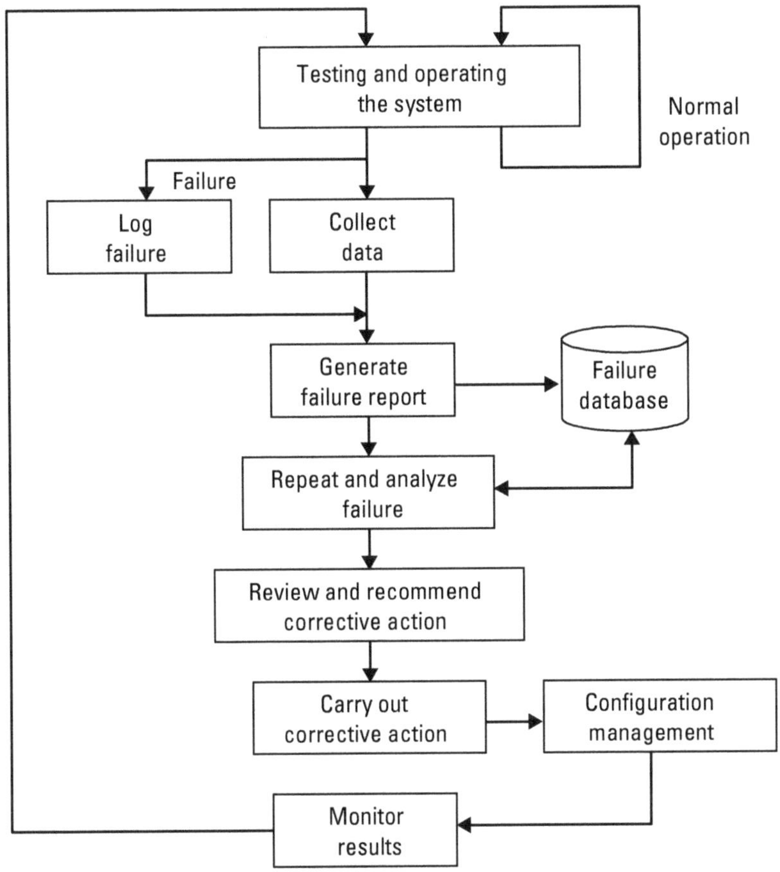

Figure 7.3 FRACAS process adapted from MIL-HDBK-2155.

this process with a database that simplifies the entry, storage, and retrieval of the information.

Once a failure report has been generated, responsibility for investigating the failure is delegated to the relevant personnel in accordance with the FRACAS process. The investigators are selected based on their experience and expertise and are delegated the necessary responsibility and authority to investigate the failure, analyze the failure, and generate a recommended corrective action report. These personnel investigate the failure using the reported data and by searching the database for other similar or related failures. Under some circumstances, it may be necessary for the analysis process to repeat (and verify) the failure, but sometimes this may not be possible especially if the failure may result in injury or death of personnel or substantial loss of equipment. The analysis process makes use of established failure analysis tools such as failure modes effect and criticality analysis (FMECA), cause and effect analysis, and fault-tree analysis; these tools are also associated with the discipline of safety systems engineering. Following the analysis and verification process, the investigators determine the possible forms of corrective action likely to rectify the cause of the failures and generate a corrective action report containing their recommendations. Corrective actions are not limited to hardware or software changes but can also include changes to the training of operators or maintenance personnel or changes to operational or maintenance procedures.

The corrective action report is reviewed and approved by a review board established in accordance with the FRACAS process. The review board is closely associated with the established configuration management process. If the recommended corrective actions involve class I or class II engineering changes, these are processed in accordance with the configuration management process. The impact of the change is monitored to confirm that the failure does not recur as the system continues operation or test.

7.5 Retirement and Disposal

The retirement phase is the final stage in the system life cycle. Functions associated with phase-out and disposal include transportation and handling, decomposition, and processing of the retiring system. A retirement concept should be developed as one of the LC during the early stages of the acquisition phase of a system (see [3] for more details on design for retirement). If considered early, disposal and phase-out issues will form some of the criteria against which the system is designed ("design for disposability"). However, it is important that system designers focus on retirement, rather than the more limiting issues of disposal; planning for disposal is important, but a system may retire from a number of life cycles before it is ultimately disposed.

As illustrated in the example in Figure 7.4, a system may be developed from conceptual design in an initial life cycle and then delivered, deployed, operated, and supported throughout that life cycle. If the organization that owns the system changes focus and moves away from the business area for which the system was acquired, the initial owners may sell the system to a second organization, which, having completed conceptual design and preliminary design in that life cycle, has decided to meet business requirements with an OTS solution, albeit secondhand. As illustrated in Figure 7.4, the

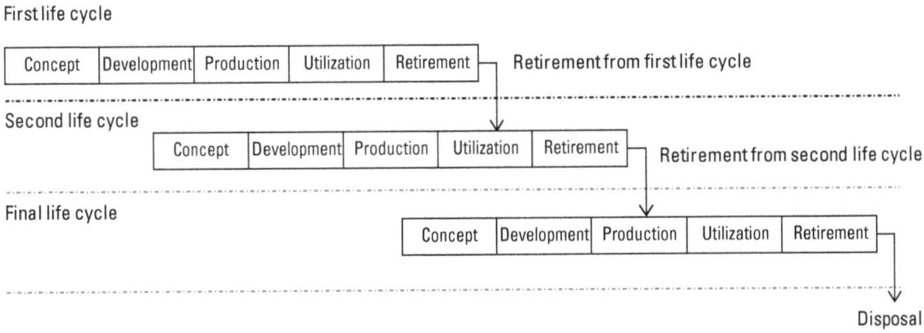

Figure 7.4 A system may be retired from a number of life cycles before final disposal at end of life.

system then enters a second life cycle, perhaps with several of its elements (the operating personnel, for example) replaced. When the second organization wishes to replace the system (perhaps because it has become costly to maintain), the system may be sold on to a third organization, which wishes to meet its systems design with a modified OTS solution based on a refurbished second-hand system; again, several system elements may be added or removed and other elements added. As illustrated in Figure 7.4, the third organization acquires the system, refurbishes it, and then fields it in a third life cycle. In this example, Figure 7.4 illustrates that the system is uneconomical to repair at the end of the third life cycle and is disposed of as scrap.

Designing for more than one life cycle is important because, if designers assume that there is only one life cycle and design just for disposal of the system products at end of that life cycle, then that disposal method may well be the only option available at the end of that life cycle (or, at least, other options may have been constrained). Yet, for minimal environmental impact, system design should address the system's existence in as many life cycles as possible (on the basis that the longer a system exists, there is lower cost and lower environmental impact resulting from the need to develop replacement systems). A good system design should therefore not only achieve longevity of use in the first life cycle of the system but should aim for longevity of use in all life cycles until disposal is the only viable option remaining.

In reality, retirement issues are often not considered until they are imminent. Very little is written about formal methods for the inclusion of retirement and disposal in the early systems engineering activities. Perhaps as environmental and ecological constraints become more stringent, retirement and disposal will begin to play a more prominent part in the system life cycle.

Indeed, environmental considerations such as emission control and material recycling are already playing an important role in constraining the design, construction, operation, and disposal in industries such as the aircraft and motor vehicle industries.

When considering the retirement aspects of the system during its design, it is useful to undertake three broad tasks: identify the reasons for potential retirement, identify potential retirement methods for the system, and then allow conceptual designers to identify design issues that may arise from the consideration of each retirement method.

- *Identify reasons for potential retirement.* There can be any number of reasons why the system may require retirement:
  - The system may have reached the end of its life and may be no longer usable and/or supportable.
  - The system may still largely be useful, but one or more critical components may be unusable or unsupportable.
  - The system may still be perfectly useful, but it (or a significant subsystem of it) is being retired because:
    - The business need for the system may have disappeared (for example, the owner's business direction may change, or the owner may be retiring and closing up the business).
    - The business need is still valid, but the business owner is forced to retire the system (due to issues such as cash flow, the operating license being revoked, scandal, public pressure, or disputes between business partners).
  - The system (or one of its critical components) may be damaged beyond economical repair (by accident, natural disaster, act of war, or vandalism).
- *Identify potential retirement methods for the system.* Once the reasons for potential retirement have been identified, each can be examined for potential retirement methods for the system (although not all may apply to all reasons for retirement):
  - Sold as a second-hand item.
  - Traded in on a replacement system.
  - Redeployed as a training aid.
  - Destroyed and disposed as waste.
  - Disassembled, that is, broken up and sold as working subsystems, assemblies, or components.
  - Scrapped, that is, broken up and sold as subsystems, assemblies, or components for their value as scrap.

    - Placed in storage, that is, mothballed because none of the other methods are currently acceptable or tenable.
- *Identify design issues that arise.* Once the reasons for potential retirement and the potential retirement methods have been identified with the stakeholders, designers can identify any design issues that may arise, such as:
    - Observance of recycling regulations;
    - Avoidance of the use of potentially hazardous/toxic materials;
    - Environmental impact;
    - Cost of disposal/destruction/uninstallation;
    - Cost of refurbishment for resale or trade-in;
    - Desirability of salvage or material recovery for resale;
    - Cost of transportation for disposal;
    - Need for specialized personnel/equipment required for disposal;
    - Observance of any caveats on disposal (such as might be placed on military equipment by international arms trade agreements, or might be placed on high technology by national sanctions);
    - Time taken to arrange disposal;
    - Availability of design data and drawings to support disposal.

7.6 A Final Thought: Systems Engineering and Sustainable Development

The World Commission on Environment and Development (WCED) defines sustainable development as development that meets the needs of the present without compromising the ability of future generations to meet their own needs [4]. The United Nations has distilled sustainable development as a concept into 17 sustainable development goals (SDG) [5], each of which can be measured using suggested metrics. Of particular interest here is SDG 12, which aims to "ensure sustainable consumption and production patterns" in order to support continued economic activity. The metrics used by the United Nations to measure SDG 12 are focused on material consumption and material footprint. This is discussed here because systems engineering, as a discipline, is well-placed to support SDG 12.

There is little doubt that sustainable consumption and production practices as promoted by SDG 12 are of critical importance as we move through the twenty-first century. For example, Primack [6] described the increasing use of global natural resources by a growing human population as a primary threat to continued biodiversity on Earth. The importance of sustainable consumption and production is further emphasized by the case studies presented by Diamond [7] in his exploration of the collapse of past civilizations caused by unsustainable consumption of finite resources.

How can systems engineering contribute to sustainable development? Collaboration and commitment between the myriads of different stakeholders involved in different phases of the systems engineering life cycle have the potential to create what is sometimes referred to as the circular economy. Design and development stakeholders in the early phases of the systems engineering life cycle can consider design concepts that incorporate raw materials produced by recycling processes associated with unrelated retired systems. System designers can also consider how their designs are able to support future reuse or recyclability. Production stakeholders should be willing to consider the use of recycled materials in their production processes and to use production processes that make their systems easier to recycle in the future. Product users should be willing to accept possible compromises around cost and performance associated with products that are fabricated with recycled material, and stakeholders associated with the retirement phase should focus on extracting valuable and reusable materials from the retired product.

The role of systems engineering in sustainable development is still emerging, but some interesting research may give a hint about the future direction. For example, Stallinger and Neumann [8] recommended updates to systems engineering standards such as ISO/IEC/IEEE 15288 to reflect circular economy concepts. Their thinking is that engineering organizations that base their design and production processes on these standards need to change their thinking (and their processes) to accommodate more sustainable practices (via the circular economy concepts). Laurent et al. [9] described life-cycle engineering (LCE) as a new "industrial culture" that can allow engineering organizations to contribute to meeting SDGs. In 2019, Schroeder et al. [10] related the concepts of LCE, SDGs, and a circular economy to discuss how engineering as a profession may be able to contribute to selected SDGs, including SDG 12.

This section has been included as the final section in the final chapter on the systems engineering process to raise awareness of sustainable development and to promote some thought on how systems engineering can play a role in ensuring that future system developments become increasingly sustainable.

7.7 Summary

The major activities during the utilization phase include operational use, system support, and modifications. The influence of systems engineering over these activities is relatively minor and is normally confined to modifications, which should be made to the system in accordance with sound configuration management procedures. Configuration management continues to play a role to ensure that the configuration of the system is managed and updated, even if modifications occur. Accurate records support the need to understand the design and capability of the system, especially if there are slight differences in configuration across a fleet of systems (for example, an aircraft fleet). Modifications made in accordance with configuration management practices enhance supportability and operations, but unmanaged modifications adversely impact training, support, safety, and operations in the long run.

In terms of the framework we introduced in Chapter 1, repeated in Figure 7.5, we have now finished our coverage of the systems engineering processes. In Chapter 8, we consider the broad topic of systems engineering management and detail the major associated activities. Some of the more common and popular systems engineering standards are introduced in Chapter 9, and their application to engineering management and process is explained. Chapter 10 explains the interrelationship between the systems engineering effort and other closely related disciplines such as project management, business analysis, quality management, and integrated logistics support management. Finally, Chapter 11 discusses the relationship between systems engineering methodologies and each of several acquisition methods.

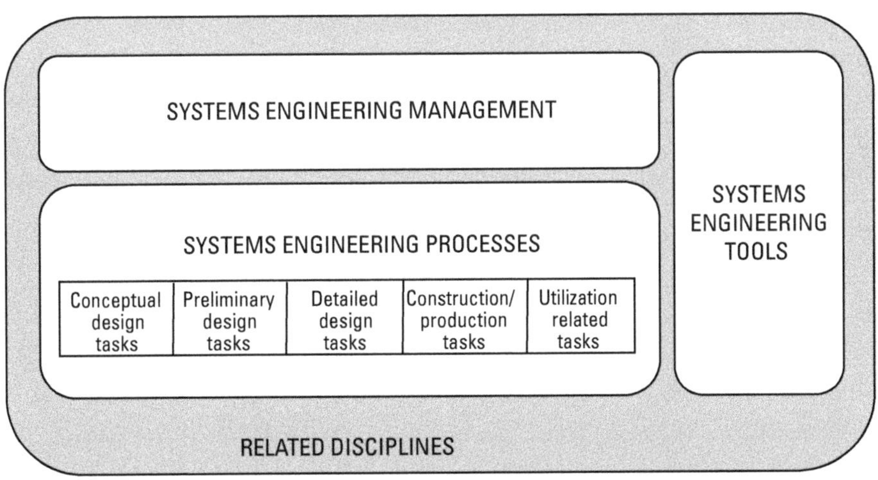

Figure 7.5 A framework for the consideration of systems engineering.

Review Questions

1. Detail the main activities associated with the utilization phase of a system's life cycle.
2. Give some examples of reasons for modifications to the system during the utilization phase.
3. Explain using an example why systems engineering could be applied during a modification exercise.
4. Briefly describe the reasons for potential retirement of a system.
5. Briefly describe the potential methods for retirement of a system.
6. Briefly describe the design issues that might arise from consideration of system retirement.

Endnotes

[1] Examples of commercially available standards that either explicitly or implicitly require a FRACAS-like process include:

ASD/AIA, *S5000F International Specification for In-Service Data Feedback*, Part 1/2, 2021.

ISO 9001:2015, *Quality Management Systems—Requirements*, International Organization for Standardization.

RTCA DO-178C, *Software Considerations in Airborne Systems and Equipment Certification.*

RTCA DO-254, *Design Assurance Guidance for Airborne Electronic Hardware.*

SAE ARP5580, *Recommended Failure Modes and Effects Analysis (FMEA) Practices for Non-Automobile Applications.*

[2] MIL-HDBK-2155, *Military Standard Failure Reporting, Analysis and Corrective Action Systems*, US Department of Defense, Washington, DC, 1995.

[3] Ryan, M. J., "Design for Retirement," *Journal of Cleaner Production*, Vol. 70, 2014, pp. 203–210.

[4] World Commission on Environment and Development, *Our Common Future*, Oxford, UK: Oxford University Press, 1987.

[5] United Nations, *The Sustainable Development Goals Report*, 2019.

[6] Primack, R. B., *Essentials of Conservation Biology*, Sunderland, MA: Sinauer Associates Publishers, 2014.

[7] Diamond, J. M., *Collapse: How Societies Choose to Fail or Survive*, New York: Penguin Books, 2011.

[8] Stallinger, F., and R. Neumann, "Enhancing ISO/IEC 15288 with Reuse and Product Management: An Add-On Process Reference Model," *Computer Standards & Interfaces*, Vol. 36, No. 1, 2013, pp. 21–32.

[9] Laurent, A., et al., "The Role of Lifecycle Engineering (LCE) in Meeting the Sustainable Development Goals—Report from a Consultation of LCE Experts," *Journal of Cleaner Production*, Vol. 230, 2019, pp. 378–382.

[10] Schroeder, P., K. Anggraeni, and U. Weber, "The Relevance of Circular Economy Practices to the Sustainable Development Goals," *Journal of Industrial Ecology*, Vol. 23, No. 1, 2019, pp. 77–95.

8

Systems Engineering Management

8.1 Introduction

Systems engineering management is responsible for planning and directing the systems engineering effort, monitoring, and reporting on that effort to the appropriate areas and reviewing and auditing the effort at critical stages in the system life cycle. The management function is also responsible for ensuring balance is struck across the systems engineering effort between striving for perfection and continuing to make adequate progress. In this chapter, we consider the major systems engineering management elements of technical review and audit management, test and evaluation, technical risk management, configuration management, the use of specifications and standards, and systems engineering management planning.

8.2 Technical Review and Audit Management

As we have discussed in the preceding chapters, the technical review and audit considerations of a project can be quite extensive and can add significantly to both project cost and schedule. Management of the review and audit program helps to ensure that the overall aim of reducing technical risks and enhancing confidence in the design process is achieved within cost and schedule constraints. In the previous chapters, we have considered the major reviews and audits in each stage of the acquisition phase. This section discusses the management considerations of technical reviews and audits.

Technical reviews and audits provide both the acquirer and developer with a measure of progress toward the goal of successfully introducing a system into service and reducing the technical risks associated with the system development. Reviews and audits achieve this by:

- Providing a formal evaluation of the design maturity;
- Measuring and reporting on planned and actual performance;
- Clarifying and prioritizing design requirements;
- Evaluating and establishing the system baselines at discrete points in the design process;
- Providing an effective means of formal communications between the stakeholders;
- Recording design decisions and rationales for later reference.

Technical reviews and audits are a vital part of the overall systems engineering effort. Reviews can range from very formal and structured reviews normally associated with a critical system milestone, to less formal reviews involving only a few people and represent only a minor component of the overall system. Regardless of the level of the review, all reviews aim to determine the ability of the design in question to meet the relevant technical requirements. Reviews tend to become more detailed as the design progresses and cover all aspects of the engineering design effort (for example, software, hardware, and integration). The acquirer normally specifies the required technical reviews and audits and their timing in the contract/agreement artifacts (based on their initial expectations tempered by the nature of the successful tenderer's proposed approach to the technical program). Once agreed, the contract/agreement must specify the length, duration, location, content, conduct, and attendees for each review. Based on these requirements, the contractor is better equipped to estimate the cost and schedule associated with the reviews. A full set of reviews in a large project can represent a sizeable portion of both time and cost for the project. The contractor organization may also decide to include additional internal reviews, at their discretion, to determine further the technical adequacy of the design.

Obviously, the number of reviews required, and their respective scope, depends on the complexity and size of the system in question, so the reviews and audits must be tailored accordingly. The technical risk associated with the system also has an impact on the number of reviews scheduled. For example, a system that makes heavy use of unmodified OTS equipment may be seen as a low-risk project compared to a pure developmental project. Fewer lower-level reviews or the use of combined reviews would be expected for the OTS system compared to the developmental project. In the case of modified OTS equipment, formal reviews may be limited in scope to cover the modifications.

It is clear that reviews and audits can provide significant assurance of the technical progress of the system development. However, these reviews and audits also take considerable time and effort to plan, prepare, and conduct. While reviews and audits are being conducted, attention is being diverted from the design and development effort. It is important that the costs and benefits associated with conducting technical reviews and audits are considered fully prior to planning and mandating the reviews and audits to ensure that an appropriate level of review and audit is placed on the system development effort.

Inadequate review and audit may add risk to the system development by allowing potential design problems to go unchecked. However, the usual tendency is to make the review and audit requirements bulletproof by over-reviewing the project. Unnecessary reviews and audits increase the cost and schedule of the system without adding any additional value in the way of technical risk reduction. The additional cost and schedule are invariably ultimately passed onto the customer.

The scheduling of reviews and audits is also very important. Reviews held too early in the design process are unable to determine technical adequacy of the design because the design is immature and subject to change. Reviews held too late in the process may miss the opportunity to avoid costly and time-consuming rectifications if the design proves inadequate. It is also useful to relate technical reviews and audits to artifact release [1]. For example, the PDR may be scheduled to coincide with the release of the hardware and software development specifications. Obviously, as the design progresses, the reviews and audits require hardware and software products as well as artifacts, but the concept has merit because design artifacts tend to be a strong indicator of design maturity.

Throughout the previous chapters, we have introduced the major formal reviews and audits referred to in MIL-STD-499B [2]. Although this standard was never released, it still provides an excellent source of reference for formal technical reviews and audits and their intended purpose. As mentioned earlier, not all reviews are applicable to all systems. To that end, some reviews and audits can be "tailored" out of the requirements to suit the individual requirements of each system. Another excellent reference source for technical reviews and audits is MIL-STD-1521B [3], which was canceled when MIL-STD-499B was released in draft. MIL-STD-1521B provides a detailed set of entry and exit criteria from each of the reviews and audits listed in this section. Several commercially available reference sources also describe the need for technical reviews and audits as part of a systems engineering process, including ISO/IEC/IEEE 15288.2:2003, which is an example of a modern standard provides a technical review and audit framework (roughly aligned to MIL-STD-1521) aimed at military systems and typical defense systems life cycles [4].

Where system size and complexity warrant it, configuration audits may be considered as a separate topic from technical reviews. MIL-STD-973 [5], MIL-HDBK-61B [6], and EIA-649-C [7] provide a comprehensive coverage of configuration audits, including the role that audits play in the overall configuration management process.

As well as specifying the reviews and audits required, the contract/agreement should also specify how the reviews and audits are to be conducted. The requirement for chairperson, agenda, data packages, and minute-taking will help in making the reviews and audits as effective as possible.

Both the supplier and the acquirer are active participants in reviews and audits. Both parties need to prepare prior to the event to ensure that the attendees have the appropriate expertise to contribute to the areas being covered in the review. The attendees must be given adequate time to review the agendas and data packages prior to the review. Each person with a role in the review must read the data packages prior to attending the review. This helps to avoid spending valuable review time covering material that has already been distributed as pre-reading. However, observers may attend the reviews as a convenient way for new staff to become familiar with the design and development progress and status.

Decisions, action items, and agreements should be noted in the minutes that are distributed following the meeting. Each action item should also be assigned to an individual who is responsible for following the action through to fruition by an agreed-upon date. Unassigned action items invariably remain unactioned.

At the conclusion of the review, the acquirer team should meet to determine the level of satisfaction with the meeting. Successful technical reviews are a good indication of project status and schedule. Significant milestone payments are often tied to the successful completion of design reviews. Therefore, technical reviews and audits are an important part of both the systems engineering and the project management effort.

There must be a willingness to make difficult decisions in the event of unsatisfactory technical reviews and audits that may result in the need to revisit some or all the preceding effort. The old adage that bad news does not improve with time certainly applies to inadequate design. While there may be pressure to meet progress and schedule targets, it is often better to stop work, go back a little in the process, and get it right before continuing. Succumbing to the pressure to continue is rarely in the interests of the project.

The technical review and audit effort is principally managed via a technical reviews and audits plan (TRAP). The TRAP records all formal reviews, detailing the entry criteria that must be met prior to the commencement of the review or audit and the exit criteria that must be demonstrated prior to approval of the review or audit. The TRAP (or the contract/agreement) normally lists the design artifacts that need to be produced for evaluation during the relevant review. The TRAP is normally drafted during conceptual design and should be finalized and approved prior to any contract/agreement being signed. It is then maintained throughout the project and updated to account for any changes that occur.

8.3 Test and Evaluation (T&E)

T&E is a systems engineering management function that ensures a coordinated and consistent testing effort is applied to the system for the entire system life cycle. By coordinating the T&E from a system perspective, the focus and emphasis of the testing can be varied with different life-cycle phases without compromising the entire T&E effort. T&E applies throughout the entire system life cycle and involves both the acquirer and the developer.

A thorough evaluation of a system involves validating the system against the original business and stakeholder needs and requirements. Obviously, this validation cannot be completed until the entire system has been designed, developed, and constructed and then operated in the intended operational environment by operational personnel; however, much can be done along the way. The aim of a useful T&E approach is to test and evaluate the system progressively as it passes through the acquisition and utilization phases in order to avoid costly and time-consuming modifications to the system design late in the life cycle. With this in mind, progressive T&E could be considered a risk mitigation measure as it provides a high degree of confidence early in the system life cycle that the design will ultimately perform as required.

Testing can be an expensive, time-consuming, and sometimes dangerous exercise that requires the use of specialized personnel, test equipment, and facilities. A formal plan is normally put into place very early in the acquisition phase to manage the system T&E effort. This plan is the test and evaluation master plan (TEMP), which forms a major deliverable from the systems engineering effort. The TEMP is discussed in more detail later in this section.

8.3.1 Verification and Validation (V&V) Methods

The entire systems engineering process aims to produce a system that is both verified against the artifacts produced during the systems engineering process and validated against the original needs, goals and objectives that initiated the system development in the first case. Often these two associated aims are combined into the term verification and validation, although compounding the two into V&V is misleading and the practice is discouraged [8] since the two activities have quite different purposes. Verification ensures that we have "built the system right" and validation ensures that we have also "built the right system" [9]. A well-managed approach to T&E aims to support the delivery of a system that is both verified and validated. Other systems engineering management functions including technical reviews and configuration audits also support the overall V&V objective.

Example 8.1: Verification Versus Validation

> *In our aircraft example, ACME Air has noted a capability gap that has been documented in business needs and requirements. These have been developed further into an StRS that includes details of all the missions and scenarios that the airline needs to be able to carry out but currently cannot perform due to limitations with their existing capabilities. This includes limitations across the entire ACME capability domain including personnel, organization, training, support, and aircraft systems. In the example in this book, ACME procurement and systems engineering specialists have defined requirements for the aircraft that flow from the StRS to the SyRS, which has then been used to drive the design and development of a new aircraft. Contractors have used that SyRS to design and develop a new aircraft for ACME Air. Naturally, the aircraft needs to be verified against the specified requirements in the SyRS. This process will likely include comprehensive ground-based activities as well as aircraft operations and flight tests of the production version of the aircraft. The personnel performing these activities are likely to be highly trained, skilled, and experienced flight-test professionals from the developer's organization. They will show that the aircraft can meet or exceed all the requirements in the SyRS. Once done, the aircraft will have been verified. However, the aircraft will not have been validated. To validate the aircraft, it needs to be combined with all the other elements of ACME's capability including ACME's line aircrew and ground crew, integrated with ACME's operational systems including maintenance and support, and operated in accordance with ACME's standard operating procedures. When all of these capability elements are combined, ACME*

will be able to judge whether they are able to achieve the business needs and requirements, missions, and scenarios. Once complete, the system is said to have been validated.

Naturally, ACME Air needs a system that is both verified (compliant with specified requirements) and validated (meets and addresses the need for the new system).

The INCOSE *Needs and Requirements Manual* contains much more detail on the various methods of verification and validation, but briefly the most common methods employed for V&V are described as [10]:

- *Inspection.* Inspection or examination is nondestructive and typically includes the use of sight, hearing, smell, touch, and taste; simple physical manipulation; or mechanical and electrical gauging and measurement. Inspection could be of the produced item such as code or artifacts that represent the system, such as drawings, parts listings, and part specifications.
- *Demonstration.* Demonstration involves the operation of the system to provide objective evidence that the required functions were accomplished under specific scenarios and operational environments. Qualitative determination is made through observation, with or without the use of special test equipment or instrumentation, to verify characteristics such as human engineering features, services, access features, and physical accommodation (compared with the test method, which uses a quantitative approach).
- *Test.* Test leverages data taken from one or more actions to provide objective evidence that the system element satisfies a need or requirement. Test uses controlled, predefined sets of inputs, data, or stimuli to exercise the system or system element whose performance is measured using special test equipment or instrumentation to obtain the required data. One or more tests (actions) can be performed using final versions of the end item or dedicated design or engineering units or platforms of varying levels of fidelity.
- *Analysis.* Analysis uses established mathematical models, simulations, algorithms, charts, graphs, circuit diagrams, or other scientific principles and procedures to provide objective evidence that design input requirements or needs have been met. Analysis also includes engineering assessments and logic usage based on previously recorded evidence. Analysis allows predictions about the expected performance of a system based on the confirmed test results of a sample set or by combining

the outcome of individual tests, demonstrations, or inspections performed on lower levels. Analysis is often used to predict the breaking point or failure of a product or system by using nondestructive tests to extrapolate the failure point.

8.3.2 Categories of T&E

There are three major categories of T&E that are applied to coincide roughly with the acquisition Phase, the transition between the acquisition and utilization phases and the utilization phase. Figure 8.1 illustrates the role and timing of each type of testing.

- *Developmental test and evaluation (DT&E).* DT&E refers to the test and evaluation activities undertaken during the acquisition phase of the system life cycle to support the design and development effort and to verify the system's elements are being developed in accordance with the relevant element specification. DT&E activities may also occur during the utilization phase to support activities such as modification development.
- *Acceptance test and evaluation (AT&E).* As DT&E completion approaches, AT&E activities become increasingly relevant. AT&E represents the formal acceptance testing conducted on the system to enable the customer to accept the system from the developer by verifying that the system meets its requirements in the contracted

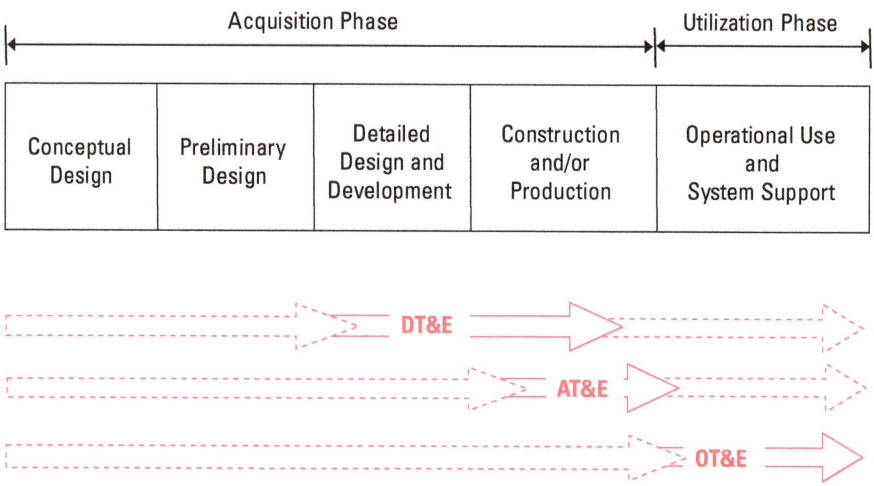

Figure 8.1 Timing of DT&E, AT&E, and OT&E.

SyRS. AT&E effectively forms the boundary or transition between the acquisition phase and the utilization phase. Unlike DT&E and OT&E, AT&E tends to be a discrete testing activity (with a defined start and a defined end).

- *Operational test and evaluation (OT&E).* OT&E is the test and evaluation effort that is focused on the functional or operational testing of the capability system and its components, conducted under realistic operational conditions including being operated by operational personnel and with all the other elements of capability. OT&E is normally conducted for a period following acceptance of the system by the customer, although limited OT&E activities are possible during acquisition, especially where a long production cycle means that some systems have been accepted prior to other systems being produced, or when concept demonstrators are being used. OT&E validates that the system meets the BNR/SNR as encapsulated particularly in the StRS.

This section describes the three forms of T&E and, where possible, details the specific test activities associated with each of the activities within the acquisition and utilization phases.

8.3.2.1 Developmental Test and Evaluation (DT&E)

As the name suggests, DT&E encompasses the verification activities conducted throughout the acquisition phase of the system life cycle and aims to highlight deficiencies in the system design as early as possible. As we have seen a number of times, the earlier problems are identified in the system life cycle, the cheaper and simpler they are to rectify. DT&E also provides confidence that the system, as designed, is likely to meet the SyRS. The designers of the system also use DT&E results to validate the design process and to minimize the risks associated with poor system design.

DT&E covers a broad range of testing levels corresponding to the different design activities conducted during the acquisition phase. At one end of the spectrum, DT&E is responsible for T&E of the lowest-level elements in the system. As such, DT&E effort may involve time on the test bench testing the initial hardware design mocked up in a development environment.

At the other end of the spectrum, full system prototypes undergo T&E as part of the DT&E program prior to the system being produced, offered for AT&E, and subsequent transitioning into the utilization phase. DT&E on system prototypes verifies that integration has been successful and provides confidence that the design meets requirements prior to moving into construction and/or production.

Responsibility for the planning and conduct of DT&E normally lies with the developer. The acquirer normally places requirements on the developer with respect to T&E and invariably wants visibility into DT&E activities. Visibility is normally provided during the design reviews, but it is not unusual for the acquirer to be invited to witness DT&E activities and be provided with copies of any DT&E reports and results. The acquirer may consider accepting some DT&E results in lieu of repeating expensive, dangerous, or time-consuming test procedures during AT&E. DT&E also offers a potential early opportunity to collect data on performance of a statistical nature such a reliability.

Although DT&E tapers off as the system nears completion and moves into construction and/or production, the development of modifications during the utilization phase revives the requirement for DT&E.

8.3.2.2 AT&E

AT&E is a discrete activity normally shared between the acquirer and supplier that aims to verify that the developed system meets the requirements of it in the SyRS that drove its development. The acquirer approves the specific acceptance test procedures to be completed as part of AT&E, but both the supplier and acquirer may jointly conduct the tests. Even if the customer does not conduct the AT&E testing, they will most likely witness the process and analyze the results. AT&E is focused on the system-level requirements contained in artifacts such as the SyRS and the contract/agreement and is designed to verify that the system meets the requirements as contained in the FBL.

The test results and any discrepancies are documented and rectified where required prior to the conclusion of AT&E. Regression testing may be required where some acceptance test procedures have failed. The supplier will perform the necessary modifications to the system and the test procedures will be repeated. Care must be taken in these circumstances to understand what additional impact system modifications may have on other aspects of system performance. Regression testing should be conducted on all aspects of the system affected by the modification, which may involve significant effort.

At the successful conclusion of AT&E, the acquirer accepts the system from the developer, and the system formally enters the utilization phase of its life cycle.

8.3.2.3 OT&E

OT&E is focused on operational functionality of the capability rather than design issues of each constituent element and is therefore commonly the responsibility of the customer organization. OT&E is normally conducted by an entity within the customer organization to assess the ability of the system and its components to meet the BNR/SNR and the StRS. The testing entity is normally independent from the procuring agency within the customer organization to ensure that some level of independent assessment is provided. Separation of OT&E responsibilities from the customer's project office assists in maintaining objectivity in the assessment.

OT&E aims to estimate the operational effectiveness and suitability of the system when it is combined with other capability system elements such as personnel, support, and training and with external systems in the desired operational environments. Potential modifications may be identified during the OT&E activities and recommendations are passed through the appropriate channels to the designers. OT&E must be conducted in as realistic an environment as possible. Whereas DT&E, and AT&E to a certain extent, can be performed in clinical conditions, OT&E must be conducted in operational conditions including the use of all the inputs to capability including operators, support personnel, maintainers, and logistics support.

In this way, OT&E supports validation of the system because the original BNR/SNR for the system should be satisfied when the customer's operational personnel operate the equipment that is supported by the customer's support personnel. These operational and support personnel will have to be trained via the customer's training system and are doing their jobs in accordance with the customer's processes, procedures, and doctrine.

There is also another important distinction between AT&E and OT&E. As illustrated in Figure 8.2, AT&E is conducted at the end of the acquisition

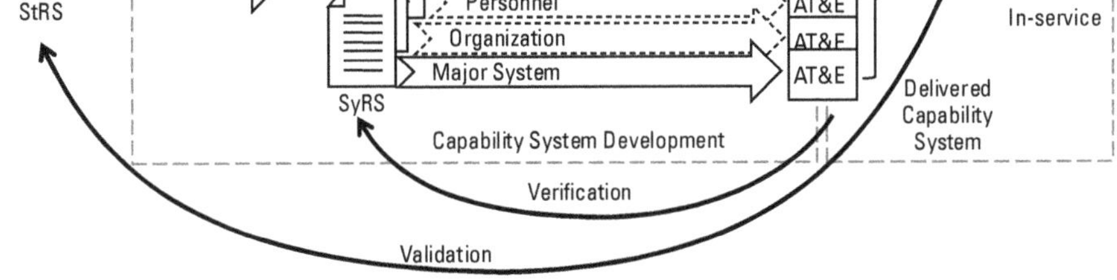

Figure 8.2 AT&E verifies that the system meets the requirements in the SyRS, and OT&E validates that the capability system meets the StRS.

phase of each of the elements of the capability system in order to ensure that the developer for each element has met the requirements of the relevant SyRS for that element; AT&E is therefore principally an act of verification to ensure that the contactor has met their obligations. OT&E is an act of validation that ensures that the elements of the capability have been integrated, and the delivered capability system meets the stakeholder requirements in the StRS (BNR/SNR).

OT&E may also be used to assist the customer in developing operational procedures for the system. The system also continually undergoes OT&E throughout its life in service, albeit often informally as operations reveal shortfalls in performance and suggest areas for improvement.

Example 8.2: Testing in Our Aircraft Example

> *In our aircraft example, the company acquiring the aircraft (ACME Air) may be organized into three functional areas: operations, maintenance and support, and procurement. It is normally the responsibility of the procurement section to manage the project associated with the procurement of the new medium-range aircraft. To that end, procurement is responsible for customer T&E. Part of their role as the customer T&E agent is to involve operations personnel in the T&E effort, particularly in AT&E and OT&E. Through their T&E efforts, the company has much greater confidence as a customer for the new aircraft.*

8.3.3 Test Management

Regardless of whether the tests are conducted by the acquirer or the developer, the developer should be responsible for managing DT&E and AT&E to ensure a coordinated approach to testing. Coordination will help maximize the effectiveness of testing and may also result in some savings in expensive testing resources such as personnel, facilities, and test equipment. Management is necessary to ensure that the distinct testing functions are mutually beneficial, avoiding unnecessary duplication of effort and other conflicts. It is also common to retain the developer to support, and sometimes manage, OT&E.

8.3.4 Testing Activities and the System Life Cycle

Having introduced the three main categories of T&E, it is instructive to investigate the focus of T&E activities in each of the stages within the acquisition phase.

8.3.4.1 T&E Activities During Conceptual Design

The major T&E activity associated with conceptual design is to plan the overall approach to T&E and the subsequent drafting of the TEMP. From an engineering management perspective, the customer typically prepares an outline of the TEMP to address the T&E requirements specified in the contract/agreement; the tenderers are required to respond with a draft TEMP that conforms to the customer's preliminary outline. As mentioned, the TEMP also is a contractual deliverable that details how the contractor intends to approach the overall test and evaluation effort during the acquisition phase of the system life cycle. Naturally, the very early versions of the TEMP produced during this phase of the life cycle lack the detail associated with a more mature artifact, but the initial TEMP is still an important part of the T&E management effort. To that end, the TEMP continues to be updated throughout the acquisition phase of the life cycle as details become available. In some of the more popular TEMP formats, room is allocated for summaries of testing conducted to date, requiring the TEMP to be updated regularly. It should be noted that, in some cases, the TEMP remains a customer/acquirer artifact and the contractor/developer T&E effort is recorded in a V&V plan; here we refer to both efforts as the TEMP for the project.

Some very early T&E may assist designers to select the most appropriate concepts and designs during conceptual design. Testing may occur on existing systems being considered for modification and reuse to meet customer requirements. Where similar systems exist elsewhere, the customer may use these systems to develop a better understanding of their expectations and requirements. This is particularly valuable when the system being proposed is completely new and the customer has no experience in its operation or capabilities. This idea of using a similar existing system to explore requirements and potential methods of employment is often called concept demonstration or technical demonstration.

8.3.4.2 T&E Activities During Preliminary Design

The activities associated with preliminary design normally involve the validation of the concepts considered during conceptual design and development of draft acceptance testing plans and procedures. These plans and procedures are ultimately used to conduct and report on formal AT&E. The acquirer has an interest in these plans and procedures, and they are normally reviewed prior to the completion of preliminary design. Design validation will inevitably involve the development of some demonstration entities. For example, concept hardware may be mocked up in a hardware development environment, software modules may be produced using some form of rapid prototyping environment, and models may be constructed using some of the many design tools available today. These models and mock-ups do not necessarily represent the final design, but they provide a valuable vehicle for validating certain design concepts.

DT&E during preliminary design may include informal tests designed to exercise these models and mock-ups and collect the results. Designers can derive information regarding the system performance from these test results and evaluate the adequacy of the overall design concept under consideration (once again, the analysis-synthesis-evaluate loop). DT&E may also be required to confirm the performance of OTS or modified OTS equipment selected for inclusion in the design.

As in conceptual design, there will be little scope for OT&E during preliminary design although, in some development approaches (particularly those involving rapid prototyping), aspects of the system development may be mature enough to allow customer input. For example, designers of systems using very complex user interfaces may opt to completely design and develop the user interface prior to moving onto the back end of the system. The user interface may be advanced enough to be tested and accepted by the acquirer prior to preliminary and detailed design of the back end.

Most testing at this stage in the life cycle can be informal and is usually conducted in the "test bench" environment. Even though this testing is informal and concentrated on early engineering models, it is vitally important as it provides the designers with the opportunity to identify any gross problems with the system design at a stage where design changes are relatively cheap and quick to implement.

In summary, T&E during preliminary design achieves the following aims to:

- Evaluate any design areas that have been identified as significant risks in the overall design;
- Help to select the preferred technical approach to the system design from the alternatives under consideration;
- Help to validate that the preferred technical approach can meet the original user requirements;
- Finalize the T&E requirements for the remainder of the acquisition phase and record this in the TEMP.

8.3.4.3 T&E Activities During Detailed Design and Development

By the later stages of detailed design and development, the design has matured to an extent where a more formal testing effort can be applied. System prototypes and mature software and hardware components are available for testing, allowing reasonably detailed and thorough testing to occur. The TPMs established and prioritized during conceptual design should provide specific testing requirements and expected results.

The DT&E effort during this stage of the acquisition concentrates on demonstrating that the design of the system is very close to complete and that all outstanding design issues have been adequately resolved. The testing must also provide some level of confidence that the design meets the requirements contained in the specifications.

Examples of some testing possible during this stage of the project include:

- *Functional testing.* This verifies that the specified functions can be performed by the system to the required level of performance under specified conditions.
- *Interface and integration testing.* This ensures that the external and internal interfaces are operating correctly. This includes the system's ability to interoperate with other relevant systems and involves verification of any human-machine interfaces.
- *Environmental testing.* This places the system in a range of environments to ensure acceptable operation under all specified conditions. Environmental testing will normally involve testing under conditions of varying temperature, humidity, terrain, and rainfall. Testing may also extend to cover less obvious environmental influences such as electromagnetic effects and nuclear, biological, and chemical factors. The physical environment also needs to be investigated, including the effects of dust, vibration, and stress.
- *Physical and configurational testing.* This verifies that the system meets requirements relating to modularity, interchangeability, and accessibility. Tests also confirm that the physical characteristics of the system (such as size, mass, and volume) are acceptable.
- *Quality factor testing.* This includes early verification of the quality factors such as useability, reliability, maintainability, and availability.

Testing during detailed design and development is normally conducted at the supplier's facility. Problems and modifications identified during this testing are then easier to incorporate. The tests qualify the system to enter construction and/or production.

There is no way of listing exhaustively a set of test categories that should be considered for a system as the testing requirements of every system is completely dependent upon that system and its intended purpose. The main point to consider is that testing needs to be detailed and thorough enough to provide the necessary confidence that the system design is ready to produce.

8.3.4.4 T&E Activities During Construction and/or Production

Once the system has successfully qualified through the tests conducted during the previous design stages, the system enters construction and/or production. This stage is focused on AT&E conducted in the later stages of this stage as systems begin to be produced. In most circumstances, AT&E is conducted by operational personnel from either the contractor or the customer. For example, by the time an aircraft project had moved through construction and/or production, production versions of the aircraft would be operating. The customer would want to see the aircraft tested in as close to an operational environment as possible. If the aircraft was a military attack aircraft, this environment may include an air weapons range including representative targets and threats. In a civilian context, other air traffic control radars and navigation aids would need to be considered as part of the environment.

Testing at this stage in the process may be the first time that all elements of the system have been fully integrated into a single entity. The indications provided by previous tests in areas such as quality factors can be verified and demonstrated throughout the testing at this stage of the development.

DT&E is now focused on the production process and procedures and ensuring that production is operating correctly. The first system produced by the production process is tested fully. If more than one system is produced, each subsequent unit needs to be tested to an extent to ensure that the production and manufacturing process is continuing to be effective.

AT&E is focused on providing the acquirer with sufficient confidence in the as-produced system, to accept the system as meeting the original requirements. AT&E can be as exhaustive as time and budget allows, but it is normal for some problems to surface after AT&E has been completed.

OT&E is now able to start ensuring that all production systems meet the original system requirements as they are accepted into service and recommending any modifications that may be required to the system. OT&E has the benefit of more time and a greater range of variables to stress and test the system than AT&E. OT&E eventually detects any problems with the production system.

8.3.4.5 T&E Activities During the Utilization Phase

Testing of the system continues throughout the utilization phase. OT&E plays a relatively minor role in the T&E program up to this point in the life cycle. It is now possible for the fielded systems to undergo testing in a true operational environment. All elements of the capability system are tested during OT&E including facilities, training, data, logistics support, and the prime equipment. OT&E will be planned and conducted on an as-required basis but should be recorded in the TEMP.

Typical T&E focus areas during the utilization phase include integration/interface with external systems, supportability issues, functional performance, and operation in the true operational environment. This type of testing is conducted by the user organization and may result in modifications to the system. It also assists in the development of operating procedures and policies regarding the delivered system.

Example 8.3: Validating a Training System

The procurement of a complex aircraft system (such as in our example) includes delivery of a training system as part of the system. The training system is responsible for training aircrew in the operation of the new aircraft, ground crew in handling procedures, and the maintenance practices necessary to keep the aircraft serviceable.

The effectiveness of the training system is only validated once all courses of instruction have been used on "real" students, and those students have demonstrated the competencies required of the training system. This validation process takes place in final training location, using delivered training aids and students from the customer's organization.

Validation of a training system may be partially achieved prior to system utilization but sufficient confidence in the training system may only be provided following a number of successful courses conducted in the field. The conduct of these courses is an example of OT&E.

DT&E will have been completed in the acquisition phase but may be required again during the utilization phase to support activities such as modification development. As discussed in Chapter 7, modifications can be considered a system in their own right; therefore, DT&E is required during the equivalent of the acquisition phase of the modification process.

8.3.5 TEMP

The TEMP is the major planning artifact for the entire T&E effort planned for the acquisition phase of the project. The TEMP is also commonly referred to as the master test plan (MTP) or the V&V plan. The TEMP is required by the contract/agreement, prepared by the contractor, and reviewed and approved by the acquirer. The TEMP should be finalized and approved during conceptual design and then reviewed as part of the technical review process. The TEMP should be in an approved state by SDR and should be maintained throughout the acquisition phase. In some cases, the acquirer may develop a TEMP to cover all phases of the system life cycle, and the contractor will be required to prepare a V&V plan for the acquisition phase.

The content of the TEMP and V&V plan will be specified by the contract/agreement, which may also mandate a template for the layout. A typical TEMP may contain the following information [11]:

- *System description.* This section includes a summary of the mission description of the system being developed and a description of the system that has been designed to meet the requirements. Major TPMs are listed in prioritized order to reinforce the mission description.
- *Program summary.* The program summary should detail the T&E responsibilities of all organizations participating in the T&E effort. The schedule of the T&E effort should also be outlined in this section of the TEMP to ensure that relevant organizations are aware of the proposed T&E activities.
- *T&E outline.* The TEMP should include references to DT&E, AT&E, and OT&E in this section. Critical technical characteristics (DT&E) and operational issues (OT&E) should be detailed, and any testing conducted to date should be recorded in the TEMP. As well as recording testing to date, the TEMP should detail the future testing (DT&E, AT&E, and OT&E). This way, the TEMP becomes a working artifact for the entire T&E effort.
- *T&E resource summary.* This section details the resources required for the entire T&E effort. Resources include personnel (both acquirer and contractor personnel), specific test and support equipment, test facilities, any simulators, models or test beds required, and any necessary personnel training required prior to testing. A schedule should also be included to show when the specific resources are required.
- *Appendixes.* Any other artifacts relevant to the T&E effort should be attached to the TEMP. The set of test procedures that will be used to test the system should either be attached to the TEMP or referenced by the TEMP. Test procedures include the setup procedure and initial conditions, a detailed set of instructions needed to run the test and provide the expected results for each of the tests. The test reports generated following testing record the results to determine whether the system has passed the test and recommend a course of action for failed tests should also be either attached to the TEMP or referenced therein.

The TEMP may also reference other plans and artifacts (such as discrepancy reports) necessary to support the T&E process.

8.4 Technical Risk Management

Risk is defined as the possibility of a loss or injury or the possibility of some disadvantage or destruction. Systems engineering management is concerned with the management of technical risk; that is, the risk associated with the technical aspects of the system life cycle. Broadly speaking, there are two major categories of risk: internal risk and external risk. Internal risks are those that are within the control of the acquirer's organization; external risks are beyond the control of the project office. An example of an internal technical risk may be the risks associated with inappropriately staffing the project office with qualified technical staff. This risk can normally be managed from within the project office. An external risk may be the introduction of some new legislation that places more stringent requirements on the system, leading to some changes to the user requirements.

The concept of risk is subjective in that risk identification and assessment is based on the perception of an individual and their interpretation of the term risk. What one individual may find risky, another individual may not. To that end, risk management must be performed by personnel with a broad and experienced knowledge of the subject area to ensure that risks are assessed realistically. Given the extremely broad nature of risk, it is unlikely that a single person will be able to perform the entire risk management function, nor is that recommended; risk management is a team effort.

8.4.1 Risk Management Steps

There are several components that combine to define the concept of risk. Perhaps one of the best structures for risk management is provided by ISO-31000 [12]. Figure 8.3 provides an overview of steps in the risk management processes, which are described further in outline in the following sections.

The broad activities contained in Figure 8.3 are:

- *Establish the context.* The first step in the risk management process involves understanding the strategic, organizational, and risk management context within which risk management is to be conducted. It is very important that risk management is not conducted without understanding the context; for example, the context for the analysis

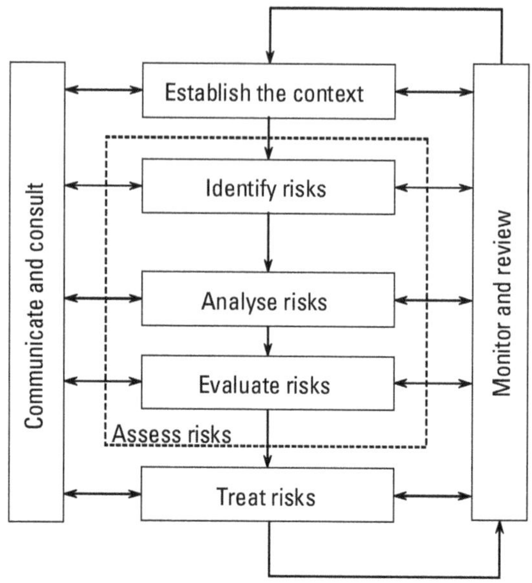

Figure 8.3 Risk management overview. (After: ISO-31000.)

of the risk of injury to employees in the workplace is a very different issue for a warehouse than it is for a football franchise or a marine battalion. This step also involves establishing the criteria against which risks will be evaluated, as well as determining the structure of the following analysis.

- *Risk identification.* The next step in the risk management process involves the identification of the potential risks relating to the system. Once the risks have been identified, they can be analyzed and managed accordingly.
- *Risk assessment.* Risk assessment consists of two parts: risk analysis and risk evaluation. Risk analysis involves determining the risk controls that already exist in the organization and then analyzing risks in terms of their level of risk should they occur. The level of risk is normally considered to result from the product of the probability, or likelihood, of the risk occurring and the consequence of its occurrence. Risk evaluation involves comparing the estimated levels of the risk to the criteria determined when establishing the risk context.
- *Risk treatment.* Risk treatment involves identifying the options available for dealing with the risks identified. Classic means of treating risks include avoidance, mitigation, transfer, and acceptance.

- *Communicate and consult.* Risk management is not an isolated exercise in either time or scope. Communication and consultation are essential to ensure that all risks are identified and treated appropriately.
- *Monitor and review.* This activity monitors and reviews the performance of the risk management system to ensure that it is always current and relevant.

Technical risk management is not a distinct systems engineering activity assigned to an individual risk manager, but rather an integrated part of a sound systems engineering management effort. This is demonstrated by the fact that many tools and concepts described in this text as systems engineering processes or management functions contribute to technical risk management. Examples include the process of TPMs, technical reviews and audits, T&E, and configuration management. Technical risk management is not a one-off exercise conducted early in the acquisition phase of the system. Technical risk management is an ongoing concern. New risks may arise, existing risks may change, and old risks may disappear throughout a system life cycle. To that end, technical risk management must continue throughout the system life cycle. Some popular systems engineering standards such as MIL-STD-499B contain guidance on risk management.

8.4.1.1 Risk Identification

Risk identification is the process of identifying the major areas of risk within a given system and recording those risks. The process should record the source of risks and the symptoms that may be evidence of an impending risk occurring. Typical sources of technical risk in projects include changes in customer requirements leading to design changes, design errors or misunderstandings leading to unsatisfactory performance, and insufficiently skilled staff leading to longer design times or substandard designs.

No foolproof set of tools exists that guarantees the complete identification and recording of all risks. Potentially the most effective means of risk identification is via a series of interviews or meetings with the appropriate groups of people. Given that both internal and external risks exist, the skill is in identifying the appropriate groups of people to interview.

Any source of information may provide insight into potential problems and risks. Sources of information include systems engineering design artifacts such as drawings and system architectures, trade-off studies, or design review artifacts. The other risk management tools such as audit results, TPM monitoring, and T&E results also provide invaluable insight into potential new risks and the performance of existing risks.

It may be helpful to consider the following major categories as potential sources of risk during the identification process:

- Design, including design process, requirements, analysis, and testing;
- Testing, including the test reports containing test results;
- Production, including the manufacturing plan and process, test equipment requirements, and subcontractor controls;
- Logistics, including facility requirements, manpower and personnel, spares and technical records;
- Management, including cost and schedule estimates, acquisition strategies, and staffing considerations.

By using these major categories, it may be simpler to assign responsibility for managing the identified risks.

Technical reviews and audits are also an excellent forum for discussing risk and potentially identifying new risks as the system moves through the acquisition phase. Strong consideration should be given to contractually requiring risk management activities in each of the major formal reviews and audits.

8.4.1.2 Risk Analysis

Once the major risks have been identified, an analysis of each risk must be carried out to determine the likelihood (probability) of the risk occurring and the consequence (impact) of that risk should it occur. This is also often called risk quantification. Based on the assessments of likelihood and consequence, the level of the risk (which is a function of both likelihood and consequence) can be calculated. By attempting to analyze aspects such as likelihood and consequence, the analysis exercise will focus on the cause and effect of the risk and, in doing so, provide some insight into the potential management of the risk.

Several tools exist to assist in the analysis of risks including classic project management scheduling tools such as Program Evaluation and Review Technique (PERT) and Critical Path Method (CPM) charts, simulation, and models that allow what-if analysis of potential impact. However, a very popular tool is expert judgment. There is benefit in using available tools to provide an input into the risk quantification, but it is easy to become overly focused on the answers given by these risk analysis tools. It is vitally important to maintain focus on the fact that risk analysis is primarily subjective in nature. There is much merit in keeping the risk analysis relatively simple. Its main aim is, after all, to provide management with a list of potential risks (in order of level of risk) and provide some indication of the causes and effects of these risks. Management will then decide on the best available means of treating these risks as discussed in the following section.

A simple but effective means of ranking the level of risk is to categorize the likelihood of occurrence into several categories. Figure 8.4 illustrates how that may be done with three categories for likelihood (highly likely, possible, and unlikely) and three categories for consequence (high, moderate, and low), so that there will be nine possible levels of risk. For practical risk analysis, it is advisable to use five categories of likelihood and consequence, resulting in 25 possible levels of risk. Excessive levels add confusion without adding greatly to the ranking exercise, so it is common to use only four levels (extreme, high, medium, and low). Because it only uses three categories of likelihood and consequence, Figure 8.4 illustrates only three levels of risk (high, medium, and low).

8.4.1.3 Risk Treatment

There are four main options for the treatment of the risks identified and analyzed in the previous two steps: risk avoidance, risk mitigation, risk transfer, and risk acceptance. The management of risk should not focus solely on the negative aspects of risk, but also the benefits that may accrue from exposing the project to the risk.

Risk avoidance is the option selected when the risk is too severe to be accepted at any level and must be avoided. Risks that may fall into this category are those that may result in loss of life or loss of business. The causes of the risk will need to be investigated to decide on how best to avoid the risk. It should be noted that, by avoiding one risk, exposure to other risks may well result. This additional exposure also must be considered when deciding on how best to avoid a risk. As an example of risk avoidance, designers may choose one technology over another to avoid the risks associated with the first technology.

		Likelihood		
		Highly Likely	**Possible**	**Unlikely**
Consequence	**High**	High	High	Medium
	Moderate	High	Medium	Low
	Low	Medium	Low	Low

Figure 8.4 Illustration of a table to identify level of risk (for three categories of likelihood, consequences, and level).

Risk mitigation may be the option selected when the risk falls in the middle ground in terms of level of risk. Mitigation basically means weakening and, in the context of risk, mitigation means reducing the risk rating. This can be achieved by reducing the consequence of the risk, reducing the likelihood of its occurrence, or preferably both. Again, the causes and effects of the risk must be considered to mitigate the risk successfully. An example of risk mitigation may be the inclusion of redundancy in the design to reduce the likelihood of failure.

In risk transfer, the identified risks can be transferred to another party through organizational structures, contracts, or insurance. In our personal lives, for example, we often transfer risk by taking out insurance. However, care must be taken because the risk still exists; it has simply been transferred. This is particularly pertinent if the risk is transferred to another party by way of contract/agreement; consequently, risk transfer is often referred to in modern standards as risk sharing.

Risk acceptance is also a valid form of risk management usually reserved for risks of low level. Risks may also be accepted when there is significant benefit to be gained should the risk not occur. The adage of risk equals return does hold true sometimes for technical risks. For example, the use of leading-edge technology may be judged to be a reasonably severe risk as the impact of it going wrong may be quite high. However, management may decide to go ahead with the use of the technology because if it does not go wrong, the benefits may be extremely attractive.

Risk treatment also considers how to monitor the risks identified. Severe risks should be monitored more closely and more often than less severe risks. However risks are treated, they must be managed continually to address the fact that circumstances are continually changing: old risks will disappear, new risks will appear, and existing risks will change in nature. Risk management is the responsibility of all parties involved in the system life cycle and must be performed continually.

8.4.2 Risk Management Records

If there appears to be significant risk associated with a given system development, there may be a requirement for a risk management plan (RMP), which is a formal artifact developed by the contracting organization and reviewed and approved by the customer. The RMP details the risk management process and the results of the application of that process to the development effort. In addition, the customer develops a more general RMP associated with all project management-related aspects of system development.

The risk management process should, as a minimum, contain the major steps outlined in this section, and the results (of the identification and quantification) will be dependent upon the nature of the system development. The RMP should detail the schedule for reviewing the risks and should contain details of the management (response and control) of each of the risks. The RMP will also provide details of more focused risk-management records such as risk logs. The plan must be periodically reviewed, since it is likely that risks will change with time.

8.5 Configuration Management

The term *configuration* can be defined as the relative disposition or arrangement of parts of something. In the context of systems engineering, the "something" is normally a configuration item (CI) that forms a part of the overall system, and the "relative disposition or arrangement of parts" is called a baseline. To that end, configuration management (CM) is the act of controlling and managing the physical and functional makeup (defined in the baselines) of the configuration items that comprise the system [13]. EIA-649-C defines configuration management as: "A technical and management process applying appropriate processes, resources, and controls, to establish and maintain consistency between product configuration information, and the product" [14].

The main aims of CM [15] are to identify the functional and physical characteristics of selected system components, designated as configuration items; to control changes to those characteristics; and to record and report on the change processing and implementation status.

An effective CM system provides an accurate snapshot of the state of each CI (and, through these, the system). CM allows an effective history of changes to the system design to be maintained including the alternatives considered and the reason for selecting the preferred alternative.

8.5.1 Establishing the Baselines

Baselines are a very important part of system acquisition as each approved baseline provides a firm position from which to move forward. Once established, each baseline will be placed under configuration control, which means that changes to the baseline can only occur through a formal change management process. This process is different for each project, but an example CM process is described in this section. The baselines used throughout system development have already been introduced a number of times in Chapters 2, 3, and 4 [16]:

- *FBL.* The FBL (also called the requirements baseline [17]) of the system has at its center the SyRS that is produced during the early stages of the acquisition phase.
- *ABL.* The ABL for the system is defined by the set of development specifications for each of the CIs making up the system. These specifications should be developed and approved during the preliminary or detailed design activities of the acquisition phase; a preliminary ABL might even exist earlier. The ABL is initially established when the set of development specifications is complete, and finalized or approved for each CI when the relevant CI-level FCA has been conducted successfully.
- *PBL.* The PBL for the system is the set of product specifications (and any associated material and process specifications) for each of the CIs comprising the system. The stage at which this baseline is first established can vary but is typically toward the end of the detailed design and development stage or very soon thereafter. The PBL is eventually confirmed or approved by the conclusion of PCA.

Generally, the FBL takes priority over the ABL and the ABL takes priority over the PBL should there be any conflict or discrepancy.

8.5.2 CM Functions

For CM to be effective, EIA-649-C outlines the following functions that must be performed throughout the life cycle [18]:

- *CM planning and management.* This is an essential CM function that is conducted over the life cycle of the system to ensure that the appropriate tasks are performed and the necessary resources are available at the right times.
- *Configuration identification.* This identifies the items (the CIs) to be placed under configuration management.
- *Configuration change management.* This ensures that changes to CIs can be conducted in accordance with an effective, systematic, and measurable change management process.
- *Configuration status accounting.* This provides accurate and up-to-date status information on the configuration of the items placed under control.
- *Configuration verification and audit.* These establish that the information relating to the CI is complete, accurate, and current; that the allocated physical, functional, and interface requirements are met by

the CI; and that an adequate process is in place to confirm the correct operation of the CM system.

Although defining largely the same set of activities, ISO 1007 uses slightly different titles for the functions: CM planning, configuration identification, change control, configuration status accounting, and configuration audit [19].

These CM functions are described further in the following sections.

8.5.2.1 CM Planning and Management

CM planning and management occur over the life cycle of the system to ensure that the CM process is effective, predictable, and repeatable [20]. Consequently, the CM process must be recorded and approved and then controlled. Additionally, the planning and management function ensures that the appropriate level of CM is undertaken at the appropriate stage in the life cycle by the appropriate element of the organization. Planning and management also ensure that personnel are trained, appropriate resources are in place, performance indicators are established, and CM information is integrated with the broader organizational processes.

8.5.2.2 Configuration Identification

Configuration identification revolves around the selection of the CIs during preliminary design. The selection of CIs is a design decision recorded through the analysis and allocation of the SyRS into lower-level specifications such as product and design specifications. Typically, systems are broken into the major categories of computer software configuration items (CSCI) and hardware configuration items (HWCI). Naturally, there are likely to be several CSCIs and HWCIs in the system design. Each CI is managed as a separate entity, including the design and T&E of the item. As discussed in Chapter 4, the selection of CIs is based on several factors: complexity, interfaces, use/function, existing items available, commonality, provided by a single supplier, criticality, and maintenance and recording needs.

Once each CI has been identified, the configuration identification function ensures that the technical artifacts required to define completely the physical and functional performance of that CI has been produced and that those artifacts are current, approved, and available for use as required. As discussed in Chapter 4, the initial identification of CIs represents the translation from functional to physical description during preliminary design.

8.5.2.3 Configuration Change Management

At various stages throughout the system development, changes may be required. The reasons for these changes will be wide and varied, ranging from a change to one of the user requirements to a change in some available technology. Requests for change (and the subsequent changes) must be managed by a formal process to ensure that the impacts of the change are investigated and handled correctly. This process is called configuration change management in EIA-649 and configuration control in ISO 10007.

Normally, a body called the configuration control (change) board (CCB) is charged with the responsibility of managing changes. The CCB would include major stakeholders and users, relevant ICWG participants, and production representatives. The CCB investigates the change and decides on the severity of the change. Major changes are sometimes called Class I changes and include changes to form, fit, or function of the system or major changes to system cost and schedule. As would be expected, the customer must be involved in the approval process of any Class I change. Minor changes, or Class II changes, include all other changes to the system configuration. Examples of Class II changes include updates to artifacts following identification of errors or changes to aspects of the design that do not impact on form, fit, or function such as material substitutions. The customer should also be involved with Class II changes but will probably be involved for information purposes only. However, the customer must be involved in the process by which a change is classified as either Class I or Class II. The customer must be very aware of the potential ramifications of the change before agreeing to a classification of Class II, because once a change is classified as a Class II change, the customer can effectively lose visibility into, and control of, that change and its impacts.

The CCB should meet on a regular basis to review the progress of changes that are currently in progress and to review other proposed changes. Sometimes, it is necessary to call a CCB meeting urgently to process a change that is considered urgent.

A change can be requested by any of the parties involved, including the customer, the design team, or a subcontractor. In the case of a customer-initiated change, the customer may request the contractor to put together a change proposal for consideration by the CCB. This is requested because the customer is not able to provide the CCB with sufficiently detailed information regarding the impact of the change.

A request for change (or change proposal) must contain sufficient information for the CCB to decide whether to approve the change. The request for change includes a statement outlining the change and the reasons that the change has been suggested. The request then details the alternatives available to implement the change and outlines the preferred alternative. The preferred alternative should be provided in sufficient detail to show that it will result in the change being implemented successfully and not introduce new problems into the system design; the change will be categorized as major (Class I) or minor (Class II). Of particular importance is the impact of the change on other CIs within the design. The impact of the change may result in changes to the interface requirements as specified in the baseline artifacts. To that end, any resultant changes to the baseline artifacts should also be included in the change proposal. The ability to assess impact relies heavily on sound traceability across the different levels of design. Finally, the change request would include details of how the change is to be implemented as well as how it proposed to verify the change.

In some cases, special interface teams are organized (comprising members from relevant CI development teams) to ensure that interfaces are handled correctly and completely. These interface teams will need to be involved with the change management process, especially if changes to the interfaces are likely. A simple change management process is shown in Figure 8.5.

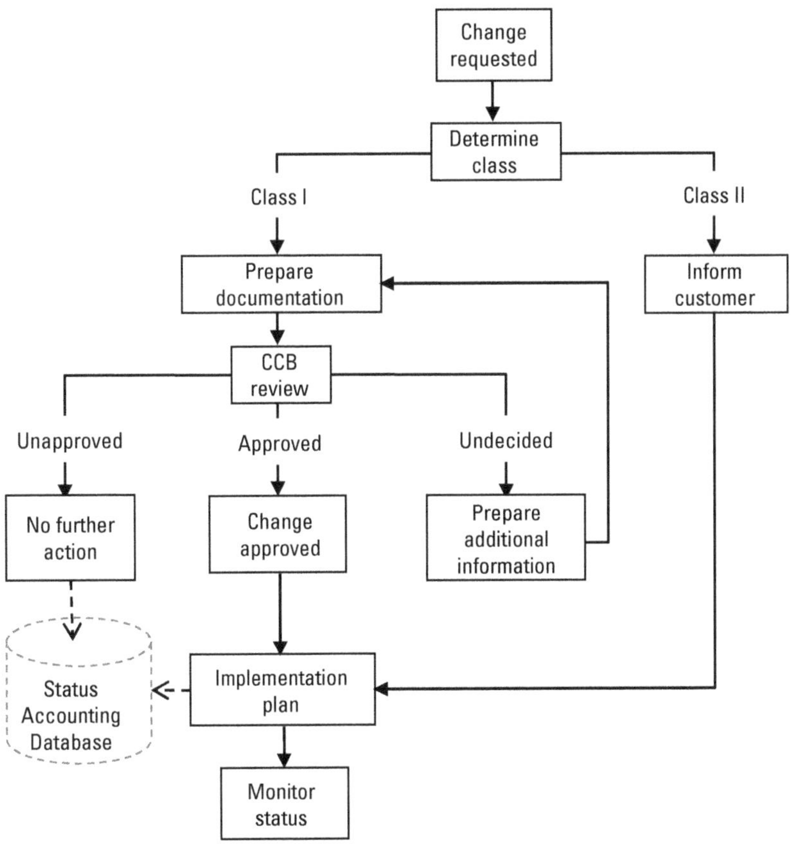

Figure 8.5 A simple change management process.

Requests for Variance

In some cases, it becomes clear during construction or production that items within the supplies are not going to comply with the relevant specification contained in the established configuration baseline. There may be any number of reasons for this occurring. Normally, these nonconformances would be rectified using the formal engineering change process; however, there are situations when departures from specifications may be accepted by the customer without the need to go through the engineering change process. CM calls these special cases requests for variance [21] (previously called deviations or waivers [22]) or concessions [23]. Note that a variance differs from an engineering change in that an approved engineering change requires corresponding revision of the CI's current approved configuration, whereas a variance does not.

Example 8.4: Request for Variance

> *Following the analysis of functional requirements, a design approach is developed that includes a computer system. The computer system is specified by the design team and is available as an OTS item. The specific OTS computer, as specified, is not available in time to support continuing integration, testing, and initial acceptance of the system. An option is to use an alternative but technically equivalent computer system that is available in the meantime. A request for variance may be sought and approved prior to the procurement of the computer system to allow the designers to use the alternative computer system (even though it is not the specified computer) until the specified computer is available. In this example, the customer would be accepting the variance for a specified period of time. There is no need to seek an engineering change because the contractors are not seeking to change the configuration baseline of the system, nor is the change permanent.*

Deviations and Waivers

In some organizations, the older terms of deviations and waivers may still be in use. Deviations and waivers are forms of request for variance, although, strictly speaking, a deviation is the closest to a request for variance.

- A *deviation* is defined as "a specific written authorization, granted prior to the manufacture of an item, to depart from a particular requirement(s) of an item's current approved configuration for a specific number of units or a specified period of time" [24].
- A *waiver* is defined as "A written authorization to accept an item, which during manufacture, or after having been submitted for government inspection or acceptance, is found to depart from specified

requirements, but nevertheless is considered suitable for use as is or after repair by an approved method" [25].

Note that the principal difference between the two types of variances relates to when the variance is authorized. Deviations are essential decisions made before the event to allow a project to continue without unnecessary delay. Waivers are granted after the event and, as tacit acceptance of failure to meet requirements, are becoming a less commonly accepted means of allowing a departure from the baseline. Note again that deviations and waivers differ from an engineering change in that the CI's current approved configuration record is not amended.

8.5.2.4 Configuration Status Accounting

Status accounting is the function that provides information about the current status of the items identified as requiring configuration control. This information includes the current state of the CIs and a history of any changes to the CIs from initial identification. Effective status accounting aids greatly in the change management process, which, in turn, allows changes to be made to the various system baselines during the design process.

8.5.2.5 Configuration Verification and Audit

Configuration audits have been covered in Chapter 6 and include FCA and PCA. FCA confirms that all necessary functionality required of the CI is included in the CI design. PCA concentrates on the physical configuration of the CI as described by the relevant artifact and displayed by the physical CI. If many copies of the CI are being produced as part of the system production process, usually the first production copy is audited followed by the auditing of an agreed sample of copies during the subsequent production process. For items that are not physical in nature such as CSCIs, the final versions of software listings are normally audited to ensure that the executable code is produced by a compilation of the approved source code.

In addition to audits associated directly with the CIs, the CM process will include audits to verify that an adequate process is in place to confirm the correct operation of the CM system.

8.5.3 CM in Acquisition and in Utilization

CM does not stop when a system is introduced into service; CM is just as important in the utilization phase as it is in acquisition (perhaps even more so, if the system is to be managed effectively over decades of operation). So, as illustrated in Figure 8.6, there are at least two configuration management activities, one conducted by the contractor during acquisition and one conducted by the customer's in-service management staff. There will therefore be two CM plans: one developed by the contractor, and one developed by the in-service support organization.

However, in service, the CIs may be slightly different to those defined in acquisition, principally because the CIs in service are chosen to facilitate through-life support and will depend on the support philosophy adopted. Nonetheless, it is useful to minimize the differences between the acquisition and utilization CIs to reduce the impact of the transition to the in-service support agency as the system is fielded. It is therefore important that the two parties engage as soon as possible during preliminary design to ensure that there is a smooth transition between acquisition CM and in-service CM.

8.5.4 CM Records

A range of records may be required to support the CM effort for system development. The need for and nature of the records are normally specified in the contractual artifacts. This section outlines some of the more typical artifacts related to CM.

8.5.4.1 CM Plan (CMP)

The CMP is one of the major systems engineering management plans and details the approach and processes involved in the CM for the project. As noted earlier, there will most likely be two CMPs for the system; the first is prepared by the contracting organization and applies to the acquisition phase of the life cycle, and the second applies to the CM of the system during the utilization phase and is written and maintained by the organization responsible

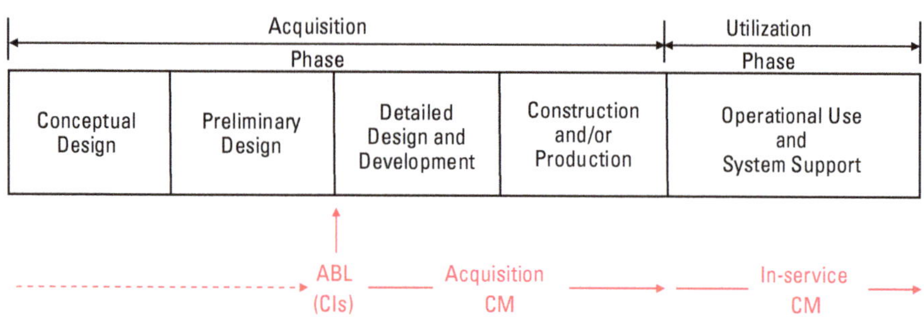

Figure 8.6 CM during acquisition and in-service.

for through-life support. It is important that the two parties engage as soon as possible during preliminary design to ensure that there is a smooth transition between acquisition CM and in-service CM.

The content and layout of the CMP vary from system to system and, for the acquisition phase, is usually determined by the customer and dictated in the contractual artifacts. An outline structure for a CMP is provided in Annex A of ISO 10007 [26] and Annex A of the ECCS-M-40B [27]. Not surprisingly, the contents of the CMP reflect the five principal functions of CM.

8.5.4.2 Change Control Forms

There will need to be some basic form that can be used during the change management process. This form is generically called the change control form, which leads the CCB through a formal change management process and will facilitate the initialization, evaluation, approval, and releasing of the changes as they occur. During the change process, the change control form provides an excellent source of status information regarding the change.

8.5.4.3 Change Proposals

There are two major types of change proposal relevant to the field of systems engineering management: engineering change proposals (ECPs), and contract change proposals (CCPs). As the names suggest, an ECP relates to the engineering design aspects of the project and a CCP relates to the contractual aspects of the project. ECPs and CCPs can exist independently of one another but often they will coexist; that is, an ECP will probably force a contract change such as a change of the cost or schedule (and therefore a CCP). This section concentrates on the ECP because it is more related to the systems engineering aspects of the project.

The content and use of the ECP are determined by the customer and stated in the contract. As a minimum, the ECP should contain sufficient information to allow the CCB to evaluate and approve the ECP. This information will include a statement of the reason for the change, the various alternative solutions considered, and the preferred solution. Details of any trade-offs should be included, as should a statement of the impact if the change is not approved and incorporated. Another important piece of information that must be contained in the ECP is an impact assessment of the change that analyzes the impact the change will have on other aspects of the design, the interfaces with other CIs, and the likely changes required to the current configuration descriptions.

ECPs may also affect the current cost and schedule for the project, in which case these impacts must also be included. It will be rare that a Class-I

ECP would be approved and incorporated with no cost and no impact on the schedule.

8.6 Specifications and Standards

One of the major outputs from the systems engineering processes is a complete set of descriptions of the designed system. It is not possible to understand the systems engineering processes without understanding how the description requirements fit into the overall system development effort. Every system development effort will have varying requirements with respect to the set of descriptions.

In addition to the description requirements, it is worth considering the role of standards in the overall design effort. These standards can take many forms including military standards, international standards, and standards from recognized institutions such as the International Standards Organization (ISO), the Institute of Electrical and Electronic Engineers (IEEE), the International Council on Systems Engineering (INCOSE), and the Electronic Industries Alliance (EIA). These standards bring together a wealth of knowledge and experience and provide the system designers with technical guidance during system development.

Both standards and artifacts requirements need to be considered carefully by the customer before stipulating their use. Although artifacts and the use of certain standards enhance the chances of project success, overdoing the requirements can add unnecessary overhead to the design effort, which further results in increased cost and longer acquisition schedules.

8.6.1 Specifications

Specifications are produced (or procured) during the acquisition process that describe the system or system components to varying levels of detail. A specification can be a document, a structured database, or another data source such as a model. Standards such as MIL-STD-961E [28] provide sound guidance on how to structure and write document-based engineering specifications. This section briefly describes the main categories of specification at the system level and below.

8.6.1.1 System Requirement Specification (SyRS)

An SyRS needs to state the requirements of the system in terms of both operational terms and maintenance and support terms and record the interfaces between the system and its environment and between major functional areas of the system. Significant design constraints should also be detailed in the SyRS. The requirements stated in the SyRS should avoid reference to products and equipment and should concentrate on explaining the requirements in mission or capability terms. In its mature and approved state, the SyRS will define the functional baseline of the system.

The SyRS is also called a system specification or, in the INCOSE *Needs and Requirements Manual*, the design-to specification. The SyRS is reviewed and approved at the SDR and, when approved, is a central element of the functional baseline of the system.

8.6.1.2 ICD

The ICD provides the mechanism for managing the integration of different system elements. The ICD is discussed in considerable detail in Chapter 4.

8.6.1.3 Development Specification

Development specifications are focused on subsystems (or CIs) that make up the system as opposed to the system. The development specification will detail the functional, performance, and interface requirements of each CI and will consist of two major parts: the requirements specification and the interface requirements specification. The development specification should be sufficiently detailed to allow designers to design and evaluate the CI design and to support the selection of OTS elements during detailed design and development. An important requirement of the development specification is that it maintains a traceability of requirements back to the SyRS.

The development specifications are normally produced during preliminary design and will be reviewed and approved at PDR. It should also be noted that separate development specifications are written for hardware, software, and facilities (and any other major developmental component of the system). Standards such as MIL-STD-490A [29] define different types of development specification including prime item development specification (PIDS) and critical item development specification (CIDS), depending on the specified item. The development specifications derived from the SyRS form the ABL for the system.

8.6.1.4 Product Specification

The product specification is aimed at any component below the system level that needs to be either developed or procured (for OTS or modified OTS). Product specifications for equipment that needs to be constructed contain the detailed design requirements for that equipment flowing from the relevant development specification. The product specification for OTS or modified OTS equipment is aimed at specifying the performance requirements for the equipment. Obviously, the functional and physical descriptions delivered with OTS items will be less detailed than that produced for developmental items and may even be limited to the small amount of tabulated data provided by the OTS manufacturer; an OTS product specification is therefore often a specification in name only as it does not contain anywhere near the information contained in the product specification for a developmental item.

The contents of a product specification depend very much on the requirements detailed in the project contract; however, a typical product specification could be expected to include [30]: a complete set of performance requirements for the product, detailed description of the interface requirements, detailed description of the parts and assemblies comprising the product, and tests and expected results to verify correct fabrication of the product.

For software products, an additional section may be added (if relevant) to detail the design and development requirements of any underlying data repository such as a database. If these requirements are insignificant, they may be included in the software design section of the product specification.

Product specifications are produced during detailed design and development although draft product specifications may be available as early as PDR. Mature product specifications are reviewed at CDR. Approval of the product specifications may be done in conjunction with the PCA. The approved product specifications for all products in the system collectively form the PBL of the system.

8.6.1.5 Process Specification

During detailed design, designers are careful to ensure that the products described by their design can be constructed or produced during manufacture or production; that is, the processes for manufacturing/producing the products are well-practiced within the developer's facilities. These processes need to be performed within the skill sets of the workforce involved and within the nominated production facility. In some cases, particularly with systems that have novel elements, a specific or unique process may have to be developed to support the manufacture of one or more products within the system. In that case, a process specification is developed to describe a process such as fabrication/production, painting, welding, soldering, or assembly. The process used to fabricate an element will contribute a great deal to the ability of that element to meet its specified requirements and must therefore be defined and proven by the engineering process and then adhered to during the production phase.

8.6.1.6 Material Specification

Material specifications are also written to support the production process and apply to raw materials such as metals and alloys, mixtures such as paints, and semi-fabricated materials such as fiber optical cable. The use of a material specification allows the selected material to be obtained reliably in production or even in modification of the system during utilization. As with fabrication processes, the material used to fabricate an element is central to that element meeting its specified requirements and must therefore also be developed, defined, and proven during the design phases and then adhered to during the production phase.

8.6.1.7 Specification Tree

Figure 8.7 shows a typical specification tree containing the specifications described in this chapter. The specification tree is used to illustrate how the lower-level specifications flow from the SyRS.

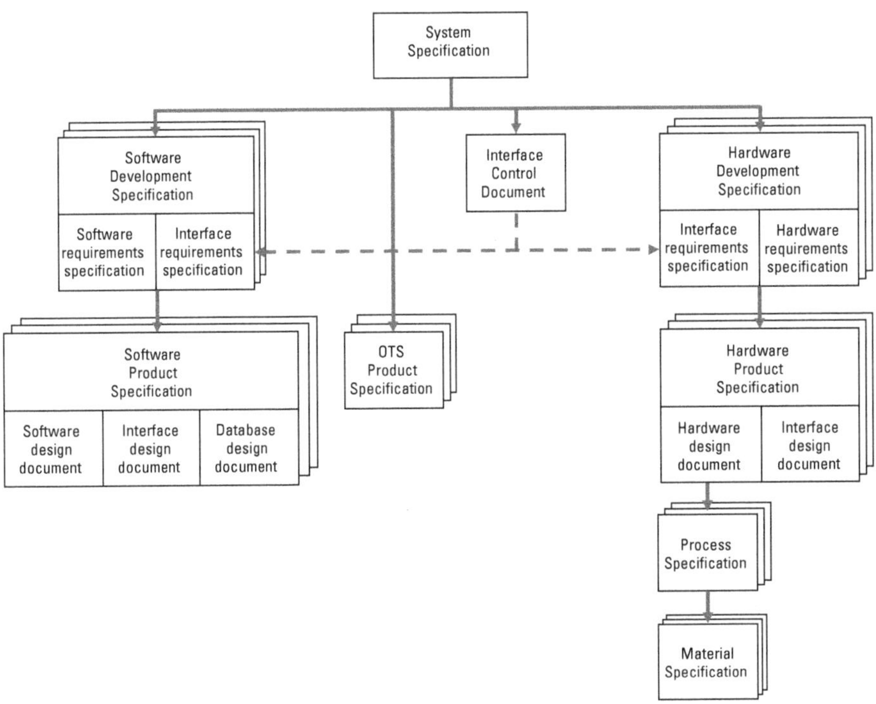

Figure 8.7 A typical specification tree.

8.6.2 Standards

An engineering standard is a generic term used to describe artifacts that establish engineering and technical requirements for processes and procedures that (when required by the contract/agreement) should be applied to a specific aspect of the acquisition phase. Standards have usually evolved over an extended period and are a refined source of significant expertise and experience. Within a typical engineering contract/agreement, standards of various types will be found in both the engineering Statement of Work (SOW) and across all levels of the specification tree (Figure 8.7). Even when standards are not contractually required, they should be considered as sources of information and potential guidance.

This section outlines some engineering management issues relating to the use of standards during the system life cycle. Chapter 9 introduces the most relevant systems engineering standards in use today.

8.6.2.1 Overspecification

There is a general tendency by customers to overspecify requirements, including requirements relating to the use of engineering standards. The reason for overspecifying may be the belief that including more stringent requirements enhances the chances of success of the system development. Often, unnecessarily specifying the use of engineering standards introduces significant overhead without necessarily reducing the risks associated with the development. The increased overhead for the contractor is passed directly onto the customer in the form of extended schedules and increased costs.

The customer needs to be certain that the standards being specified in the contract are required and not merely repeated there because "that standard has always been used in our contracts." This requires that the customer has read and understood the standard prior to specifying it.

8.6.2.2 Tailoring

A problem related to over specification is that of the "blanket specification" of engineering standards. This means specifying the entire standard without consideration of the relevance (or lack thereof) of each of the sections within that standard. Standards are normally designed to be tailored to meet the specific requirements of each acquisition project. For example, certain technical reviews and audits may not be necessary in a particular project. If this is the case, the unnecessary audits and reviews should be "tailored out" of the standard that has been used to specify the review and audit requirements of the project. Without tailoring, unnecessary (and therefore expensive) activities and artifacts may be undertaken by the contractor organization.

Tailoring should be considered by the customer before the tendering process begins for the acquisition project. This gives the tendering (and eventual contractor) organizations the ability to respond to the tender with accurate estimations of effort, cost, and schedule. In some circumstances, it may be acceptable to leave the tailoring until conceptual design and tailor the standards in a joint effort between the customer and the contractor.

Regardless of when or how the tailoring occurs, all specified standards must be considered in their entirety prior to the beginning of the acquisition phase. Unnecessary requirements must be tailored out before the requirements start needlessly impacting the acquisition effort. Effective tailoring can only be done by individuals who are conversant with the standard, the overall systems engineering discipline and the specific system in question.

8.6.2.3 Unintentional Invocation of Standards

Standards sometimes reference other standards where necessary rather than repeating information that has been recorded elsewhere. For example, standard A may reference the requirements in standards B, C, and D. Standard B may reference the requirements in standards E, F, and G, and so on. By specifying the use of standard A, the customer may also, by implication, invoke the requirements of standards B through to G. This implication inherent in some standards may result in an exponential explosion in the number of standards and requirements referred to in the contract. The customer can avoid these problems by carefully selecting and tailoring the standards required by the contract.

8.7 Systems Engineering Management Planning

Systems engineering is a broad subject and, to cover the entire systems engineering effort, a systems engineering management plan (SEMP) is formulated to detail the required effort. The contract SEMP is normally constructed by the contractor in accordance with the requirements in the contract and reviewed and approved by the customer. Once approved, the SEMP becomes the governing plan controlling the entire systems engineering effort and all technical aspects of the project. Changes to the SEMP must be reviewed and approved as they occur, and the SEMP is normally reviewed at each of the formal design reviews.

In addition to the contract SEMP, the customer may also have developed a SEMP for any work performed within that organization, and the contractor will also require subcontractors to develop a SEMP that conforms to the main contract SEMP. Once the contract has been awarded, it is useful to reconcile the customer and the contractor (and any subcontractor) SEMPs to make sure that the technical processes described are seamless across the contractual boundaries.

The SEMP should cover all the major systems engineering functions. It may do so by referring to other subordinate plans such as the CMP, the RMP, the TRAP, and the TEMP, specification lists/trees, and CI lists. Internal company design plans and processes should also be referenced by the SEMP if applicable. In this way, the SEMP completely defines the engineering management and processes to be applied to the project.

In addition to engineering management and processes, the SEMP should also detail positions of responsibility within the design team, including chief designers, software and hardware team leaders, project managers, and testing personnel. Part of the SEMP approval process should include an assessment of the skills and qualifications of these key personnel. Naturally, if this information is found in other management plans, the SEMP needs to refer to that plan.

The content of the SEMP should be maintained throughout the system design and development effort. Changes to the SEMP must be approved by the customer organization as this ensures visibility into changes that may expose the project to unexpected risks (such as a change in key personnel).

Due to its coverage of the entire systems engineering effort for the project, the SEMP is a fundamental source of information when evaluating the ability and capacity of a contractor to perform the technical activities associated with the system development (as well as providing a good indication of their understanding of the system and what is needed to bring it into being). Therefore, it is common for customers to request a draft SEMP to be submitted as part of the solicitation process for a project.

A number of systems engineering standards suggest contents for the SEMP. These standards detail the systems engineering requirements for systems development projects and all view the SEMP as the tailoring of the requirements of the standard to meet the specific systems engineering needs of the project. The most popular of these standards are reviewed in Chapter 9. Their descriptions of the SEMP are summarized here:

- *MIL-STD-499B.* The SEMP is seen as the key planning artifact containing details of the intended systems engineering processes to be adopted. The following topics for inclusion in the SEMP: TPM planning, technical review planning, technical integration planning, technology transition planning, and relating TPM and SEMS/SEDS to schedule and cost performance.

- *DI-S-3618/S-152, Data Item Description.* In support of MIL-STD-499, the US Department of Defense provides a data item description (DID) providing a template and suggested contents for the SEMP [31].
- *EIA/IS-632.* The content of the SEMP as required by EIA/IS-632 is largely the same as the MIL-STD-499B SEMP. Appendix B of EIA/IS-632 describes the contents of the SEMP with major sections describing the SE process, organizational responsibilities and roles, transitioning critical technologies, integration of SE effort, and SE activities. The SEMP is described as used by the customer to evaluate the tenderers' SE approaches, to make technical risk assessments, and to gauge progress and as used by the contractor to manage and integrate all technical activities.
- *EIA-632.* In transitioning from the interim standard, the explicit definition of SEMP content was omitted, and the full standard simply makes reference to the SEMP in Annex D as the engineering plan, which is listed as one of the SE planning artifacts.
- *IEEE-STD-1220.* IEEE-STD-1220 also refers to the SEMP as the engineering plan and provides a very useful template in Annex B. The content of the IEEE-STD-1220 engineering plan is designed to allow the tailoring of the standard to suit the project under consideration. The topics to be included in the engineering plan are identical to those required by both MIL-STD-499B and EIA/IS-632. IEEE-STD-1220 is a more useful source in that it provides a detailed lower-level breakdown of each section and also provides more guidance on completion of the sections of the template.
- *ISO/IEC/IEEE 15288.* ISO/IEC/IEEE 15288 is focused on the system life cycle and takes a considerably broader view than EIA-632 and IEEE-STD-1220, which are system-focused. ISO/IEC/IEEE 15288 therefore does not concern itself with the details of a SEMP but is supported by a range of guides (ISO/IEC/IEEE 24748-1 to 24748-8).
- *ISO/IEC/IEEE 24748-4 [32].* In its support as a guide to ISO/IEC/IEEE 15288, ISO/IEC/IEEE 24748-4 addresses systems engineering planning, referring explicitly to the SEMP and providing a very detailed description of the required SEMP clauses. Upper-level headings include process definition, technical project planning, technical project execution and control, technical baseline management, supporting processes, and specialty engineering activities and plans. Annex B to the guide also provides guidance on how the SEMP should be

modified depending on the life-cycle stage or stages that the project encompasses. Annex C provides guidance on tailoring policies.

- *SAE 1001.* As discussed in Chapter 9, SAE 1001 provides systems engineering detail not provided by ISO/IEC/IEEE 15288, providing the next level of tasking detail performed by a project, focusing on tasks, inputs, and outputs to show they are interconnected (integrated) to form project processes for systems engineering. That is particularly useful for a small organization that does not want the full set of processes as defined in ISO/IEC/IEEE 15288 (particularly the high-level organizational project-enabling processes) but just wants to focus on the project level. SAE 1001 provides considerable guidance on project and SE planning but does not prescribe an SE template itself, preferring to rely on ISO/IEC/IEEE 24748-4 as a normative reference (and consequently calling up the SEMP outline in that guide).
- *NASA Systems Engineering Handbook [33].* The NASA SEMP (Appendix J to the *NASA Systems Engineering Handbook*) includes three general sections: (1) technical program planning and control, which describe the processes for planning and control of the engineering efforts for the design, development, test, and evaluation of the system; (2) systems engineering processes, which include specific tailoring of the systems engineering process as described in the NASA Procedural Requirements, implementation procedures, trade study methodologies, tools, and models to be used; and (3) engineering specialty integration, which describes the integration of the technical disciplines' efforts into the systems engineering process and summarizes each technical discipline effort and cross-references each of the specific and relevant plans.

8.8 Summary

Systems engineering management is responsible for planning and directing the systems engineering effort, monitoring and reporting on that effort to the appropriate areas, and reviewing and auditing the effort at critical stages in the system life cycle. In this chapter, we considered the major systems engineering management elements of technical review and audit management, T&E, technical risk management, CM, the use of specifications and standards, and systems engineering management planning. In the next chapter, we will examine the major systems engineering standards.

Review Questions

1. Explain why technical reviews and audits are vital to systems engineering management and describe the major formal reviews and the audits that may occur prior to construction and/or production.
2. List and describe the three categories of T&E and describe the primary focus of each of them.
3. Briefly describe the major steps in the ISO 31000 risk-management process.
4. List and explain the four main options for the treatment of risk and explain when each option may be preferred over the others.
5. Why is risk management an ongoing concern?
6. CM involves five major functions. List these functions and briefly describe them.
7. Explain the role of the CCB in the context of CM.
8. Explain the major role and likely content of the following specifications:
   (a) System requirement specification;
   (b) Development specification;
   (c) Product specification;
   (d) Process specification;
   (e) Material specification.
9. Explain the common management errors and likely impacts associated with the use of engineering standards. Explain how these errors (and their impacts) can be avoided.
10. Briefly describe the role and content of the SEMP.

Endnotes

[1] Defense Systems Management College, *Systems Engineering Management Guide*, Washington, DC: US Government Printing Office, 1990, p. 12-3.

[2] MIL-STD-499B, *Military Standard—Systems Engineering—Draft*, Washington, DC: US Department of Defense, 1994.

[3] MIL-STD-1521B, *Technical Reviews and Audits for Systems, Equipments, and Computer Software*, 1985.

[4] Examples of commercially available standards and guides that contain information on technical reviews and audits include:

ISO/IEC/IEEE 15288, *Systems Engineering—System Lifecycle Processes*, International Standards Organisation, 2023.

ISO/IEC/IEEE 15288.2, *Application of Technical Reviews and Audits for Defense Programs, IEEE, 2014.*

International Council on Systems Engineering (INCOSE), INCOSE-TP-2003-002-05, *Systems Engineering Handbook*, Fifth Edition, New York: Wiley, 2023.

[5] MIL-STD-973, *Military Standard—Configuration Management*, Washington, DC: US Department of Defense, 1992.

[6] MIL-HDBK-61B, *Military Handbook—Configuration Management Guidance*, Washington, DC: US Department of Defense, 2020.

[7] EIA-649-C, *Configuration Management Standard*, Warrendale, PA: SAE International, 2019.

[8] Wheatcraft, L. S., M. J. Ryan, and T. E. Katz, *Needs and Requirements Manual: Needs, Requirements, Verification, Validation Across the Lifecycle*, Hoboken, NJ: Wiley, 2024.

[9] Boehm, B. W., "Verifying and Validating Software Requirements and Design Specifications," *IEEE Software*, Vol. 1, No. 1, January, 1984, pp. 75–88.

[10] Wheatcraft, L. S., M. J. Ryan, and T. E. Katz, *Needs and Requirements Manual: Needs, Requirements, Verification, Validation Across the Lifecycle*, Hoboken, NJ: Wiley, 2024.

[11] Defense Systems Management College, *Systems Engineering Management Guide*, Washington, DC: US Government Printing Office, 1990.

[12] ISO 31000:2009, *Risk Management—Principles and Guidelines*, International Standards Organisation, 2009.

[13] CM is a well-described process. Some very useful sources of detailed information on traditional and modern CM practices can be found in:

EIA-649-C, *Configuration Management Standard*, Warrendale, PA: SAE International, 2019.

EIA-836, *Configuration Management Data Exchange and Interoperability*, 2002.

GEIA-HB-649, *Implementation Guide for Configuration Management*, 2005.

ISO 10007, *Quality Management Systems—Guidelines for Configuration Management*, 2017.

MIL-HDBK-61A(SE), *Military Handbook—Configuration Management Guidance*, Washington, DC: US Department of Defense, 2001.

MIL-STD-973, *Military Standard—Configuration Management*, Washington, DC: US Department of Defense, 1992.

[14] EIA-649-C, *Configuration Management Standard*, Warrendale, PA: SAE International, 2019.

[15] Defense Systems Management College, *Systems Engineering Management Guide*, Washington, DC: US Government Printing Office, 1990, p. 11-1.

[16] EIA-649-C, *Configuration Management Standard*, Warrendale, PA: SAE International, 2019.

[17] EIA-649-C, *Configuration Management Standard*, Warrendale, PA: SAE International, 2019.

[18] EIA-649-C, *Configuration Management Standard*, Warrendale, PA: SAE International, 2019.

[19] ISO 10007, *Quality Management Systems—Guidelines for Configuration Management*, 2017.

[20] EIA-649-C, *Configuration Management Standard*, Warrendale, PA: SAE International, 2019.

[21] EIA-649-C, *Configuration Management Standard*, Warrendale, PA: SAE International, 2019.

[22] Other sources of information on deviations and waivers include:

MIL-STD-973, *Military Standard—Configuration Management*, Washington, DC: US Department of Defense, 1992.

MIL-HDBK-61A(SE), *Military Handbook—Configuration Management Guidance*, Washington, DC: US Department of Defense, 2001.

MIL-STD-481B, *Military Standard Configuration Control—Engineering Changes (Short Form), Deviations and Waivers*, US Department of Defense, Washington, DC, July 15, 1988.

[23] ISO 10007, *Quality Management Systems—Guidelines for Configuration Management*, 2017.

[24] MIL-STD-973, *Military Standard—Configuration Management*, Washington, DC: US Department of Defense, 1992.

[25] MIL-STD-973, *Military Standard—Configuration Management*, Washington, DC: US Department of Defense, 1992.

[26] ISO 10007, *Quality Management Systems—Guidelines for Configuration Management*, 2017.

[27] ECCS-M-40C:2008, *Space Project Management—Configuration Management*. https://ecss.nl/, 2008.

[28] MIL-STD-961E with Change 1, *Department of Defense Standard Defense and Program-Unique Specifications Format and Content*, Washington, DC: US Department of Defense, April 2, 2008.

[29] MIL-STD-490A, *Military Standard—Specification Practices*, Washington, DC: US Department of Defense, 1985.

[30] Defense Systems Management College, *Systems Engineering Management Guide*, Washington, DC: US Government Printing Office, 1990, p. 10-2.

[31] DI-S-3618/S-152, Data Item Description, *Systems Engineering Management Plan (SEMP)*, Department of Defense, February 9, 1970.

[32] ISO/IEC/IEEE 24748-4, *Systems and Software Engineering—Lifecycle Management—Part 4: Systems Engineering Planning*, 2018.

[33] NASA, *NASA Systems Engineering Handbook*, NASA SP-2016-6105 Rev2, 2016.

9

Systems Engineering Standards

9.1 Introduction

Systems engineering standards are available to assist both systems engineering management and the systems engineering processes, as well as assisting related disciplines such as project management. This chapter introduces a number of standards that may assist management. Finally, we summarize the major guidance available from the International Council on Systems Engineering (INCOSE).

9.2 Evolution of Systems Engineering Standards

This section introduces some of the more prominent systems engineering standards and summarizes their contents. Repeating sections of text from the standards has been avoided, and readers should refer to the standard itself if detailed content information is required.

While there has been considerable movement toward a harmonious suite of standards relating to systems engineering over the last decade, there is some way to go and there are still a number of standards guiding system development in organizations around the world. Although ISO/IEC/IEEE 15288 and its attendant handbooks and guides are being rapidly adopted, there are many projects still making use of older standards such as EIA 632 and IEEE-STD-1220. There are also many organizations whose internal processes have been developed based on an older standard; there is considerable inertia to rapid change.

Consequently, we have briefly covered all the major systems engineering standards in this chapter for two reasons. First, they are all in use to various degrees in projects around the world so the reader will most likely be interested in their use of a heritage standard as well as the content of the modern standards, handbooks and guides. Second, we want to illustrate that, while there are clear differences, they all provide a framework for the establishment of useful systems-engineering processes and procedures in an organization. Additionally, any selected standard will almost always have a scope greater than the project and system under consideration, so tailoring is essential, not only because of the scope of the standard but also because of the need to temper the selected standard's requirements with the culture and structure of the adopting organization.

The evolution of systems engineering standards is shown in Figure 9.1 to illustrate the history and previous standards upon which current standards are based. In the following sections, we briefly describe the standards from MIL-STD-499B to the current standards most recently revised in 2023.

9.3 MIL-STD-499B—Systems Engineering (Draft)

MIL-STD-499B was written within the US Department of Defense (DoD) and released in draft form to replace MIL-STD-499A, which was dated May 1, 1974. The Bravo version of 499 was to be ratified as a standard in 1994, but it was never released as anything other than draft form. In 1994, the US government, through the Perry Initiative, directed that a number of US military

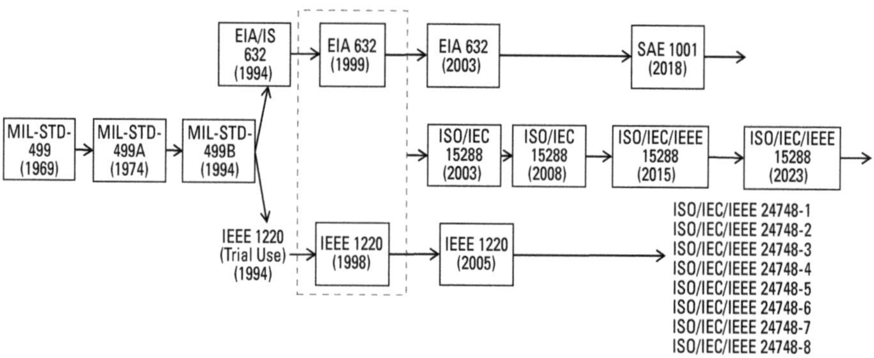

Figure 9.1 Evolution of systems engineering standards.

standards were to be canceled. Another standard reviewed in this section (EIA/IS-632) was created to continue the MIL-STD-499 linage. Accordingly, EIA/IS-632 could be considered to be a demilitarized version of MIL-STD-499B.

9.3.1 General Standard Content

MIL-STD-499B emphasizes that it focuses on explaining what to do rather than describing how to go about doing it. The standard attempts to assist with defining, performing, managing, and evaluating the systems engineering efforts required in each system development or acquisition project. The SEMP is seen as a central systems engineering artifact and MIL-STD-499B recommends that both the customer/acquirer (referred to as the tasking activity) and contractor/developer (referred to as the performing activity) should produce SEMPs to describe their individual systems engineering efforts. The SEMP is seen as an effective way to tailor the requirements in MIL-STD-499B to meet the specific requirements of the individual project.

The standard has been written with a systems approach in mind to make the standard relevant to a wide range of projects, from total systems development projects, through modifications and upgrades, to rectification projects. It is considered applicable to large or small projects, using new or existing technology, producing single or multiple units, dominated by either hardware or software. Certainly, MIL-STD-499B offers an excellent basis for all those project types, but, again, the importance of tailoring the requirements must be emphasized. For example, the entire set of formal reviews and audits may be appropriate in a large, risky project but inappropriate in a small project making use of mature technology.

MIL-STD-499B avoids the problems associated with the invocation of additional standards by avoiding references to other standards when specifying requirements. MIL-STD-499B uses different terms for the systems design and development stages of the acquisition phase. The stages used in MIL-STD-499B correspond well to the life-cycle stages used throughout this book in terms of activities and products. The stages used in MIL-STD-499B and the corresponding stages used in this book are shown in Table 9.1.

In the general requirements section, the standard calls for an integrated approach to the technical effort to best meet the customer's needs. This integrated effort must be reflected in the SEMP. MIL-STD-499B also requires schedule information to be contained in the SEMP and calls for two scheduling artifacts: the systems engineering master schedule (SEMS) and the systems engineering detailed schedule (SEDS). The SEMP (and the SEMS and SEDS) must be written in accordance with the tailored requirements of

Table 9.1
Relationship Between MIL-STD-499B Life-Cycle Stages and Those Used in This Book

MIL-STD-499B Stage	Corresponding Phase/Stage in This Book
Preconcept	Preacquisition
Concept exploration and definition	Conceptual design
Demonstration and validation	
Engineering and manufacturing development	Preliminary design, detailed design, and development
Production and deployment	Construction and/or production
Operation and support	Operational use and system support
Disposal	Retirement

MIL-STD-499B, but, once approved, the SEMP becomes the key systems engineering artifact for the project.

9.3.1.1 Systems Engineering Process

MIL-STD-499B sees systems engineering as having a series of inputs, which proceed through the systems engineering processes (with a management overlay) and result in a series of outputs. As illustrated in the Figure 9.2, the process starts with requirements analysis, which is aimed at taking the customer's needs, objectives, and requirements and translating them into requirements. The requirements include functional and performance requirements and relevant constraints for each of the system's primary functions.

A functional analysis/allocation follows, which breaks the functions identified in the requirements analysis down to lower-level functions that are allocated to the functional architecture. Traceability between low-level functions and the parent function is emphasized. Performance requirements should be defined for each high-level function so that measurable performance requirements can be allocated to each of the low-level functions. The systems engineering processes chapters in this book describe a basic process of analysis-synthesis-evaluation. Requirements analysis and functional analysis/allocation collectively form the analysis stage of this basic process.

Next in the MIL-STD-499B process is synthesis, which corresponds directly with our use of the term in this book. Basically, solutions need to be designed and described for each logical group of functional and performance requirements previously defined. This design must be integrated to form a physical architecture. Following this design and integration task, a verification function must be performed to verify that the physical design does meet the

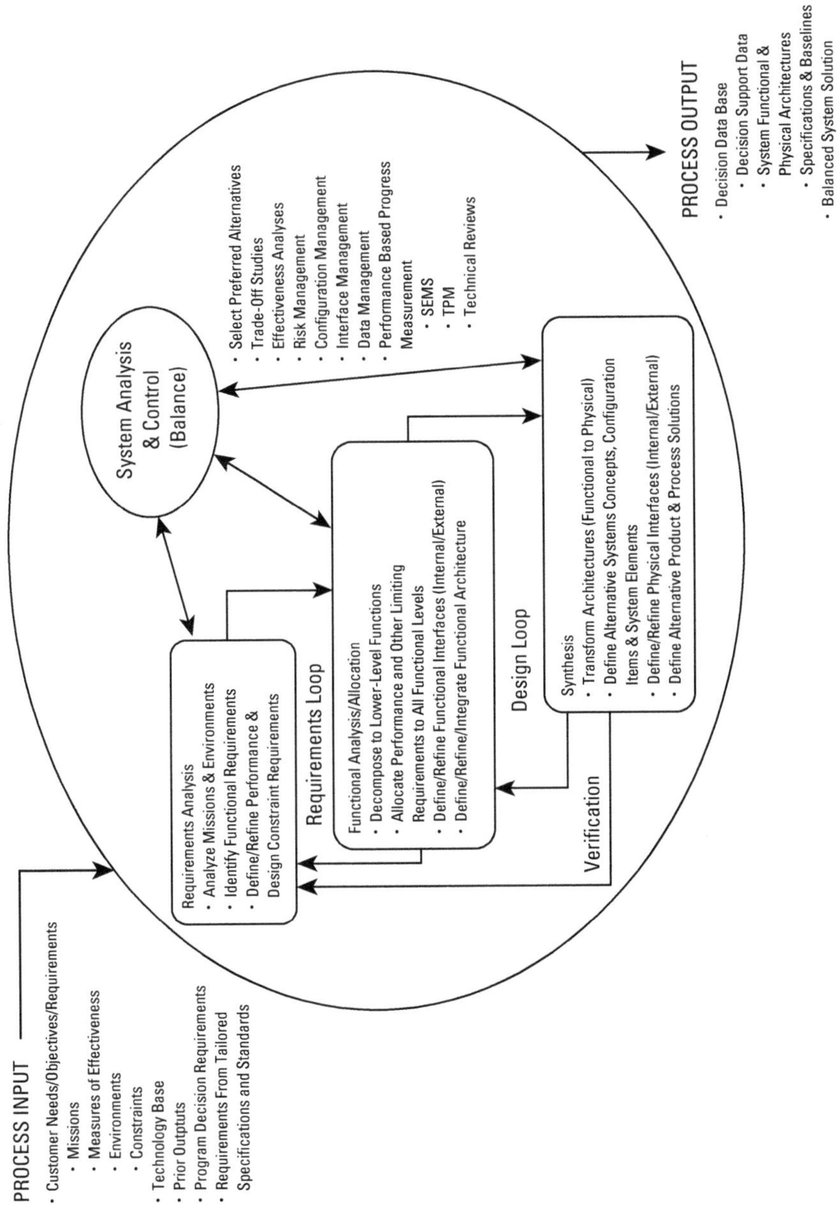

Figure 9.2 MIL-STD-499B systems engineering process.

original functional and performance requirements. Verification forms part of the synthesis function in MIL-STD-499B [1] but better describes the evaluation function described in Chapter 1.

Finally, MIL-STD-499B describes a process of analysis and control that can be described as a management overlay over the engineering process. Analysis and control are responsible for measuring progress, considering alternatives, and recording the design decisions made. It includes functions such as CM, risk management, technical reviews and the use of TPMs.

9.3.1.2 Content of the MIL-STD-499B SEMP

MIL-STD-499B talks of systems engineering planning as consisting of SEMP, SEMS, and SEDS. The SEMP is again seen as the key planning artifact containing details of the intended systems engineering processes to be adopted. MIL-STD-499B also lists the following topics for inclusion in the SEMP:

- *TPM planning.* MIL-STD-499B states the importance of identifying technical performance measures that are "critical indicators of technical progress and achievement." The SEMP should detail how these are identified, how often they are updated and reviewed, the level of detail that TPM tracking covers, and the recovery planning process relating to TPMs.
- *Technical review planning.* The technical review planning should consider both tasking activity reviews (formal reviews) and performing activity reviews (informal). All the tasks associated with each of the reviews should be specified including responsibilities, venues, procedures, and entry/exit criteria. Although MIL-STD-499B does not refer to a separate plan, it may be more appropriate to include these details in a separate plan, the TRAP, and merely reference the TRAP from the SEMP. This may be particularly suitable if the review and audit process is complex and likely to change, in which case the TRAP can be updated without the need to update the SEMP also.
- *Technical integration planning.* The SEMP should detail how multidisciplinary teams and subcontractors will work together during the development effort and how the many different products will be integrated. Again, an overview of the integration process with reference to other more detailed plans may be appropriate here.
- *Technology transition planning.* MIL-STD-499B refers to emerging technologies and the processes involved with transitioning this leading-edge technology from development to production and utilization.

If this is applicable, MIL-STD-499B requires details of the transition planning in place to make the transition a success.

- *Relating TPM and SEMS/SEDS to schedule and cost performance.* MIL-STD-499B requires the performing activity to relate the risk management tools of TPM and SEMS/SEDS to schedule and cost performance of the project. In this way, the information provided by the TPM and SEMS/SEDS can be more readily related to the project management aspects of the project.

9.3.1.3 Additional Information and Requirements

MIL-STD-499B lists and explains a number of functional tasks that need to be considered in a successful systems engineering effort. These tasks may be considered less obvious and may be potentially overlooked in the technical planning effort. The functional tasks listed in MIL-STD-499B are listed in Table 9.2.

All these tasks need to be considered during the technical planning process. In some projects, all areas may be relevant, but, in most cases, some of the tasks will not be applicable. Each project must be considered individually as emphasized during the discussions on standards tailoring.

The concept of leverage options is introduced, which relates to design options for reducing the risks and costs associated with design. Performing activities need to consider options such as the use of NDI, dual-use technology, open system architectures (OSA), and reuse during the planning stages. Dual-use technology is defined as "technology that can be used in either the commercial or government sectors or both" [2]. OSA basically entails the use of architectures that are widely used and have become industry standards. By requiring the use of OSAs, MIL-STD-499B avoids proprietary solutions that may become obsolete, difficult, and expensive to support. The exploration of these options should be recorded in the SEMP.

Table 9.2
Functional Tasks Listed in MIL-STD-499B

Reliability and maintainability	Survivability
Safety and health	Human factors
Electromagnetic compatibility	Security
Producibility	Test and evaluation
Integrated logistics support	Diagnostics
Transportability	Infrastructure

MIL-STD-499B highlights (in a section entitled “Pervasive Development Considerations”) some related support issues that may need addressing in the SEMP.

The computer resources needed to support the design and development effort need to be considered and planned for. Issues such as the processor capabilities, development tools, and development environments need to be considered to ensure that the hardware can support the mature version of the system software.

Parts, materials, and processes needed to support the development effort are critical and need planning. This section needs to address inventory and supply support issues and must take LCC into consideration. Intended use of simulation and prototyping to support the design process should also be detailed in the SEMP.

The use of digital data, computer-aided design tools, and other design technologies must be specified. This requirement may be showing the age of MIL-STD-499B, as these techniques were only emerging in the early 1980s when the standard was initially drafted, whereas computer-aided design is commonplace today. Even so, there may be a requirement for the performing activity to include design tool details in the SEMP.

MIL-STD-499B emphasizes that cost-effectiveness should be considered throughout the project. The standard requires that a range of analyses is considered and recorded by the performing activity with a view to improving the cost-effectiveness of the project. The analysis tasks contained in MIL-STD-499B are listed in Table 9.3.

A large section of MIL-STD-499B is focused on technical reviews and audits. The section provides a summary of the formal reviews and audits that can be conducted during development, and it provides details such as when the reviews and audits should be conducted and what should be achieved from the reviews. The review responsibilities are also listed in terms of tasking and performing activities. The most critical (and commonly used) reviews and audits from MIL-STD-499B have been listed and described in the appropriate section in Chapter 8.

Table 9.3
Analysis Tasks Listed in MIL-STD-499B

Manufacturing	Verification	Deployment
Environmental	Supportability	Training
Life-cycle cost	Operational	Disposal

MIL-STD-499B also provides a section of notes, which contains some very important tailoring and application guidance covering the use of the standard in different situations. The importance of thorough and thoughtful tailoring has been emphasized throughout this text. The notes section also contains a list of suggested military standards for related topics such as configuration management and software engineering. This list is out of date and contains standards that are no longer used or supported, but the list does provide an interesting summary of topics related to systems engineering management.

There are also three appendixes to MIL-STD-499B, namely a glossary of terms, a list of acronyms and abbreviations, guidance, or technical reviews, and some guidance on the development of the SEMS.

9.3.1.4 Summary

MIL-STD-499B is out-of-date and was never issued or supported by the US DoD. Despite this, the standard is an excellent source of information and guidance about systems engineering management and technical planning. The contents of MIL-STD-499B reflect a vast and varied range of experience in the field and provide both customer and contractor organizations with learned advice on engineering management. MIL-STD-499B remains a highly recommended source of systems engineering information.

For defense projects, MIL-STD-499 has been replaced by ISO/IEC/IEEE 15288.1 Application of Systems Engineering for Defense Programs and ISO/IEC/IEEE 15288.2 Application of Technical Reviews and Audits for Defense Programs.

9.3.2 EIA/IS-632—Systems Engineering

When it became evident that MIL-STD-499B was not going to be released, the EIA had an established committee that undertook the adaptation of MIL-STD-499B for use as an industry standard. The adapted standard was called EIA Interim Standard (IS) 632 (EIA/IS-632) and was released in December 1994. Work on the updated standard was completed in 1999 with the release of EIA-632.

A number of organizations were involved with the EIA during the development of EIA/IS-632, including the US DoD, the National Council on Systems Engineering (NCOSE), American National Standards Institute (ANSI), and the IEEE. NCOSE has since become an international organization and is now called the International Council on Systems Engineering (INCOSE).

9.3.2.1 General Standard Content

The result of the work carried out by the EIA during the development of EIA/IS-632 is an industry standard that represents a demilitarized version of MIL-STD-499B. The differences in general content between MIL-STD-499B and EIA/IS-632 are minor. By scanning the respective tables of content, the similarities become obvious.

9.3.2.2 Systems Engineering Process

The systems engineering processes detailed in EIA/IS-632 are identical to those contained in MIL-STD-499B. EIA/IS-632 has added some additional requirements in the requirements analysis and functional analysis and allocation areas. The additional requirements involve a verification function as part of each stage to ensure traceability of both requirements and functions to either systems engineering inputs (customer requirements and needs) or to a higher-level requirement identified earlier in the process. Note that both EIA/IS-632 and MIL-STD-499B require verification as part of the synthesis process; however, MIL-STD-499B does not mention verification as part of requirements analysis or functional analysis and allocation.

The addition of the verification function in all the engineering processes adds a degree of rigor to the requirements contained in EIA/IS-632 and ensures adequate traceability in all the systems engineering processes.

9.3.2.3 Content of the EIA/IS-632 SEMP

The content of the SEMP as required by EIA/IS-632 is largely the same as the MIL-STD-499B SEMP. Recall that MIL-STD-499B required the SEMP, SEMS, and SEDS to form the basis of the systems engineering planning effort. To this list, EIA/IS-632 additionally requires the work breakdown system (WBS) to play a role in the scope planning process.

9.3.2.4 Other Information and Requirements

The remaining information and requirements contained in EIA/IS-632 are identical to those contained in MIL-STD-499B. The review and audits section of both standards refers to the same set of formal and informal reviews and provides the same guidance on issues such as roles, responsibilities, expected outcomes, and entry/exit criteria.

EIA/IS-632 contains a notes section (as in MIL-STD-499B) that includes important information on the tailoring and application of the standard for different projects.

The appendixes in EIA/IS-632 are slightly different from those in MIL-STD-499B in that EIA/IS-632 provides an appendix detailing the expected structure and content of the SEMP, including details such as preparation and numbering conventions. This appendix provides useful guidance and will help to ensure uniform standards and content across SEMPs formulated by different organizations and different projects. EIA/IS-632 also provides an additional useful appendix containing information about the SEDS, which is not contained in MIL-STD-499B.

9.3.2.5 Summary

EIA/IS-632 is almost identical to MIL-STD-499B. Some military-specific phrases and examples have been altered to make the EIA/IS-632 standard less military-focused. These changes are largely cosmetic in nature. The other changes (as detailed in this section) are very minor and do not make EIA/IS-632 substantially different from MIL-STD-499B.

As with MIL-STD-499B, EIA/IS-632 contains excellent information, advice, and guidance. It is worth remembering that MIL-STD-499B was only ever a draft and was never officially released for use by the US DoD. Therefore, the major advantage of EIA/IS-has over MIL-STD-499B is that EIA/IS-632 was officially released by the EIA. For this reason, the use of EIA/IS-632 over MIL-STD-499B was recommended until it was replaced by EIA-632 [3] when released in 1999.

For defense projects, EIA/IS-632 has been replaced by ISO/IEC/IEEE 15288.1 Application of Systems Engineering for Defense Programs and ISO/IEC/IEEE 15288.2 Application of Technical Reviews and Audits for Defense Programs.

9.3.3 IEEE-STD-1220—IEEE Standard for Application and Management of the Systems Engineering Process

The IEEE started work on IEEE-STD-1220 [4] in 1989. Early work on the standard focused on parallel effort with the development of MIL-STD-499B to ensure consistency between industry and military/government systems engineering direction. When released in 1994, IEEE-STD-1220 was the first commercial systems engineering standard so IEEE-STD-1220 was released as a trial use standard. The IEEE updated the trial use standard into a full standard in December 1998 but retained the designation IEEE-STD-1220. In 2005, the IEEE released another revision of IEEE to include a mapping of the standard to a systems engineering life-cycle standard called the ISO/IEC/IEEE 15288 (reviewed later in this chapter). There are only minor differences between the 1994, 1998, and 2005 IEEE-STD-1220 standards.

IEEE-STD-1220 has the same scope as both MIL-STD-499 and EIA/IS-632 in that it is designed to be applied to the technical effort of a wide range of projects covering development, modification, or rectification of equipment. It is applicable to large or small projects, using new or existing technology, producing single or multiple units, being dominated by either hardware or software. IEEE-STD-1220 was intended to be applicable to commercial, government, or military projects.

The systems engineering planning requirements contained in IEEE-STD-1220 center around the engineering plan, which is the result of the tailored application of the standard. Tailoring is again emphasized as an important part of the application of the standard to different situations and projects.

IEEE-STD-1220 is focused on a system paradigm in that everything within the system (including products, subsystems, and processes) can be dealt with as a smaller system within the main system. This is possible because products, subsystems, and processes all have boundaries, relationships and interfaces with the other products, subsystems and processes within the main system. In this way, the requirements of IEEE-STD-1220 can be applied to each individual system within the main system, ensuring a consistent approach to the technical effort behind the project.

IEEE-STD-1220 presents a very familiar systems engineering process. The process is consistent with the processes adopted by MIL-STD-499B and EIA/IS-632 and detailed in this book. The process is applied at each stage of the system life cycle and at each level of system decomposition. The IEEE-STD-1220 process is summarized in Figure 9.3 [5].

The first step in the systems engineering process is the analysis stage and this is represented in IEEE-STD-1220 as the requirements analysis and the functional analysis. The synthesis function exists in both IEEE-STD-1220 process model and the systems engineering process model used in this text. The final stage of the systems engineering process is evaluation. IEEE-STD-1220 emphasizes the evaluation function throughout the entire process. Accordingly, evaluation functions occur at each stage in the process and are called requirements baseline validation, functional verification, and physical verification.

9.3.3.1 General Standard Content

The engineering plan plays a central role in technical planning when using IEEE-STD-1220. In addition to the engineering plan, the master schedule, detailed schedule, and various technical plans are also referenced. IEEE-STD-1220 explains that typical plans produced to support the engineering plan include risk management, technical reviews, configuration management,

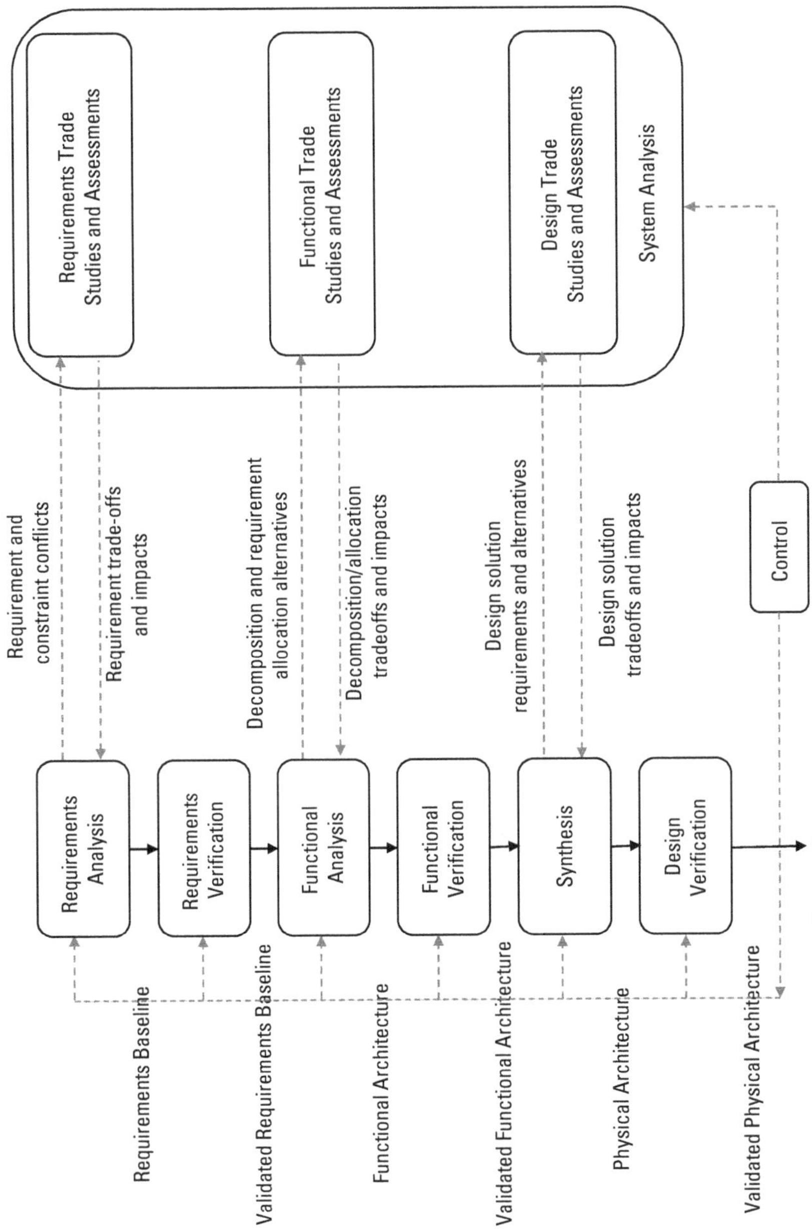

Figure 9.3 IEEE-STD-1220 systems engineering process.

verifications, computer resources, manufacturing, maintenance, training, security, and human systems engineering.

The importance of design data is emphasized in IEEE-STD-1220 through the requirement for an integrated database to capture and track design data. An integrated data package is also required to record the architecture and design information, as necessary to support manufacturing and through-life support. IEEE-STD-1220 lists and briefly describes the specifications and drawings that are expected in a typical integrated data package.

IEEE-STD-1220 refers to a tool called the system breakdown structure (SBS), which takes on the same role as the WBS in the other standards. The roles and importance of quality assurance, systems integration, and technical reviews are emphasized in the general requirements section.

9.3.3.2 IEEE-STD-1220 Life-Cycle Model

IEEE-STD-1220 dedicates an entire section to the application of systems engineering across the different phases of the system life cycle and provides a comprehensive description of the activities within each stage of the life cycle.

The IEEE-STD-1220 life-cycle stages (and the corresponding stages in this book) are described in the following sections:

- *System development.* System development in the IEEE-STD-1220 life-cycle model best corresponds to conceptual design in the model in this book. This phase of the life cycle (according to IEEE-STD-1220) concentrates on the conceptual design of the system, the development of project and technical plans, consideration of major interfaces (including human/system interfaces), identification of major risks, and the development of system-level specifications. The system baseline will result from the system development effort. The system baseline corresponds with the more generally used term *functional baseline.* The phase concludes with the selection of the preferred concept and a review that ensures that the system-level requirements are completely defined.
- *Subsystem definition.* System definition in IEEE-STD-1220 is the major development phase of the system life cycle and consists of a number of subphases.
- *Preliminary design.* The IEEE-STD-1220 preliminary design activity corresponds directly to preliminary design as presented here. Subsystem identification, definition, and design are considered during this stage of the development, including interfaces, risks, and specifications. The project and technical planning outputs will also be reviewed and updated where necessary. The result of the preliminary design phase

is a design to baseline that corresponds to the more generally accepted allocated baseline. The phase is concluded with the conducting of a design review to ensure that the preliminary design is complete and satisfactory.

- *Detailed design.* The detailed design phase of the IEEE-STD-1220 life cycle corresponds to the detailed design and development phase as presented here. In this phase of the development, the subsystem design is further broken down to the lowest component level, thus completely defining the subsystem design. Component definitions and risks are recorded in the form of detailed drawings, parts lists, and specifications. The work in this phase of the development results in the build-to baseline that we have called the PBL. Technical reviews are focused on ensuring component level definition and design is satisfactory. The design reviews completed following detailed design may also revise the system and subsystem designs if any changes have occurred.
- *Fabrication, assembly, integration, and test (FAIT).* FAIT activities are the beginning of the construction and/or production phase. System integration and testing occur in this phase where the components and subsystems are integrated and analyzed to ensure that the higher-level functionality and interfaces are satisfied. Where problems are identified, these are fixed and retested. The main reviews conducted during this phase include the TRR and the FCA. Successful completion of this phase provides approval to move the system into production.
- *Production and customer support.* This final phase of the IEEE-STD-1220 life-cycle model encompasses construction and/or production and the utilization phase. Production-related activities such as inventories, production control, and integration occur. A PCA is conducted on the initial production system to ensure compliance with the technical artifacts. Customer support and after-market products are provided as required.

9.3.3.3 Systems Engineering Process

IEEE-STD-1220 details the tasks that need to be performed in each of the process steps in an analysis-synthesis-evaluation loop. Each of the processes is summarized using a flow diagram in the standard.

- *Requirements analysis.* The requirement analysis activity is designed to establish what the system is required to accomplish in quantitative and measurable terms. The analysis must also define the environmental constraints within which the system will be required to operate. This information forms the requirements baseline upon which the remainder of the development effort will be based. Sixteen subclauses are detailed in IEEE-STD-1220 pertaining to conducting the requirements analysis effort. These subclauses provide direction and guidance to those responsible for performing the requirements analysis effort. Clause 6.1.16 describes the requirements baseline, which is the product of the analysis and is recorded with three views: operational, functional, and design. The operational view is written from a user's perspective and describes what the system must be capable of and how it will be used. The functional view describes what the system products must do to meet the operational requirements. The design view concentrates on the physical interfaces, the human engineering aspects and the physical form of the system.

- *Requirements validation.* Requirements validation concentrates on the requirements baseline produced during the requirements analysis. It basically ensures that the baseline represents the users' requirements within the identified constraints, and that the baseline is a complete representation of those requirements. Five subclauses are detailed in IEEE-STD-1220 to describe the validation process. There may be a requirement to revise and update the requirements baseline during the validation process if deficiencies are identified.
- *Functional analysis.* Functional analysis takes the requirements in the requirements baseline and further defines the problem in clearer detail. This is done by translating each requirement into functional terms and into a functional architecture. Decomposition of the requirement into functions and subfunctions (including interfaces and performance requirements) continues until the description is complete. Functional analysis does not result in a design solution. Grouping of logical subfunctions will occur during the synthesis stage to produce a design solution. IEEE-STD-1220 provides eight subclauses to direct the analysis effort.
- *Functional verification.* Functional verification investigates the completeness of the functional architecture in line with the requirements baseline and the suitability of the architecture as an input into the synthesis stage. The importance of upward traceability from the functional architecture to the system-level requirements and constraints is emphasized. Also, downward traceability from the system level to

the functional level will help to ensure that all requirements have been addressed.

- *Synthesis.* Synthesis defines the product solutions and subsystems that will meet the functional architecture requirements. First, the functions are grouped together and a range of design solutions considered. The most suitable of these alternatives is selected with associated artifacts verifying the selection. A range of considerations is used during this selection. Synthesis results in a design solution defining integration arrangements of subsystems and the internal and external interfaces. IEEE-STD-1220 provides 18 subclauses to direct the synthesis effort.
- *Design verification.* Physical verification is primarily aimed at ensuring traceability from the lowest-level physical architecture to the functional architecture verified during the functional verification stage. This traceability should also be continued through to the requirements baseline. There are a number of possible approaches to verification, including inspection, analysis, and demonstration. The method of verification will need to be defined prior to the exercise.
- *Management layer.* The management layer identified in IEEE-STD-1220 consists of systems analysis and control. Systems analysis includes activities such as trade-off studies to consider alternatives, risk assessments, and LCC analysis. Control covers issues such as design data management, technical management, configuration management, risk management, interface management, and project and technical planning.

9.3.3.4 Content of the IEEE-STD-1220 Engineering Plan

IEEE-STD-1220 also refers to the SEMP as the engineering plan and provides a very useful template in Annex B. The content of the IEEE-STD-1220 engineering plan is designed to allow the tailoring of the standard to suit the project under consideration. The topics to be included in the engineering plan are identical to those required by both MIL-STD-499B and EIA/IS-632; IEEE-STD-1220 is a little more useful in the lower-level breakdown of each section and also provides more guidance on completion of the sections of the template.

9.3.3.5 Additional Material and Requirements

Annex A to IEEE-STD-1220 is an informative summary of the role of systems engineering within an enterprise. This annex does not add additional requirements to the IEEE standard; it merely provides the reader with some background information.

9.3.3.6 Summary

IEEE-STD-1220 is a detailed and complete systems engineering standard. It is widely used today and provides some excellent systems engineering advice and guidance. Unlike EIA/IS-632, IEEE-STD-1220 is not simply a close replica of MIL-STD-499B, but rather a comprehensive and independent coverage of systems engineering. It provides details of the system life cycle and how the different phases are covered by systems engineering. In addition, it presents a slightly more involved systems engineering process than has been considered in this book and provides the reader with some guidance on the content of the SEMP.

IEEE-STD-1220 was replaced by ISO/IEC/IEEE 24748-4: Systems and Software Engineering—Lifecycle Management—Application and Management of the Systems Engineering Process, which replaced IEEE-STD-1220 as well as ISO/IEC 26702 Systems Engineering—Application and Management of the Systems Engineering Process.

9.3.4 EIA-632—Processes for Engineering a System

EIA-632 [6] was developed jointly by the EIA and INCOSE and released in 1999. Martin [7] provided an excellent overview of EIA-632 and its heritage. The standard represents a significant change from the interim standard and focuses on the enterprise and its policies and procedures rather than focusing narrowly on a particular system development being undertaken by that enterprise. This broad focus sets EIA-632 apart from previous systems engineering standards. So great is the change that people may not consider EIA-632 to be a systems engineering standard at all; it certainly does not define systems engineering or even use the term.

EIA-632 introduces 13 processes required to successfully engineer a system. Within these 13 processes, there are 33 requirements set out in the standard. Not all of the 33 requirements will be relevant or required in every situation; therefore, EIA-632 is expected to be tailored to individual development projects.

The standard has been designed to help enterprises develop policies and procedures relevant to their role in engineering systems within their industries. Projects being conducted by the enterprise are then responsible for further development of company policies and procedures into project-specific plans, schedules and so on. By operating at the enterprise level, EIA-632 avoids being a prescriptive standard and concentrates on what must be done to successfully and completely engineer a system, as Figure 9.4 [8] illustrates.

The organization and approach of EIA-632 provides a very detailed framework upon which systems engineering enterprises can base their company policies and procedures.

9.3.4.1 EIA-632 Processes

Figure 9.5 [9] shows the 13 processes contained in EIA-632 grouped into the five broad categories described in the standard. The categories and processes are arranged into a flow diagram that effectively describes the EIA-632 logic. An interesting point to note is that the analysis-synthesis-evaluation loop exists only as a subset of the overall EIA-632 logic. This is because EIA-632 is an enterprise-wide standard and must address more than the classic systems engineering processes.

A majority of the standard is focused on describing the 13 processes and the 33 requirements that flow out of those processes. Notes and examples are used throughout to clarify the intent of the standard.

EIA-632 focuses on process rather than method. The methods for achieving the requirements are expected to come from enterprise policy and procedures. Each of the process areas shown in Figure 9.5 is dealt with separately by considering the inputs and outputs of the process and the requirements within the process.

9.3.4.2 EIA-632 Requirements

Guidance and explanation for each of the 33 requirements are provided in the standard, along with an example wording of the requirement. The reader is expected to examine each requirement for relevance and tailor the requirements where necessary. Annex C provides the reader with a process task outcome table that details the representative tasks that will need to be performed for each of the requirements and the expected outcomes from those tasks. This annex represents a prescriptive description of the 33 requirements but aids in the understanding of the requirements. Readers gain an excellent insight into

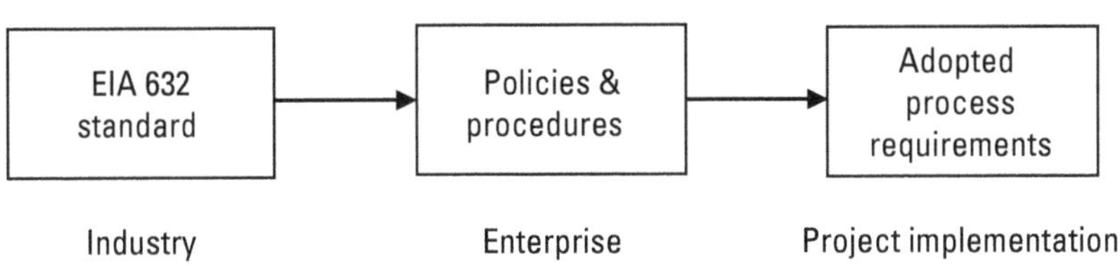

Figure 9.4 Application of EIA-632.

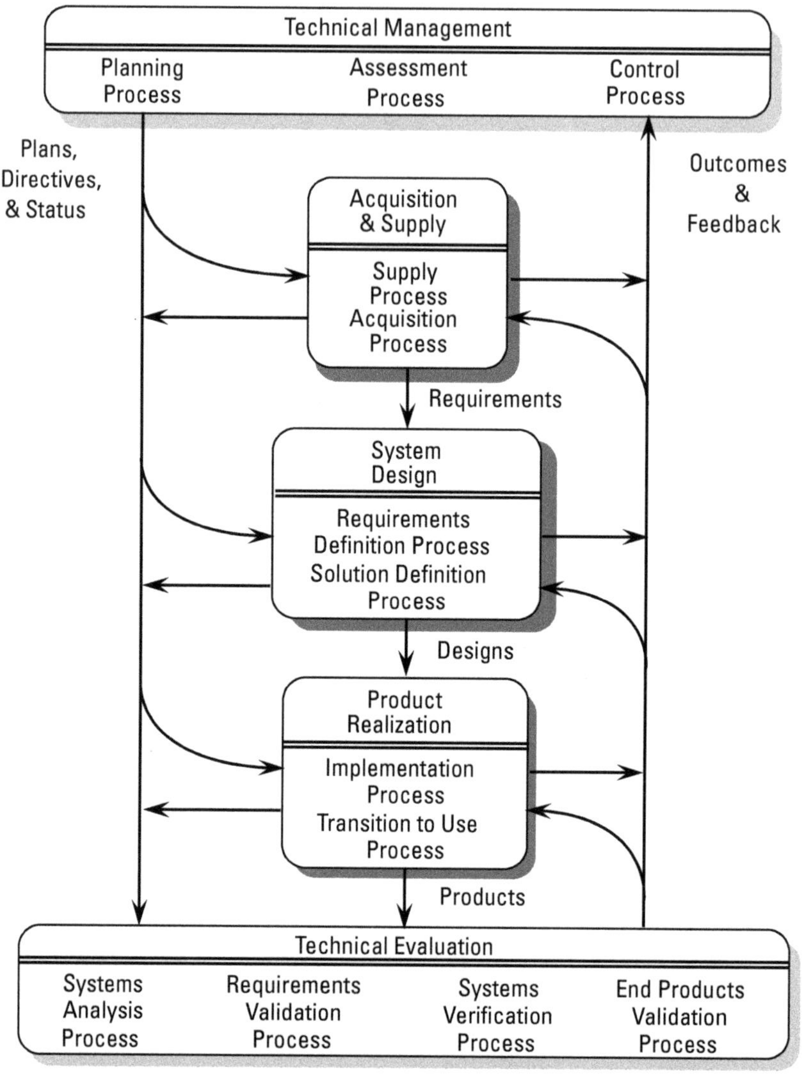

Figure 9.5 EIA-632 processes.

how the requirements might be satisfied by reading this annex. The following lists the 33 requirements by process category [10]. Further details are available in the full standard.

- Supply Process Requirements
    1. Product supply

- Acquisition Process Requirements
  2. Product acquisition
  3. Supplier performance
- Planning Process Requirements
  4. Process implementation strategy
  5. Technical effort definition
  6. Schedule and organization
  7. Technical plans
  8. Work directives
- Assessment Process Requirements
  9. Progress against plans and schedules
  10. Progress against requirements
  11. Technical reviews
- Control Process Requirements
  12. Outcomes management
  13. Information dissemination
- Requirements Definition Process Requirements
  14. Acquirer requirements
  15. Other stakeholder requirements
  16. System technical requirements
- Solution Definition Process Requirements
  17. Logical solution representations
  18. Physical solution representations
  19. Specified requirements
- Implementation Process Requirements
  20. Implementation
- Transition to Use Process Requirements
  21. Transition to use
- Systems Analysis Process Requirements
  22. Effectiveness analysis
  23. Trade-off analysis
  24. Risk analysis
- Requirements Validation Process Requirements
  25. Requirement statements validation
  26. Acquirer requirements validation
  27. Other stakeholder requirements validation
  28. System technical requirements validation
  29. Logical solution representations validation
- System Verification Process Requirements

30. Design solution verification
31. End product verification
32. Enabling product readiness

- End Products Validation Process Requirements

33. End products validation

9.3.4.3 EIA-632 Concepts

EIA-632 defines a system as comprising hardware, software, people, facilities, data, materials, services, and techniques. In addition to this definition, a system is proposed to consist of the end product that is the classic deliverable, but also the enabling products that allow the end products to be produced. The categories of enabling products and some examples provided in EIA-632 are listed in Table 9.4.

A major concept introduced in EIA-632 is that of the building block that we discussed in Chapter 1. This shows that each end product (of which there may be many per system) can be broken into subsystems. Each subsystem can be considered a system in its own right and dealt with in accordance with EIA-632. This forms the basis of the top-down development concept supported by EIA-632, which breaks the end products into increasing levels

Table 9.4
EIA-632 Enabling Products and Examples

Enabling Products	Examples
Development	Plans, schedules, policies, procedures, tools, models, interfaces, and development personnel
Production	Plans, schedules, policies, procedures, facilities, tools, equipment, materials, and production personnel
Test	Plans, schedules, policies, procedures, mock-ups, test equipment, test facilities, simulations, models, and test personnel
Deployment	Plans, schedules, policies, procedures, packaging, storage, handling, installation, transportation, sites, and installation personnel
Training	Plans, schedules, policies, procedures, simulators, training aids, courses, facilities, and trainers
Support	Plans, schedules, policies, procedures, tools, repair equipment, support facilities, maintenance manuals, diagnostic equipment, and repair personnel
Disposal	Plans, schedules, policies, procedures, disposal sites, special equipment, and disposal personnel

of details from layer 1 (system level) down to as many layers as required by the development process. Once the system has been completely broken down in this way, design decisions can be made regarding how each subsystem is designed. Figure 9.6 [11] shows the building-block concept.

EIA-632 then goes on to describe the bottom-up realization concept, which details how the very low-level layers are progressively integrated and tested until the system-level end product is produced. The top-down development and bottom-up realization concepts are not unique to EIA-632 and form the basis for most popular systems engineering thinking on system development (as described in Chapter 1).

Table 9.5 [12] shows the enterprise-based life cycle as detailed in Annex B to EIA-632 and compares it with the system life cycle used in this book.

9.3.4.4 EIA-632 Annexes

EIA-632 provides a number of useful annexes. Annex A is a standard inclusion listing the glossary of terms used throughout. Annex B provides additional detail on the engineering lifecycle used throughout the standard. Annex C provides a very useful description of each of the requirements in terms of representative tasks needed to achieve the requirements and the expected outcomes from each of the representative tasks.

Annex D provides a description of the artifacts that may be required from a system development project, divided into the following categories:

- *Source documents.* These documents include concepts, such as maintenance concepts, operational concepts, and disposal concepts.
- *Technical artifacts.* This section provides a comprehensive list of technical artifacts that could be expected to flow from the system development effort. These documents are listed alphabetically and include the systems engineering management documents and the systems engineering process documents described in this book.
- *Enterprise or project documents.* These documents provide guidance and constraints for the development effort. These documents typically include enterprise policies and procedures.

Annex E provides details of the technical reviews and audits that could be expected throughout the system development effort. This detail is provided relative to the engineering life-cycle activities. The description of the reviews and audits provides details of what each review and audit should aim to achieve.

Annex F describes the activities needed to fully define the end products. The activities depend on whether or not the end products are already known.

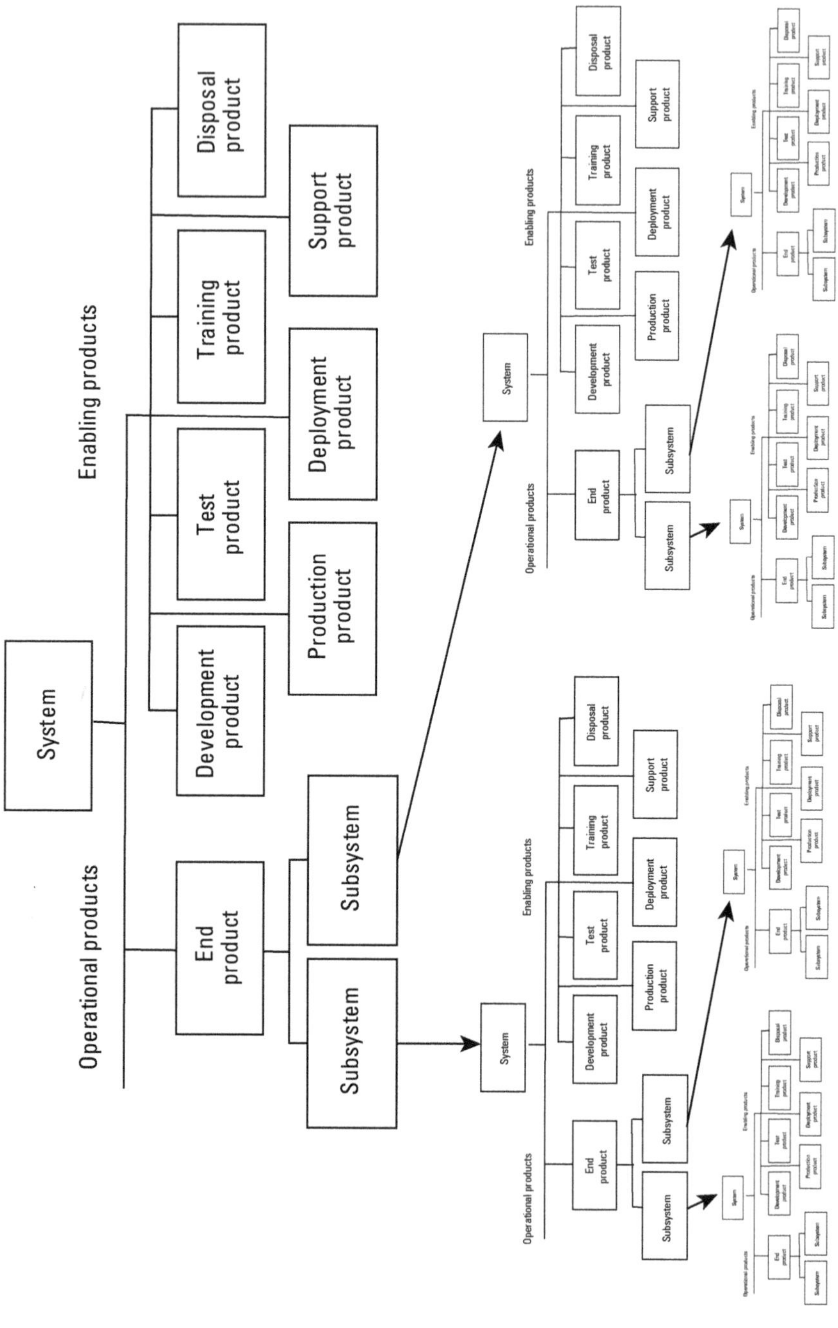

Figure 9.6 EIA-632 building-block concept.

Table 9.5
EIA-632 Life-Cycle Activities Compared with Those Used in This Book

EIA-632 Life-Cycle Activities	Life-Cycle Activities in This Book
Assessment of opportunities	Preacquisition
Investment decision	
System concept development	Conceptual design
Subsystem design and pre-deployment	Preliminary design
	Detailed design and development
	Construction and/or production
Deployment, operations, support, and disposal	Utilization Retirement

The end products will be known if OTS equipment is to be used. However, end products that are to be designed require a slightly different approach to definition.

Annex G provides an explanation of the relationship between the many sources of requirements and the eventual solution. The sources of requirements include acquirer requirements, technical requirements, and derived technical requirements. These are then assigned to logical and physical representations that eventually lead to a set of specified requirements.

9.3.4.5 Summary

EIA-632 is an excellent systems engineering document that has progressed significantly since the release of the interim standard. The scope of EIA-632 has grown to incorporate the entire enterprise rather than just the system development project. The standard is based on five process categories that include 13 individual processes. EIA-632 breaks these 13 processes into 33 requirements that must be performed to develop a product. These requirements are designed to be incorporated into enterprise policies and procedures. Depending on the specific nature of the system development effort, some of the 33 requirements may be tailored out of the standard. EIA-632 provided a very solid framework upon which systems engineers could base system development efforts.

When the EIA disbanded in 2011, its standards were transferred to other standards organizations. The Society of Automotive Engineers now maintains SAE/EIA-632A-2021, which has been stabilized and is not expected to change in the near future.

Then EIA-632A was superseded by SAE1001 Integrated Project Processes for Engineering a System [13] and replaced EIA-632 when it was published on October 9, 2018. Although SAE1001 is a successor in one way to EIA-632, the name is changed to reflect that the standard is significantly different. SAE1001 is discussed in more detail after the discussion of ISO/IEC/IEEE 15288 in the next section.

9.3.5 ISO/IEC/IEE 15288 System and Software Engineering—System Life-Cycle Processes

ISO/IEC/IEEE 15288 was ratified as an international standard in 2002 and was subsequently revised in 2008, 2015, and 2023. Consideration of ISO/IEC/IEEE 15288 began in the mid-1990s with the intent of developing a standard that provided a common framework within which to address technical (both hardware and software) and social aspects of a system (in short, it is meant to be able to be applied to any man-made system). The new standard was to establish common terminology and to integrate across the organization/enterprise all necessary disciplines, technical processes, and project management. Arnold and Lawson [14] provided an excellent summary of the role of ISO/IEC/IEEE 15288 from a business management perspective. While earlier standards tended to focus on the design and development aspects of a system life cycle, ISO/IEC/IEEE 15288 defines a set of 30 tailorable processes that covers the system life cycle from a whole-of-business perspective. Again, as with EIA-632, ISO/IEC/IEEE 15288 cannot be considered to be a systems engineering standard at all; it certainly does not define systems engineering. As its name implies, the standard is focused on the life cycle of the system, not the project that brings the system into being.

As shown in Figure 9.7 [15], the 30 processes of ISO/IEC/IEEE 15288 rightly have an emphasis on the acquisition phase of the system life cycle, but the whole life cycle is addressed. Figure 9.8 [16] lists the processes in four groups: agreement processes, organizational project-enabling processes, technical management processes, and technical processes.

ISO/IEC/IEEE 15288 establishes a common framework to describe the typical processes associated with the life cycle of any human-made system, which includes anything that "may be configured with one or more of the following: hardware, software, data, humans, processes (for example, processes for providing service to users), procedures (such as operator instructions), facilities, materials, and naturally occurring entities" [17].

The standard is intended to apply to organizations, or groups of organizations, in their roles as acquirers or suppliers. It is also meant to apply to "a wide variety of systems in terms of their purpose, domain of application,

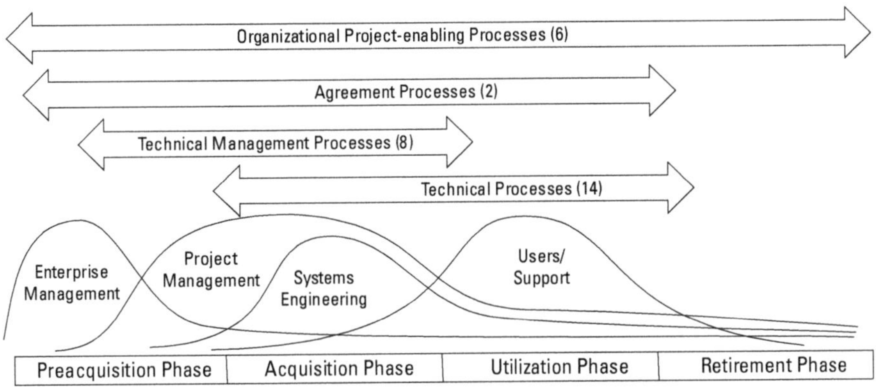

Figure 9.7 The 30 ISO/IEC/IEEE 15288 processes mapped onto the system life cycle.

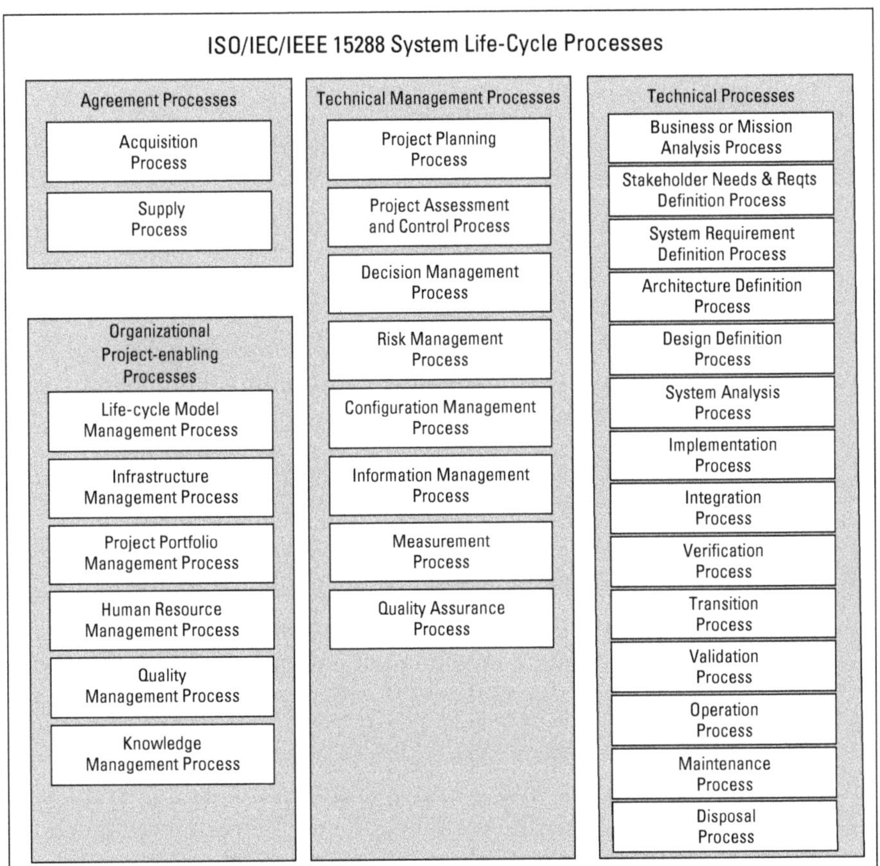

Figure 9.8 ISO/IEC/IEEE 15288 system life-cycle processes.

complexity, size, novelty, adaptability, quantities, locations, life spans, and evolution . . . to one-of-a-kind systems, mass-produced systems and customized, adaptable systems . . . complete stand-alone systems and to systems that are embedded and integrated into larger and more complex and complete systems" [18]. In other words, the standard is intended to apply to any system created by humans.

ISO/IEC/IEEE 15288 has a broad application and has been accepted to be the foundational basis of a number of other standards and extensions:

- ISO/IEC/IEEE 12207: Systems and Software Engineering—Software Lifecycle Processes.
- ISO/IEC/IEEE 15288.1: Application of Systems Engineering for Defense Programs.
- ISO/IEC/IEEE 15288.2: Application of Technical Reviews and Audits for Defense Programs.
- ISO/IEC/IEEE 21839: System of Systems Consideration in the Lifecycle Stages of a System.
- ISO/IEC/IEEE 21840: Systems and Software Engineering—Guidance for the Utilization of ISO/IEC/IEEE 15288 in the Context of System of Systems.
- ISO/IEC/IEEE 21841: Systems and Software Engineering—Taxonomy of System of Systems.
- ISO/IEC/IEEE 24641: Systems and Software Engineering—Methods and Tools for Model-based Systems and Software Engineering.
- ISO/IEC 90005: Guidelines for the Application of ISO 9001 to Systems Lifecycle Processes.
- SAE1001: Integrated Project Processes for Engineering a System.

ISO/IEC/IEEE 15288 is supported by a number of guides:

- ISO/IEC/IEEE 24748-1: Systems and Software Engineering—Lifecycle Management—Guidelines for Lifecycle Management.
- ISO/IEC/IEEE 24748-2: Systems and Software Engineering—Lifecycle Management—Guidelines for the Application of ISO/IEC/IEEE 15288.
- ISO/IEC/IEEE 24748-3: Systems and Software Engineering—Lifecycle Management—Guidelines for the Application of ISO/IEC/IEEE 12207.
- ISO/IEC/IEEE 24748-4: Systems and Software Engineering—Lifecycle Management—Application and Management of the Systems

Engineering Process (formerly the combination of ISO/IEC 26702: Systems Engineering—Application and Management of the Systems Engineering Process and IEEE-STD 1220: Management of the Systems Engineering Process).

- ISO/IEC/IEEE 24748-5: Systems and Software Engineering—Lifecycle Management—Software Development Planning.
- ISO/IEC/IEEE 24748-6: Systems and Software Engineering—Lifecycle Management—System Integration Engineering.
- ISO/IEC/IEEE 24748-7: Systems and Software Engineering—Lifecycle Management—Application of Systems Engineering on Defense Programs.
- ISO/IEC/IEEE 24748-8: Systems and Software Engineering—Lifecycle Management—Technical Reviews and Audits on Defense Programs.

9.3.6 SAE1001: Integrated Project Processes for Engineering a System

SAE1001: Integrated Project Processes for Engineering a System [20] began as Revision A to EIA-632 and superseded that standard when published on October 9, 2018. Although a successor in one way to EIA-632, the name was changed to reflect that the standard is significantly different.

The current version of SAE1001 is compatible with ISO/IEC/IEEE 15288 (with a mapping provided in Annex B); it uses ISO/IEC/IEEE 15288 terminology and lists ISO/IEC/IEEE 15288 as a normative standard, pointing to definitions and clauses in the higher standard. Although the two standards are ostensibly similar by virtue of their focus on systems engineering processes, SAE1001 is different in that it is focused on a set of "integrated" processes that inform a given project how to implement project processes for engineering a system, providing ". . . an integrated set of fundamental technical processes to aid a project in the engineering or re-engineering of a system." SAE1001 therefore contains details not provided by ISO/IEC/IEEE 15288, providing the next level of tasking detail performed by a project, focusing on tasks, inputs, and outputs to show they are interconnected (integrated) to form project processes for systems engineering. That is particularly useful for a small organization that does not want the full set of processes as defined in ISO/IEC/IEEE 15288 (particularly the high-level organizational project-enabling processes) but just wants to focus on the project level.

SAE1001 is therefore targeted at organizations inexperienced with systems engineering who will most likely have immature processes and developing systems engineering personnel so that they are dependent on other organizations for such support. Consequently, the goal of SAE1001 is to assist organizations to improve their systems engineering processes into a cohesive integrated set and to develop their systems engineering staff.

9.3.7 Other Useful Standards

The following section lists and describes some additional sources of specific information relating to systems engineering topics. Most of the documents listed are available on the internet following a simple search using their document number.

9.3.7.1 Other Systems Engineering Standards

ECSS-E-10A [21] is part of a set of standards issued by the European Cooperation for Space Standardization and tailored for use in space applications. It provides a useful coverage of the requirements engineering process, configuration management, systems engineering products and interfaces with other disciplines. The standard is written as a set of statements of what must be done in systems engineering rather than describing how to do it.

In 2004, ECSS-E-10A was superseded by ECSS-E-10 Part 1B, reflecting updates and refinements in system engineering practices within the space sector. On March 6, 2009, ECSS-E-10 Part 1B was replaced by ECSS-E-ST-10C, which represented a significant shift by presenting only requirements and deferring informative material to a new handbook, ECSS-E-HB-10. This change aimed to streamline the standard and provide clearer guidance for its application.

There are many other standards related to systems engineering, including the following:

- ISO/IEC/IEEE 15026-1: Systems and Software Engineering—System and Software Assurance—Part 1: Concepts and Vocabulary.
- ISO/IEC/IEEE 15026-2: Systems and Software Engineering—System and Software Assurance—Part 2: Assurance Case.
- ISO/IEC/IEEE 15026-3: Systems and Software Engineering—System and Software Assurance—Part 3: Integrity Levels.
- ISO/IEC/IEEE 15026-4: Systems and Software Engineering—System and Software Assurance—Part 4: Assurance in the Lifecycle.
- ISO/IEC/IEEE 15289: Systems and Software Engineering—Content of Lifecycle Information Products (Documentation).
- ISO/IEC/IEEE 24765: Systems and Software Engineering—Systems and Software Engineering Vocabulary.

- ISO/IEC/IEEE 16326: Systems and Software Engineering—Life Cycle Processes—Project Management.
- ISO/IEC/IEEE 16085: Systems and Software Engineering—Life Cycle Processes—Risk Management.
- ISO/IEC/IEEE 42010: Software, Systems and Enterprise—Architecture Description.
- ISO/IEC/IEEE 42020: Software, Systems and Enterprise—Architecture Process.
- ISO/IEC/IEEE 42030: Software, Systems and Enterprise—Architecture Evaluation Framework.
- ISO/IEC/IEEE 29148: Systems and Software Engineering—Requirements Engineering.
- ISO/IEC/IEEE 15939: Systems and Software Engineering—Measurement Process.
- ISO/IEC/IEEE 26560: Systems and Software Engineering—Reference Model for Product Line Engineering and Management.
- ISO/IEC/IEEE 26580: Systems and Software Engineering—Models and Tools for the Feature-based Approach to Software and Systems Product Line Engineering.
- ISO 10303-233: Systems Engineering Data Interchange Standard.
- ISO/IEC 29110: System and Software Engineering Standards and Guides for Very Small Entities.
- ISO/IEC 31000: Risk Management.
- ISO/IEC 31010: Risk Management—Risk Assessment Techniques.
- ISO/IEC 33060: Process Assessment—Process Assessment Model for System Life Cycle Processes.
- NIST SP 800-160 Vol. 1: Systems Security Engineering: Considerations for a Multidisciplinary Approach in the Engineering of Trustworthy Secure Systems.
- NIST SP 800-160 Vol. 2: Developing Cyber-Resilient Systems: A Systems Security Engineering Approach.
- CMMI: CMMI® V2.0.
- OMG SysML: OMG Systems Modeling Language.

9.3.7.2 Technical Reviews and Audits

MIL-STD-1521B [22] is an old standard that was canceled by the US DoD. Some of its content has been summarized and divided between MIL-STD-499B and MIL-STD-973. Despite this, it remains an excellent reference source for technical reviews and audits; a quick scan will recognize all the reviews and audits we have mentioned here: SRR, SDR, PDR, CDR, TRR, FCA, and PCA. It provides a thorough coverage of a comprehensive range of reviews and audits and links the reviews and audits to the systems engineering life cycle presented in the systems engineering standard of the day.

MIL-STD-1521B provides useful descriptions of the types of issues to be covered at each of the reviews and the criteria that should be met prior to entering and exiting each review. In addition, it provides checklists and templates that may be of assistance during the technical review and audit program.

In defense programs, the requirements for conducting reviews and audits are contained in IEEE 15288.2-2014 IEEE Standard for Technical Reviews and Audits.

9.3.7.3 Configuration Management

EIA-649-C [23] and MIL-STD-973 [24] are recognized standards that describe configuration management practices and processes described in Chapters 3 to 6. MIL-STD-973 is generally regarded as the classic configuration management standard and remains an excellent reference source, although it is becoming a little dated in some regards and was canceled with the release of EIA-649-C. More recently, ISO 10007 [25] provided guidelines for configuration management.

MIL-HDBK-61B [26] is a military handbook on configuration management. Like other military handbooks, MIL-HDBK-61B is not designed to be contractually invoked but, instead, is written as a guidance document. Regardless of what configuration management standard is being used, MIL-HDBK-61B is a valuable source of information about modern configuration management practice.

9.3.7.4 Specification Standards

MIL-STD-490A [27] has been canceled and replaced by MIL-STD-961E [28]. However, as is often the case, MIL-STD-490A remains an excellent source of information on specifications. The standards introduce a hierarchical set of specifications from the A-level system specification to the E-level materials specification. The relationship of these standards to functional (A-level specifications), allocated (B-level specifications) and product baselines (C, D, and E-level specifications) is also explained. The standard also describes the many different types of development and product specifications including prime item and critical item development and product specifications. The document provides content information for each of the specifications in an appendix.

MIL-STD-961E is the current US DoD standard on specifications and is written to describe the content and requirements of generic specifications rather than content of specific specifications as outlined in its predecessor. With that in mind, reference to both of the specifications can assist in an understanding of the types and content of technical specifications required on given system developments.

9.3.7.5 WBS

Military handbook MIL-HDBK-881A [29] is an excellent source of information on work breakdown structures and represents a conversion of the earlier military standard MIL-STD-881C, Work Breakdown Structures for Defense Materiel Items. MIL-HDBK-881A effectively describes the relationship between the program WBS (PWBS) and the contract WBS (CWBS). It also contains detailed information on the purpose, structure, and content of a range of example programs including an aircraft system that proved useful in this text.

9.4 Useful Guidance from INCOSE

The International Council on Systems Engineering (INCOSE) is a not-for-profit membership organization founded to develop and disseminate the transdisciplinary principles and practices that enable the realization of successful systems [30]. In addition to connecting systems engineering professionals with educational, networking, and career-advancement opportunities, INCOSE is also focused on producing state-of-the-artwork products that support and enhance systems engineering, providing standards, handbooks, and guidance for professionals across industries. There are many such products available in the INCOSE store (free to INCOSE members and at a small cost for non-members) that conform to the overarching ISO/IEC/IEEE 15288; this section briefly describes the major handbooks and guides.

- *Systems Engineering Handbook* [31]. The INCOSE Systems Engineering Handbook is an essential reference for systems engineering practitioners and engineers in other disciplines looking to perform or understand the discipline of systems engineering. The handbook elaborates on the key systems lifecycle processes described in ISO/IEC/IEEE 15288 with chapters covering key systems engineering concepts, system life-cycle processes and methods, tailoring and application considerations, systems engineering in practice, and more. It is the basis for certification of the systems engineering discipline.

- *Guide to the Systems Engineering Body of Knowledge (SEBoK)* [32]. The SEBoK is an online resource that provides a guide to the key knowledge sources and references of systems engineering organized and explained to assist a wide variety of individuals. It is a living product, accepting community input continuously, with regular updates.
- *Needs and Requirements Manual (NRM)* [33]. This IBCOSE manual is intended for those whose role is to perform needs, requirements, verification, and validation activities throughout the system life cycle and is addressed to practitioners of all levels of experience. Those new to project management and system development should find the detailed guidance useful. Those more experienced should be able to find new insights across all phases of the life cycle, particularly in terms of a data-centric perspective to which they may not have been exposed. The NRM is supported by a number of guides:
    - Guide to Needs and Requirements [34];
    - Guide to Writing Requirements [35];
    - Guide to Verification and Validation [36];
    - Guide to Security Needs and Requirements [37].
- Other useful INCOSE artifacts:
    - Integrated Data as a Foundation of Systems Engineering;
    - Systems Engineering Tools Database (SETDB);
    - A Complexity Primer for Systems Engineers;
    - Digital Engineering Measurement Framework;
    - Guide for the Application of Systems Engineering in Large Infrastructure Projects;
    - SE Competency Framework;
    - Project Managers Guide to SE Measurement for Project Success;
    - Systems Engineering Agility Primer;
    - Systems Engineering Measurement Primer;
    - Systems of Systems Primer;
    - Systems Engineering Principles;
    - Technical Measurement Guide.

9.5 Summary

In this chapter, we reviewed a number of relevant standards that are available to assist both systems engineering management and the systems engineering processes, as well as assisting related disciplines such as project management. We also summarized the major systems engineering guidance available from INCOSE, which is freely available to their members.

While there has been considerable movement toward a harmonious suite of standards in ISO/IEC/IEEE 15288 and its attendant handbooks, there are still a number of organizations around the world making use of older standards such as EIA 632 and IEEE-STD-1220. There are also many organizations whose internal processes have been developed based on an older standard; there is considerable inertia to rapid change, particularly if the governing contract/agreement invokes the older standard.

Having said that, while there are clear differences in the various standards, there are many more similarities and they all provide a framework for the establishment of useful systems engineering processes and procedures in an organization. In any case, any selected standard will almost always have a scope greater than the project and system under consideration, so tailoring is essential, not only because of the scope of the standard but also because of the need to temper the selected standard's requirements with the culture and structure of the adopting organization.

In the next chapter, we introduce some of the related disciplines that will interact with the systems engineering effort throughout the conduct of a project and the system's life cycle.

Review Questions

1. Explain the systems engineering processes used by MIL-STD-499B, IEEE-STD-1220, and EIA/IS-632.
2. Explain how systems engineering standards could be used to assist a systems engineer during the planning of stages of a developmental project.
3. Explain how systems engineering standards could assist a customer in defining systems engineering requirements in a contract/agreement.
4. Compare the life-cycle models in MIL-STD-499B, IEEE-STD-1220, and EIA-632.
5. Briefly explain the differences between the earlier systems engineering standards (such as MIL-STD-499B, IEEE-STD-1220, and EIA/IS-632), and later standards such as EIA-632, ISO/IEC/IEEE 15288, and SAE1001.

Endnotes

[1] MIL-STD-499B, *Military Standard—Systems Engineering—Draft*, Washington, DC: US Department of Defense, 1994.

[2] MIL-STD-499B, *Military Standard—Systems Engineering—Draft*, Washington, DC: US Department of Defense, 1994.

[3] EIA-632, *Processes for Engineering a System*, Washington, DC: Electronic Industries Association (EIA), 2003.

[4] IEEE-STD-1220, *IEEE Standard for Application and Management of the Systems Engineering Process*, New York: IEEE Computer Society, 2005.

[5] IEEE-STD-1220, *IEEE Standard for Application and Management of the Systems Engineering Process*, New York: IEEE Computer Society, 2005.

[6] EIA-632, *Processes for Engineering a System*, Washington, DC: Electronic Industries Association (EIA), 2003.

[7] Martin, J., "Processes for Engineering a System: An Overview of the EIA-632 Standard and Its Heritage," *Systems Engineering*, Vol. 3, No. 1, 2000, pp. 1–26.

[8] EIA-632, *Processes for Engineering a System*, Washington, DC: Electronic Industries Association (EIA), 2003.

[9] EIA-632, *Processes for Engineering a System*, Washington, DC: Electronic Industries Association (EIA), 2003.

[10] EIA-632, *Processes for Engineering a System*, Washington, DC: Electronic Industries Association (EIA), 2003.

[11] EIA-632, *Processes for Engineering a System*, Washington, DC: Electronic Industries Association (EIA), 2003.

[12] EIA-632, *Processes for Engineering a System*, Washington, DC: Electronic Industries Association (EIA), 2003.

[13] SAE1001, *Integrated Project Processes for Engineering a System*, SAE International, 2018.

[14] Arnold, S., and H. W. Lawson, "Viewing Systems from a Business Management Perspective: The ISO/IEC 15288 Standard," *Systems Engineering*, Vol. 7, No. 3, 2004, pp. 224–242.

[15] ISO/IEC/IEEE 15288, *Systems Engineering—System Lifecycle Processes*, International Organization for Standardization, 2023.

[16] ISO/IEC/IEEE 15288, *Systems Engineering—System Lifecycle Processes*, International Organization for Standardization, 2023.

[17] ISO/IEC/IEEE 15288, *Systems Engineering—System Lifecycle Processes*, International Organization for Standardization, 2023.

[18] ISO/IEC/IEEE 15288, *Systems Engineering—System Lifecycle Processes*, International Organization for Standardization, 2023.

[19] SAE1001, *Integrated Project Processes for Engineering a System*, SAE International, 2018.

[20] SAE1001, *Integrated Project Processes for Engineering a System*, SAE International, 2018.

[21] ECSS-E-10A, *Space Engineering—System Engineering*, Noordwijk, The Netherlands: European Cooperation for Space Standardization, 1996.

[22] MIL-STD-1521B, *Military Standard—Technical Reviews and Audits for Systems, Equipments, and Computer Software*, Washington, DC: US Department of Defense, 1985.

[23] EIA-649-C, *Configuration Management Standard*, Warrendale, PA: SAE International, 2019.

[24] MIL-STD-973, *Military Standard—Configuration Management*, Washington, DC: US Department of Defense, 1992.

[25] ISO 10007, *Quality Management Systems—Guidelines for Configuration Management*, 2017.

[26] MIL-HDBK-61B, *Military Handbook—Configuration Management Guidance*, Washington, DC: US Department of Defense, 2020.

[27] MIL-STD-490A, *Military Standard—Specification Practices*, Washington, DC: US Department of Defense, 1985.

[28] MIL-STD-961E with Change 1, *Department of Defense Standard Defense and Program-Unique Specifications Format and Content*, Washington, DC: US Department of Defense, April 2, 2008.

[29] MIL-HDBK-881A, *Department of Defense Handbook—Work Breakdown Structure*, Washington, DC: US Department of Defense, July 30, 2005.

[30] www.incose.org.

[31] International Council on Systems Engineering (INCOSE), INCOSE-TP-2003-002-05, *Systems Engineering Handbook*, Fifth Edition, New York: Wiley, 2023.

[32] SEBoK, Guide to the Systems Engineering Body of Knowledge (SEBoK), 2024, https://sebokwiki.org/wiki/Guide_to_the_Systems_Engineering_Body_of_Knowledge_(SEBoK).

[33] Wheatcraft, L. S., M. J. Ryan, and T. E. Katz, *Needs and Requirements Manual: Needs, Requirements, Verification, Validation Across the Lifecycle*, Hoboken, NJ: Wiley, 2024.

[34] International Council on Systems Engineering (INCOSE), INCOSE- TP-2021-003-01, *Guide to Needs and Requirements*, 2022.

[35] International Council on Systems Engineering (INCOSE), INCOSE- TP-2010-004-04, *Guide to Writing Requirements*, 2024.

[36] International Council on Systems Engineering (INCOSE), INCOSE-TP -2021-006-01, *Guide to Verification and Validation*, 2022.

[37] International Council on Systems Engineering (INCOSE), INCOSE- TP-2024-146, *Guide to Security Needs and Requirements*, 2024.

10

Related Disciplines

10.1 Introduction

This chapter introduces some of the related disciplines that will interact with systems engineering effort throughout the conduct of a project and the system life cycle. Particular attention is given to the project management discipline since systems engineering in most projects is seen as being subordinate during the acquisition phase to project management.

This chapter is not intended to treat any of the related disciplines in a comprehensive manner since each discipline is a large subject area. Rather, the intention is to show how each of the major disciplines is related to systems engineering.

10.2 Project Management

The principal role of project management is to meet project requirements (the project scope) to the desired quality within a given schedule and budget. Project management encompasses an extremely large body of knowledge that is recorded comprehensively in a range of texts, so the detail is not replicated here. This section simply introduces the discipline of project management and details the strong relationship between systems engineering and project management.

The Project Management Institute's (PMI) Project Management Body of Knowledge [1] defines a project as "... a temporary endeavor undertaken to create a unique product, service or result" and project management as "... the application of knowledge, skills, tools and techniques to project activities to meet the project requirements." Although there are any number of unique products, services, or results that could be the output of a project, we are interested here in projects that bring systems into being in order to solve a business problem or to take up a business opportunity. Consequently, we are interested in the relationship between project management and systems engineering.

The strong relationship exists between systems engineering and project management primarily because every technical decision made as part of the systems engineering processes has an impact on project management activities and vice versa. One way to introduce project management and investigate its relationship with systems engineering is through the project management knowledge areas defined in the PMI's Project Management Body of Knowledge, a summary of which is provided next, including a brief note on the role of systems engineering in the activities defined within each knowledge area.

- *Project integration management.* This ensures that all elements of the project are properly planned and coordinated. The coordination involves making trade-offs between conflicting alternatives to meet that the needs and requirements of the business and of the key project stakeholders. The keys to integration management are the development of the project charter and then the project plan, which result from the project planning processes and form a governing approach to control and execution. The plan must be cognizant of the business constraints and organizational policies as well as the outputs from the project planning processes. Once the plan is in place, the project integration management function ensures that the plan is executed. Systems engineering contributes significantly to the development of the activities upon which the project plan is developed as well as to the subsequent integration effort.
- *Project scope management.* This is primarily concerned with ensuring that the project includes all the work required to complete the project successfully without including any additional or unnecessary work. The first stage in scope management is to define a formal scope as the basis of the subsequent effort. The Guide to the Project Management Body of Knowledge [2] references the use of systems engineering as fundamental to the completion of this task; almost all the activities we discuss in this book are focused on developing and managing the scope

of the system. Once the scope has been developed and approved, the project must be broken into smaller, more manageable components. This breakdown is commonly performed using a WBS. The scope and the WBS must then be formalized through an acceptance process to ensure that all parties are committed to the same scope and the same work; we will see the important role that systems engineering has in defining the system requirements and then allocating them to the elements in the project WBS. Once approved, the scope management activity must take control of any changes to the approved project scope.

- *Project schedule management.* This is a key project management activity that is sometimes referred to as time management. To manage the project's schedule, the project manager must start with a comprehensive list of the work activities that make up the project. Tools such as the WBS assist in this identification and, again, systems engineering provides an essential contribution. These activities will normally be performed in a particular sequence that is dictated often by the systems engineering aspects of the project. For example, the activity "hardware/software integration" can only be performed following the activities "hardware test and acceptance" and "software test and acceptance." Other complications with respect to time management eventuate when activities have specific start and end dates that cannot be altered. Once the activities and their sequences have been determined, the durations for each of the activities need to be estimated. Based on these pieces of information, the project manager can determine the schedule for the entire project. It then becomes an exercise in schedule control as activities are added, deleted, or modified throughout the project. The project manager is assisted in the time management function by tools ranging from sophisticated, automated software packages to a simple reliance on expert judgment to predict activity duration. Systems engineering activities are the major source of information regarding WBS activities, in terms of their formation, progress, and completion.
- *Project cost management.* This ensures that the project is completed within the approved budget. The first task in determining the budget, called resource planning, determines the resources required to complete the project activities. Resources include personnel, materials, and equipment. Based on the resources required for each activity, the project manager can estimate the costs associated with each activity and the total cost of the entire project. Changes in the project scope will invariably impact on the cost aspects of the project; those impacts need to be monitored and assessed on an ongoing basis. Tools are

available to assist with cost management, but a great deal of the estimation and control effort relies on expert judgment and experience. Systems engineering provides critical input for the development of the WBS but also provides input to cost-benefit trade-offs as well as investigations to determine the impact of any proposed change on project cost and schedule.

- *Project quality management.* This is described in Section 10.3 as a discipline related to project management. The project manager must lead in the quality assurance effort to ensure that quality is taken seriously by all involved. The first stage in the process is to plan the quality management effort by identifying which quality standards are to be applied to the project and how these will be satisfied. The overall project performance must be evaluated regularly to provide users with confidence that the selected standards are being satisfied. The systems engineering efforts in establishing work packages, as well as in review and audit (particularly configuration audits), provide essential input to project quality management.
- *Project resource management.* This is not exclusive to project management but is, nonetheless, a vital part of the project management function to ensure the effective use of resources across the project. Human-resource management ensures the most effective use of the project personnel. Often, one of the first tasks of the project manager is to acquire and organize the personnel required to perform the activities associated with the project. Once this has been done, the project manager must maintain the unity and direction of those involved with the project, which involves sound leadership and management skills such as negotiation, communication, motivation, and team building. In addition to human resources, the project manager is responsible for all other resources used or consumed to complete the project. Again, the extent, nature and timing of the use of these resources is largely defined by systems engineering endeavors.
- *Project communications management.* This is a fundamental activity for project managers. Communication is required to ensure that relevant project information is collected and disseminated to the appropriate personnel in a timely manner. The project manager must determine the information requirements of business managers and the key stakeholders in the project, and the preferred format for the information to be provided to them. The project manager then must ensure that the information is distributed as planned. A particularly important part of communications management is the report generation associated

with status reporting and phase or project closure. The role of communications management is worthy of emphasis here as a project management to which systems engineering provides great value. The systems engineer is, in some respects, part management staff and part technical staff. Systems engineers, like a business analysist, are therefore largely conversant with both technical and management languages and can act as very effective translators between the management layers within organizations. Systems engineers can also assist with horizontal communication between different design teams, to assist with technical issues such as design space consideration, interfacing issues, and technical conflicts. The major systems engineering reviews also provide a significant contribution to the communication of project information.

- *Project risk management.* This is covered in some detail in Chapter 8. The systems engineering effort focuses on assessment of the technical risks associated with the project, whereas the project manager broadens the coverage of risk to include areas such as schedule, cost, and the other elements of project management as defined in sources such as the Project Management Body of Knowledge. Systems engineering supplies input to the risk management process in concert with the project manager. Project risk management involves the same steps as technical risk management. The main source of information for the task of risk management is intuition and experience.
- *Project procurement management.* This ensures timely supply of goods and services from outside the immediate project environment in support of the project. The initial step is to plan what products and services are required to complete the project and when they will be needed. These details should then be recorded. The project manager should seek quotes or offers from the suppliers and select the best of the offers. In some cases, the procurement will be a one-time purchase but often one or more standing offers or contracts will result. The project manager will be responsible for managing the resulting contracts. Once all aspects of the procurement are completed, the project manager is responsible for the closure of any extant contracts.
- *Project stakeholder management.* This is the process by which stakeholders are identified, and an appropriate management plan is developed, implemented, and controlled. The PMI's Project Management Body of Knowledge defines a stakeholder as any individual, group, or organization that is affected by or can affect the system of interest. Since stakeholders can influence the system development, it is essential that

they are identified, and appropriate mechanisms are established to engage with them throughout the project to ensure that their needs are met. From a systems engineering perspective, engagement with stakeholders is essential to the establishment of appropriate business and stakeholder needs and requirements as well as their transformation into system requirements.

10.2.1 Systems Engineering and Project Management

The strong relationship between project management and systems engineering is demonstrated by considering the 10 project management knowledge areas summarized above. Kauffman [3] investigated this relationship and demonstrated a significant intersection between the two professions by referring to the Project Management Body of Knowledge for project management, and IEEE-STD-1220 and EIA/IS-632 for systems engineering when those standards first appeared in draft form. A summary of Kauffman's findings is shown in in Figure 10.1, which has been augmented for systems engineering to include the current versions of both EIA-632 and IEEE-STD-1220 as well as ISO/IEC/IEEE 15288.

In the earlier standards, EIA/IS-632, IEEE-STD-1220, and EIA-632, there are mostly areas that overlap between the two disciplines but there are notable exceptions in activities traditionally considered the sole province of project management: cost management, human resource management, and procurement management. Yet there are still strong relationships even in those aspects. For example, project management is responsible for management of cost and schedule, although this can only be meaningful with input from systems engineering in terms of the development of the WBS.

However, Figure 10.1 illustrates [4] that, over the last two decades, there has been a considerable change in thinking associated with the discipline of systems engineering. Note, in ISO/IEC/IEEE 15288 (the current premier systems engineering standard), the interrelationship between all aspects of both disciplines has been strengthened even further because the standard not only refers to every aspect of the Project Management Body of Knowledge but also covers all life-cycle aspects of systems development that are outside the acquisition phase and therefore outside the discipline of project management.

This also reflects major changes in relevant standards and systems approaches to the system life cycle. As illustrated in Figure 10.1, IEEE-STD-1220 viewed systems engineering as that subset of project management activities that focused on design and development of the system in support of acquisition; in quite a considerable contrast, ISO/IEC/IEEE 15288 shows that project management is that subset of the systems life-cycle processes that is focused on the acquisition phase. Systems engineering focuses on the remainder of the life-cycle processes as well as the acquisition phase, that is, the entire life cycle. Figure 10.2 illustrates in another way the changes in focus with the systems engineering over time.

Whichever way they are viewed, however, Figures 10.1 and 10.2 are useful in reinforcing that project management and systems engineering are inextricably intertwined, particularly in the acquisition phase.

Since a project is a temporary endeavor (that is, it has a defined starting point and a defined end point), a project can only exist for a portion of the system life cycle. Consequently, we should note that the principal responsibility of project management during the system life cycle is to bring the capability system into being so that it can be introduced into utilization. However, as illustrated in Figure 10.3, we will need to establish a number of projects throughout utilization in order to undertake major activities such as modifications to the system.

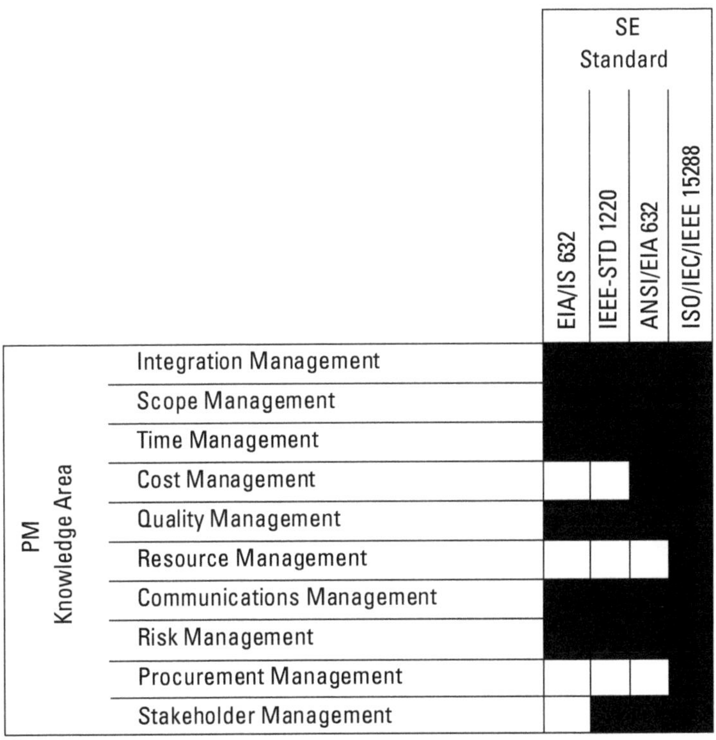

Figure 10.1 Relationship between the 10 knowledge areas of project management and systems engineering as defined in the major systems engineering standards.

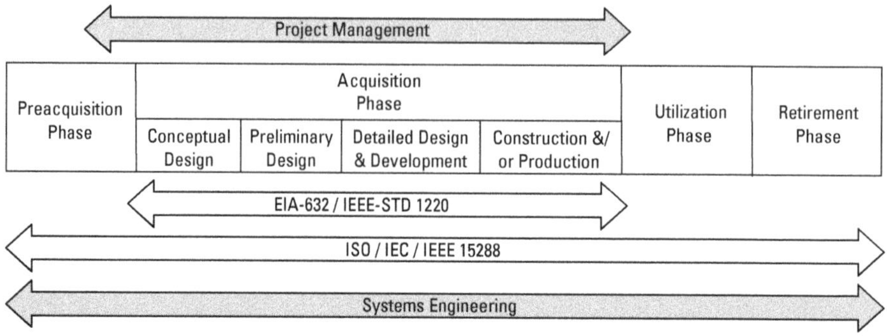

Figure 10.2 Relationship between project management and systems engineering standards throughout the system life cycle.

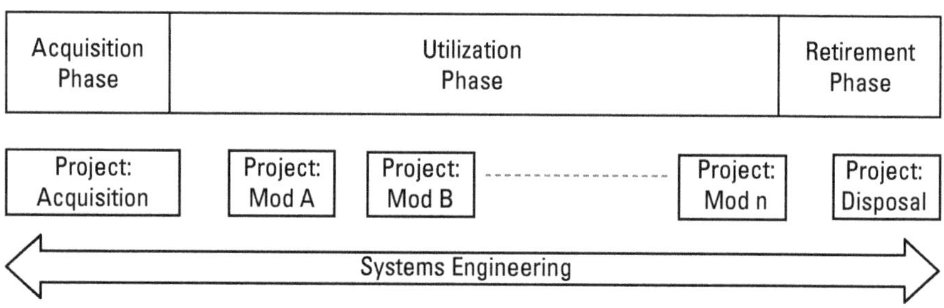

Figure 10.3 The role of project management and systems engineering throughout the system life cycle.

10.3 Business Analysis

A business analyst (BA) plays a crucial role in organizations by bridging the gap between business needs and technological solutions. Their principal role is to understand business objectives, analyze data, identify inefficiencies, and recommend improvements to processes, systems, and strategies. Their primary contribution to project management and to systems engineering is to ensure that business requirements are accurately captured and translated into actionable solutions that align with an organization's strategic objectives. In that sense, business analysis contributes most to the preacquisition phase and to the conceptual design activity in that phase where they would take carriage of the development of BNR, SNR, and the SyRS.

The most widely recognized guide outlining best practices and standards in business analysis is *Business Analysis Body of Knowledge* [5], published by the International Institute of Business Analysis (IIBA). The IIBA also provides an *Agile Extension to the Business Analysis Body of Knowledge* [6] to assist BAs working in Agile environments by detailing their role in Scrum and other Agile methodologies.

BAs operate across various industries, including finance, healthcare, information technology, government, and retail. They can work within internal business teams, as part of a consulting firm, or alongside project managers, developers, and stakeholders to ensure that solutions deliver maximum value.

BAs perform several key functions within an organization:

- *Requirements gathering and analysis.* BAs are responsible for eliciting, recording, and validating business requirements from stakeholders. They use various techniques such as interviews, workshops, surveys, and artifact analysis to gather relevant information.
- *Process improvement and optimization.* By analyzing existing business processes, BAs identify inefficiencies, redundancies, and opportunities for improvement. They use methodologies such as Lean [7], Six Sigma [8], and Business Process Modeling [9] to recommend optimized workflows.
- *Stakeholder management.* Since BAs interact with different levels of an organization, including executives, IT teams, and end-users, they act as a liaison between these groups. They must ensure clear communication and alignment on project goals.
- *Data analysis and decision support.* BAs often analyze data to provide insights into business performance and trends. They may use data visualization tools, reporting software, and statistical models to support decision-making.
- *Solution evaluation and implementation support.* Once a solution is proposed, BAs help to assess its feasibility, work with technical teams on implementation, and validate that the solution meets business needs. They also play a role in acceptance testing and post-implementation reviews.
- *Artifacts and communication.* BAs create and maintain artifacts such as business needs and requirements, functional specifications, process diagrams, and user stories to ensure clarity and alignment on project objectives.

BAs collaborate closely with project managers and systems engineers to ensure that project objectives continuously align with business needs and provide value. The BA, project manager, and systems engineer all work with stakeholders to define requirements, which the project manager and systems engineer ensure are implemented effectively. In Agile environments, BAs frequently work as product owners or Agile BAs, defining user stories and working with development teams to ensure continuous delivery of business value.

10.4 Quality Assurance

The term *quality* as used in the context of quality assurance means "the totality of features and characteristics of a product or service that bears on its ability to satisfy stated or implied needs" [10].

With this definition in mind, quality assurance aims to ensure that the products and services required from a system development effort are delivered at the required quality level. This requires a planned managerial effort to perform the following:

- Identify the quality assurance objectives and standards.
- Focus on preventative measures rather than cures.
- Collect relevant data and use the data to analyze results.
- Establish and monitor quality performance measures.
- Perform quality audits.

A major role of quality assurance is to ensure that appropriate plans and procedures are in place to perform the design and development tasks. These plans must be well recorded, approved by the appropriate personnel, and maintained and updated when necessary. However, it is not satisfactory to merely have the plans and procedures in place. The organization must operate in accordance with those plans and procedures. The quality assurance role will perform audits periodically to ensure that the systems engineering effort is progressing in accordance with the formal process. The quality assurance representative requires evidence of the plans and procedures being followed.

10.5 Integrated Logistics Support

The main aim of integrated logistics during the acquisition phase is to impact the design of the system to maximize supportability over the entire system life cycle. During utilization, the logisticians will also be interested and involved in maintaining and, where possible, improving system reliability throughout the system life cycle. The FRACAS was introduced in Chapter 7 as a process aimed at doing just that. Both logisticians and systems engineers must combine skills and expertise to make the FRACAS process as effective as possible.

A definition of logistics support is "to assure the effective and economical support of a system throughout its programmed lifecycle" [11]. To meet this definition, a number of support services are provided (by what is sometimes called the support system) during the utilization phase. The constituent capabilities that provide those support services are acquired during the acquisition phase.

- *Operating support capability.* This capability encompasses the support elements of: operating facilities system operators, support equipment, operator manuals and technical data, operating support procedures, and operating support information systems.
- *Engineering support capability.* Engineering support encompasses the support elements of: engineering facilities, engineering personnel, engineering support and test equipment, engineering technical data, engineering processes, engineering information management system, and software support.
- *Maintenance support capability.* Maintenance support is the capability that includes the support elements necessary to develop, establish, and integrate a maintenance support system capable of sustaining a system throughout its life. The capability comprises the elements of: maintenance facilities, maintenance personnel, maintenance support and test equipment, maintenance technical data, maintenance processes, and maintenance information management system.
- *Supply support capability.* Supply support encompasses the following support elements: supply facilities, supply personnel, supply support equipment, supply technical data, supply processes, supply information management system, spares, and packaging.
- *Training support capability.* The training support capability includes the support elements: training facilities training personnel, training equipment, training materials and other technical data, training processes, and training information management system.

Although the support capabilities have been dealt with independently in this section, it must be stressed that the elements are strongly interrelated. It is the role of the logistician to ensure that the elements are considered during the systems engineering processes and throughout the system life cycle.

10.6 Operations

Operations is the generic name sometimes given to the actions of the users and support personnel who will operate, support, and maintain the system when it is in service. Operations personnel are vitally important to the project, and they play differing roles as the project progresses through the acquisition phase. For example, in conceptual design and preliminary design, the operators are likely to be heavily involved in the correct and complete definition (and interpretation) of the stakeholder requirements. This is vital because these interpretations will flow through the entire systems engineering effort and result in specific system performance. During the latter stages of the acquisition phase, the operators will be involved in various testing activities. Depending on the nature of the system under development, aspects of the system's operation (and therefore testing) will only be possible using appropriately trained personnel. Consider, for example, the testing of an attack helicopter's control system, or the operation of an air defense radar system capable of tracking 150 targets simultaneously. While this can be verified by the contractor during AT&E, the customer will want to validate the requirement when the trained pilot is on board.

10.7 Software Engineering

Software engineering is defined as: "the technological and managerial discipline concerned with systematic production and maintenance of software products that developed and modified on time and within cost estimates" [12]. Software engineering has a set of processes, management requirements, tools, and related disciplines that are very similar to systems engineering.

As discussed in Chapter 5, ISO/IEC/IEEE 12207 [13] is the primary international standard for that defines the processes required for developing and maintaining software. The 2017 publication of ISO/IEC/IEEE 12207 has been harmonized with ISO/IEC/IEEE 15288 and adopts the ISO/IEC/IEEE 15288 process model with one slight change in the name of one process; the "System Requirements Definition" process is renamed the "System/Software Requirements Definition" process.

The *Software Engineering Body of Knowledge (SWEBOK)* defines 18 knowledge areas [14]:

- *Software requirements.* This introduces the requirements engineering process including the sub-areas of requirements engineering process, requirements elicitation and requirements analysis, software requirements specifications, requirements validation, requirements management, and requirements tools.

- *Software architecture.* This introduces software architecture fundamentals, description, process, and evaluation.
- *Software design.* This is concerned with the process of defining the architecture, components, interfaces, and other characteristics of the software.
- *Software construction.* This is the underlying software engineering principles by which software is crafted.
- *Software testing.* This is the process through which defects and problems are identified and software quality is improved.
- *Software engineering operations.* This introduces software engineering operations fundamentals, planning, delivery, and control.
- *Software maintenance.* This sustains software products throughout the life cycle to ensure that problems are identified and resolved, training is conducted, and modifications are implemented in a controlled manner.
- *Software configuration management.* This introduces the configuration management of software in a manner similar to that described in Chapter 8.
- *Software engineering management.* This describes the organizational, project management, and measurement aspects of software engineering.
- *Software engineering process.* This encompasses the technical and managerial processes performed during acquisition, maintenance, and disposal of software products, as well as the definition, implementation, measurement, management, change, and improvement of the software engineering process itself.
- *Software engineering models and methods.* In a manner similar to that used in this text to define tools, models, and methods associated with systems engineering process and management, the SWEBOK includes both the software development environments and software development methods.
- *Software quality.* This is concerned with software quality, in particular with software quality assurance and verification and validation.
- *Software security.* This introduces software security fundamentals, management and organization, engineering and processes, security engineering for software systems, software security tools, and domain-specific software security.
- *Software engineering professional practice.* This is concerned with the knowledge, skills, and attitudes that must be possessed by software

engineers to practice software engineering in a professional, responsible, and ethical manner.

- *Software engineering economics.* This is concerned with making decisions regarding software engineering in a business context.
- *Computing foundations.* This is concerned with the development and operational environment (the computer and its various components) in which software evolves and executes.
- *Mathematical foundations.* This helps software engineers to comprehend well-understood, unambiguous logic upon which software is based, which, in turn, is translated into programming language code.
- *Engineering foundations.* This is concerned with the engineering foundational skills and techniques that are useful for a software engineer.

Software products pass through a development process that is tied into the system life cycle leading up to the identification of CIs, and the described HWCIs and CSCIs. Each CI is managed individually and follows an individual development cycle.

Software engineering has some unique challenges that must be addressed jointly by the software and systems engineering specialists. Software is inherently difficult to test due to the number of variables often associated with its successful operation. The measurement of software reliability is also difficult. Software maintenance and support are often overlooked as many assume that software does not require maintenance and support once fielded. Software integration with host hardware brings about its own set of challenges as does the issue of firmware as a specialized category of software. These challenges mean that traditional approaches to design, development, integration, testing, and support derived from hardware-dominated environments are not always satisfactory in a software environment. To that end, software engineering is recognized as a specialty engineering discipline in this book, and systems engineers should treat it accordingly.

10.8 Hardware Engineering

Hardware engineering processes proceed in a similar manner to those of software engineering following the selection of HWCIs. The HWCIs are progressively broken into components and units in a similar way to the CSCs and CSUs.

Hardware units are designed and developed and then tested via unit-level testing. Once successful, the units are integrated and retested as they form the hardware components. This process is repeated to form and test the HWCIs. Finally, the HWCIs and CSCIs are combined, and system-level testing is conducted.

Hardware and software engineers must work closely together to ensure that interfacing issues are identified and addressed. Interfacing issues are sometimes addressed via an Interface Control Working Group (ICWG), which meets regularly during the system development process.

As with software engineering, recording and management of the development process will be managed on an individual HWCI basis. This allows different approaches to be used for each HWCI. For example, HWCI A may make use of OTS equipment with minor modifications, while HWCI B may require a bespoke design and development effort.

10.9 Summary

This chapter introduced some of the related disciplines that will interact with systems engineering effort throughout the conduct of a project and the system life cycle. Particular attention is given to the project management discipline since systems engineering in most projects is seen as being subordinate during the acquisition phase to project management.

So far, to best explain the activities and artifacts of systems engineering, we have assumed that the life-cycle activities are undertaken sequentially (what is generally referred to as the waterfall approach). However, there are several other development approaches, which we review in the next chapter.

Review Questions

1. The Project Management Body of Knowledge details 10 project management knowledge areas that are applicable to that discipline. Briefly describe each of these knowledge areas and explain how systems engineering contributes to the management associated with those areas.
2. Explain the role of business analysis in a project and describe how the discipline relates to systems engineering.
3. Explain the role of quality assurance in a project and describe how the discipline relates to systems engineering.
4. Explain the role of logistic support in a project and describe how the discipline relates to systems engineering.
5. Briefly describe each of the five constituent support capabilities of logistics support.

6. The Software Engineering Body of Knowledge details 18 knowledge areas that are applicable to that discipline. Briefly describe each of these knowledge areas and explain how each relates to the discipline of systems engineering.

Endnotes

[1] Project Management Institute Standards Committee, *A Guide to the Project Management Body of Knowledge*, Upper Darby, PA: Project Management Institute, Version 6, 2017.

[2] Project Management Institute Standards Committee, *A Guide to the Project Management Body of Knowledge*, Upper Darby, PA: Project Management Institute, Version 6, 2017.

[3] Kauffman, D., "Project Management and Systems Engineering: Where the Professions Intersect—Generate Synergy Not Conflict," *Proceedings of the Eighth Annual International Symposium*, INCOSE, 1998.

[4] Based on the original assessment published in: Kauffman, D., "Project Management and Systems Engineering: Where the Professions Intersect—Generate Synergy Not Conflict," *Proceedings of the Eighth Annual International Symposium*, INCOSE, 1998.

[5] International Institute of Business Analysis (IIBA), *A Guide to the Business Analysis Body of Knowledge (BABOK Guide)—Version 3*. International Institute of Business Analysis, 2015.

[6] International Institute of Business Analysis (IIBA), Agile Alliance, and Project Management Institute (PMI), *Agile Extension to the BABOK Guide—Version 2*, International Institute of Business Analysis, 2017.

[7] Womack, J. P., and D. T. Jones, *Lean Thinking: Banish Waste and Create Wealth in Your Corporation*, New York: Simon & Schuster, 1996.

[8] Pande, P. S., R. P. Neuman, and R. R. Cavanagh, *The Six Sigma Way: How to Maximize the Impact of Your Change and Improvement Efforts*, New York: McGraw-Hill, 2000.

[9] Dumas, M., et al., *Fundamentals of Business Process Management*, New York: Springer, 2018.

[10] Kerzner, H., *Project Management—A Systems Approach to Planning, Scheduling, and Controlling*, New York: Van Nostrand Reinhold, 2009.

[11] Blanchard, B., *Logistics Engineering and Management*, Upper Saddle River, NJ: Prentice Hall, 2003.

[12] ISO/IEC/IEEE 12207, *Systems and Software Engineering—Software Life Cycle Processes*, 2017.

[13] Fairley, R., *Software Engineering Concepts*, New York: McGraw-Hill, 1985.

[14] Washizaki, H., (ed.), *Guide to the Software Engineering Body of Knowledge (SWEBOK Guide)*, Version 4.0, IEEE Computer Society, 2024; www.swebok.org.

11

Systems Engineering and Development Approaches

11.1 Introduction

Throughout this book, we have described systems engineering without discussing in detail the development and acquisition context within which it might be undertaken. To best explain the activities and artifacts of systems engineering, we have assumed that the life-cycle activities are undertaken sequentially. In doing so, we have assumed what is generally referred to as the *waterfall approach* to system development. However, there are several other development approaches, each of which has strengths and weaknesses depending on the nature of the system under development. The selection of a suitable development approach is a critical activity early in a system life cycle. In this chapter, we review the major development approaches and discuss their applicability.

While waterfall development is often used successfully in system development, the main reason that it was used in this book is that it provides a logical, sequential flow of activities and deliverables that support teaching and explaining systems engineering. Additionally, waterfall development is generally considered to be the basic building block upon which alternative approaches such as incremental, evolutionary, and spiral developments are built [1, 2]. A solid understanding of waterfall development is therefore very useful. Further, in a 2004 survey [3], approximately one-third of software development projects reported using a waterfall development approach. Such an approach is still valid in many circumstances.

Like most aspects of systems engineering, it is simply not possible (nor desirable) to state categorically that a given approach to development is either universally acceptable or universally useless. The urge to develop a universally applicable model while dismissing other approaches promotes a one-size-fits-all mindset that can lead to an inappropriate "cookbook" approach to systems engineering. There are simply too many variables in play in today's complex project environment. By definition, projects to produce unique outputs [4] generally require a bespoke systems engineering and acquisition approach to be developed and applied. Successful systems engineering relies on intelligent, well-informed, and experienced professionals determining the most appropriate way to execute a technical program.

To help this decision-making process, this chapter investigates the characteristics of the principal approaches to project development. We examine waterfall development, incremental development, evolutionary development, and spiral development and describe when these alternatives may be considered a more pragmatic approach to development and acquisition.

11.2 Waterfall Development

Waterfall development, a variation of which is illustrated in Figure 11.1, was established as a response to the emerging software engineering discipline [5] but has since been adopted by the wider engineering community. The approach gets its name from the fact that it relies on the development of a complete set of requirements at the system level, which influences and cascades into the subsystem requirements, which, in turn, drives the requirements for the assemblies and components. At each transition, a baseline tends to be established, commonly, the FBL, the ABL, and the PBL. This idea gives rise to the visualization of requirements cascading down a waterfall as shown in Figure 11.1. Because of its nature, waterfall development is also often called the linear sequential model.

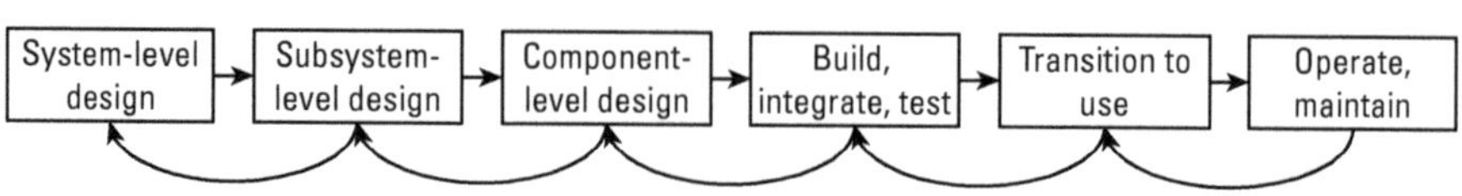

Figure 11.1 Cascading activities in the waterfall approach.

While the "flow" of a waterfall model is in the forward direction, there is a recognition that a mechanism is needed by which experience at the lower levels can be used to revisit the activities at the higher level. Although often forgotten, being able to manage changing requirements that are driven by discoveries in later systems engineering stages is a critical element of a pragmatic systems engineering process. While the waterfall model allows this feedback, it is often difficult to implement and limited to feeding back only to the preceding stage if the artifacts produced by that stage require correction. Consequently, the waterfall model is often criticized as being largely a one-shot process.

So, despite the accommodation of limited feedback, the waterfall approach to development is most effective when a complete and accurate set of requirements exists at the system level and that those requirements remain relatively stable throughout the subsequent acquisition process. This leads to the approach sometimes being referred to, rather disparagingly, as a grand design approach.

The classic waterfall development approach has been adapted into an equally well-known construct called the VEE diagram, which was first presented to the systems engineering community in 1991 [6]. The VEE diagram, as used in this book, applies to taking a system from conceptual design through to the end of detailed design and development and uses the concepts pioneered by the waterfall approach but arranged in the shape of a VEE as illustrated in Figure 11.2, in order to reflect the systems engineering philosophy of top-down design with bottom-up development.

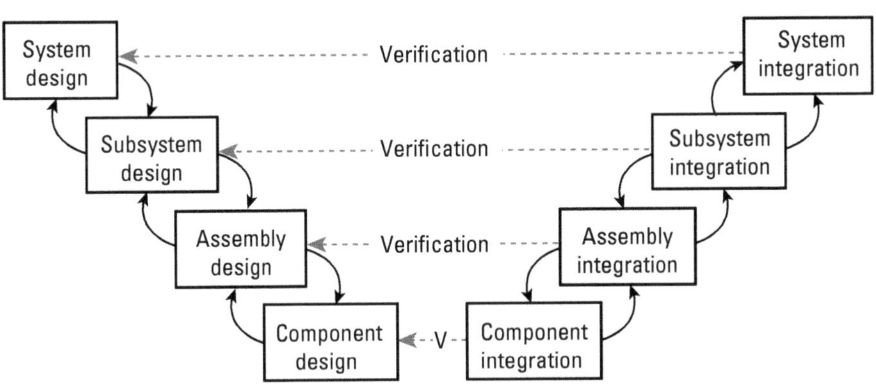

Figure 11.2 VEE representation of the waterfall approach.

Whether shown in the cascading construct of Figure 11.1, or in the shape of the VEE in Figure 11.2, waterfall development is characterized by an effort to understand and record the complete scope of the system before undertaking subsequent engineering design and acquisition effort.

In some instances, the use of the waterfall model is appropriate; if we are building a bridge across a wide channel, for example, we cannot start building until we have established all requirements, completed all detailed design, and so on. Additionally, waterfall development is almost universally applied in fixed-price contracting, which requires the development of a complete work breakdown structure (that can then be scheduled and costed) before the contract can be signed.

However, in many cases, it is not possible to establish requirements that remain stable over the system's life cycle. A large and complex system development requiring integration with other (external and changing) complex systems is a situation in which establishing accurate and stable requirements up-front may be difficult (and may be a liability). An accurate and complete set of requirements is also unlikely when the capability being introduced is completely new to the stakeholders, particularly when it is a unique system. In these cases, the stakeholders simply may not be able to articulate accurately their requirements.

Difficulties associated with developing a comprehensive set of system-level requirements up front have been well recorded in many other places. Yet weaknesses in the requirements management process are a major cause of project failure; we know that, if we do not get the requirements right up-front, there is a significant risk of project failure.

In addition to the requirements challenges, the acquisition process is generally related to some form of contract/agreement, binding the acquirer and the supplier of the system. The waterfall approach relies on sufficient money and time being available to complete the entire development effort in a single iteration. As with requirements, it is not always possible to commit the necessary time or money to large, single development efforts.

It is not only system-level requirements that tend to change over time. A characteristic of modern technology is the pace with which it changes. As the waterfall approach is based on the establishment of progressively more detailed baselines, it may be difficult in large projects to take advantage of opportunities presented by advances in technology. Design decisions and approaches tend to be made early in waterfall development, which may make design change prohibitive. Other areas of change on typical system developments that may cause problems with the waterfall approach include changes in any of the external systems with which the system under development must be integrated; changes in the operating or support environments; and changes to any of the constraints under which the system must be designed, produced, operated, or supported.

However, we must conclude by noting that waterfall development is very useful:

- In that it places strong emphasis on early elicitation of requirements;
- When the requirement set is small and well understood;
- When used as the basis of smaller builds in other development methods;
- When fixed-price contracting is used.

11.3 Incremental Development

In many cases, there may be valid reasons why the complete project scope cannot be developed and delivered in a single iteration:

- *Limited capability early.* Sometimes, some degree of capability is required by the users as soon as possible, without having to wait for the full development time.
- *Insufficient funds.* The project budget is an example of a project constraint that needs to be considered early in any system development. A smaller than required project budget may preclude a classic waterfall approach (addressing the entire scope of the system) from being employed.
- *Insufficient time.* In some circumstances, stakeholders may require a minimal set of functionalities to be delivered urgently, meaning that there is insufficient time to design, construct, and deliver the entire project.
- *Incomplete understanding of requirements.* It is not possible to completely bound the scope of requirements early in the program. Consequently, a classic waterfall approach becomes either a gamble (based on a best-guess scope) or a program plagued by continual change as the understanding of scope or functionality matures.
- *Risk management.* Perhaps all the above issues could be couched as risk-management issues. Incremental development assists in the management of risk by choosing small, well-defined, manageable work packages.

Incremental development starts with a good understanding of scope and level of functionality desired from the system and allocates different aspects of the system scope to a series of builds or increments that will be designed and delivered over time. Normally (as illustrated in Figure 11.3), the first increment introduces a kernel of basic functionality, which is then augmented over subsequent builds with progressive enhancements to scope until the entire system functionality is delivered [7].

Incremental development tends to deliver partial capability more quickly than if a traditional waterfall approach had been adopted. This reduces the likelihood of changes occurring to requirements, constraints, or external systems or the environment during an increment of the development.

Each incremental build may be executed using waterfall development where the top-level requirements of the first build drive the design of that build. If users would like the opportunity to develop or clarify their requirements between builds, the first build may be fielded for a period of use and exploration to collect data to inform subsequent builds. The top-level requirements of subsequent builds are refined to take account of the feedback to improve on the first build. In this way, it is possible to consider incremental development as a series of smaller waterfall developments, as illustrated in Figure 11.4.

The builds (and therefore the waterfalls) may be separated by considerable time (as illustrated in Figure 11.4) or may overlap to some degree. Overlapping builds may occur in instances where requirements sets are frozen, or baselined, to support an initial build, but continue to mature after they have been frozen. Requirements that change after the initial baseline has been frozen can be collected and fed into the baseline of the second build, and so on. This may give rise to a degree of overlapping as illustrated in Figure 11.5 where successive

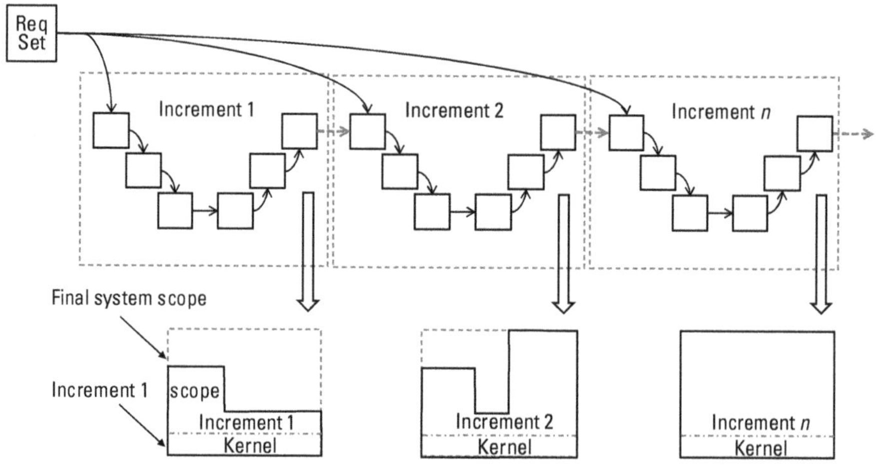

Figure 11.3 Incremental development building on system functionality.

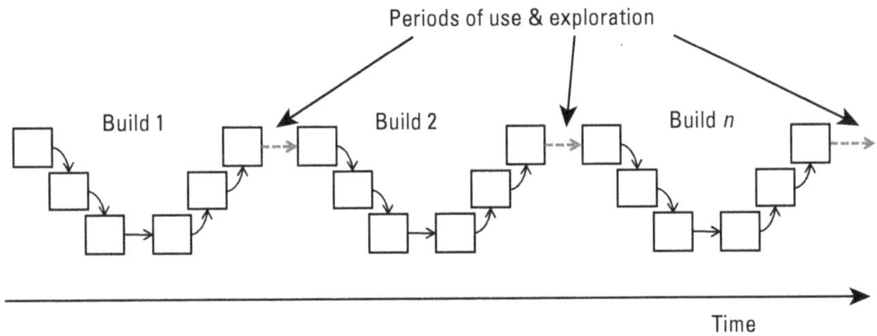

Figure 11.4 A series of waterfall builds.

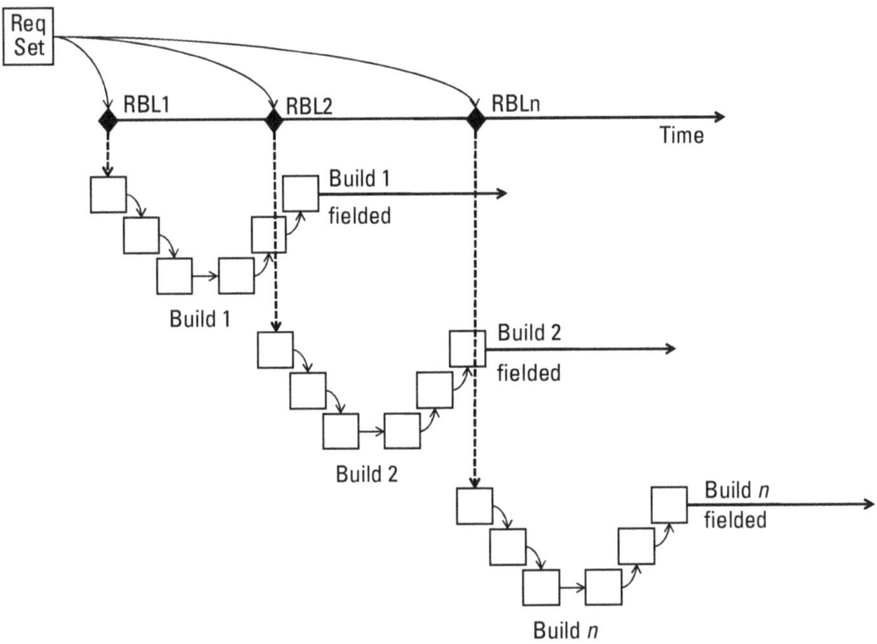

Figure 11.5 Overlapping incremental builds.

requirements baselines (RBLs) are shown as a function of time. Either way, the establishment of a number of builds that are clearly defined at the outset not only allows the use of the waterfall approach for each build, but it subsequently allows fixed-price contracting for each build.

If incremental development is being considered, special consideration should be given to including requirements in the initial build that address long-term design goals such as expandability and upgradability. This will help to ensure that subsequent builds are able to maximize use of the initial build without having to replace or upgrade significant percentages of the system each time. For example, if the system being developed is a combination of hardware and software, consideration should be given to provision of spare computer capacity; spare communication bandwidth; and spare mass, space, and power allowances in the initial build. Even when expandability and upgradability are properly considered in the initial build, incremental development strategies must recognize that design decisions made during the initial build will constrain the design of subsequent builds.

The contractual mechanism to be used to bind the customer and contractor needs to be considered carefully when employing incremental development. While the scope, timeframe, and costs associated with the initial build may be well understood and easy to translate into a contract, this may not be the case of the subsequent builds. Both the customer and the contractor need to be interested in how to broker a contractual agreement encapsulating incremental development.

Finally, this book emphasized the importance of stakeholder engagement in the systems engineering process, including the need to manage stakeholder expectations. With incremental development, management of stakeholder expectations becomes critical. The stakeholders must understand and agree to the level of functionality being delivered in each of the planned increments. This guards against stakeholder disappointment or even rejection when initial builds fail to deliver the entire scope that is ultimately being aimed for. The aim of the incremental development is to eventually deliver complete functionality progressively as illustrated in Figures 11.3 and 11.4.

11.4 Evolutionary Development

Evolutionary development is very similar to incremental development [8] in that it aims to arrive at the desired level functionality and scope over time using a number of design iterations. Although incremental development and evolutionary development are very similar in their aims, the defining difference is that incremental development is a series of discrete builds with deliberate augmentation of functionality, whereas evolutionary development is a more continuous evolution towards the final system scope including the development of some functionality that is later discarded [9]. Additionally, the presumption in evolutionary development is that not only are the requirements not completely known in advance, but the development process is used to elicit unforeseen requirements and to identify weaknesses in early sets.

The two approaches have different applications because stakeholders sometimes have little idea about what they want from the solution at the outset of the project but can provide a meaningful critique of proposed solutions when they see them. In these situations, the stakeholders are not able to describe the complete set of requirements for the system that they need or agree on how that will be delivered as a series of incremental builds. An evolutionary approach may be able to assist. In other situations, stakeholders are keen on some level of solution to at least help them with their problems and acknowledge that they are willing to accept an imperfect initial solution.

The initial stage of the evolutionary development therefore focuses on delivering something of use, mindful of the fact that significant variation in functionality is allowed until the series of builds iterates towards the final system scope. This is illustrated in Figure 11.6. As with incremental development, executing an evolutionary technical program may be thought of as a progression of builds in time or, as illustrated in Figure 11.7, evolutionary builds may be overlapped in time to speed up the delivery of functionality to stakeholders.

Given the similarity in the concepts of both incremental and evolutionary development, the same set of considerations discussed with incremental development needs to be applied to evolutionary developments. Stakeholders must be involved and appraised of the strategy, and the constraining nature of design decision made in early evolutions needs to be carefully considered.

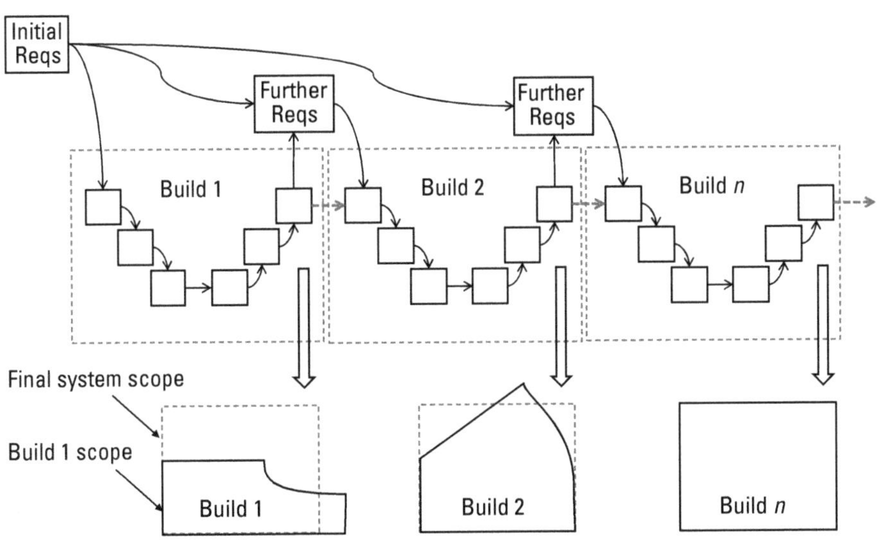

Figure 11.6 Evolutionary development of functionality.

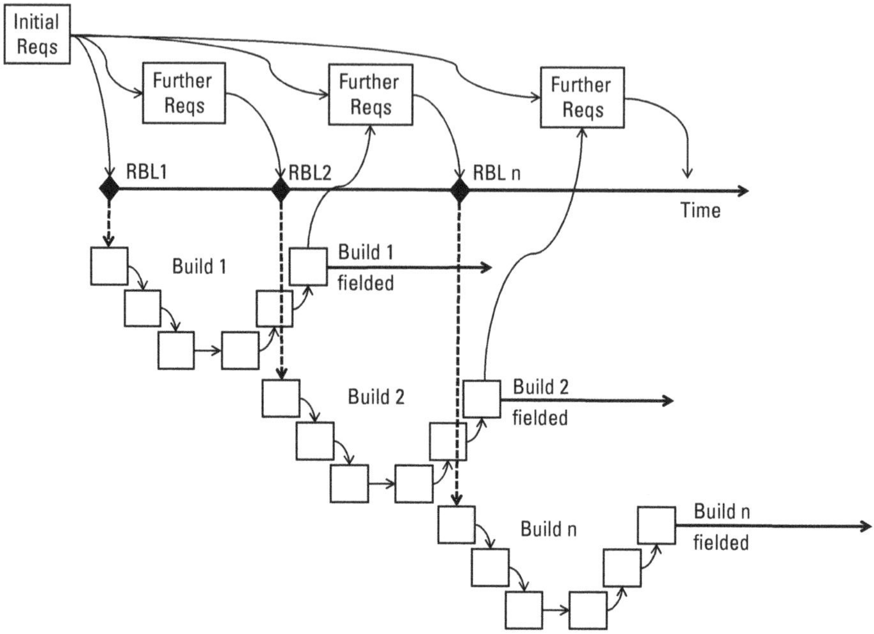

Figure 11.7 Overlapping evolutionary builds.

11.5 Prototyping and System Development

Prototyping is a pragmatic way of exploring different aspects of system concepts and requirements before making final decisions and committing to a particular acquisition path. Prototyping continues to be an important part of many design and development efforts as they proceed from system to subsystem to component-level design.

At a system level, a prototype may be produced to provide some limited functionality to key groups of stakeholders to facilitate exploration of their requirements for the mature system.

Prototyping is not the only way in which to gain a better understanding of the capability requirements before embarking on an acquisition. Another popular tool or activity is called concept exploration and demonstration. In this case, stakeholders can access or experience systems that deliver capability that is similar in scope or functionality to the capability that they desire. These similar systems are not prototypes because they are generally already in service with other companies or organizations. The stakeholders observe and, in some cases, use these systems to explore and mature their understanding of their own requirements for the capability. Acquisition is then able to proceed from a basis of knowledge rather than one of guesswork.

In this way, prototyping and concept exploration and demonstration assist in the early requirements analysis task. A range of other tools can also be used to help with requirements analysis but are not discussed specifically here, including scale models, mock-ups, functional models, and simulations.

The key difference between a prototype and the early builds in either incremental or evolutionary developments is that a prototype is provided with minimum formal process simply to assist the stakeholders in exploring their requirements. The requirements from the prototyping phase are then used as the key input in a development phase that employs a more formal methodology such as waterfall, incremental, or evolutionary acquisition. The prototype does not generally go on to become a fielded system but may continue to play a role in the requirements determination process, especially if incremental or evolutionary methodologies are employed.

Another difference between a prototyping process and early incremental or evolutionary builds is that a prototype is used to explore what the stakeholders do not know about their requirements, whereas early incremental or evolutionary builds aim to deliver what the stakeholders do know about their requirements.

Developing the prototype, although less formal than a complete system development, should be done using a basic process of some sort. After all, even prototypes can cost significant sums of money, and their acquisition and use need to be controlled. The process may be as simple as determining the need for the prototype by agreeing on its purpose, designing/building or acquiring the prototype, and then using and refining it until the prototyping phase has been completed. This basic process is illustrated in Figure 11.8.

In some cases, the prototype may eventually develop into the final system via an incremental or evolutionary development process. This strategy is sometimes called evolutionary prototyping [10]. If this is not the strategy, the prototype should be disposed of when no longer required to ensure that it is not misused.

The use of prototyping as an exploration tool does not end at the system level illustrated in Figure 11.8. Prototyping is an important tool or technique that gets used at all levels of the system development to discover and confirm requirements and constraints. For example, designing and agreeing on user interfaces during the preliminary design or detailed design and development phases commonly uses prototyping and key stakeholders to confirm the design adequacy of the proposed solutions before finalizing the design artifacts.

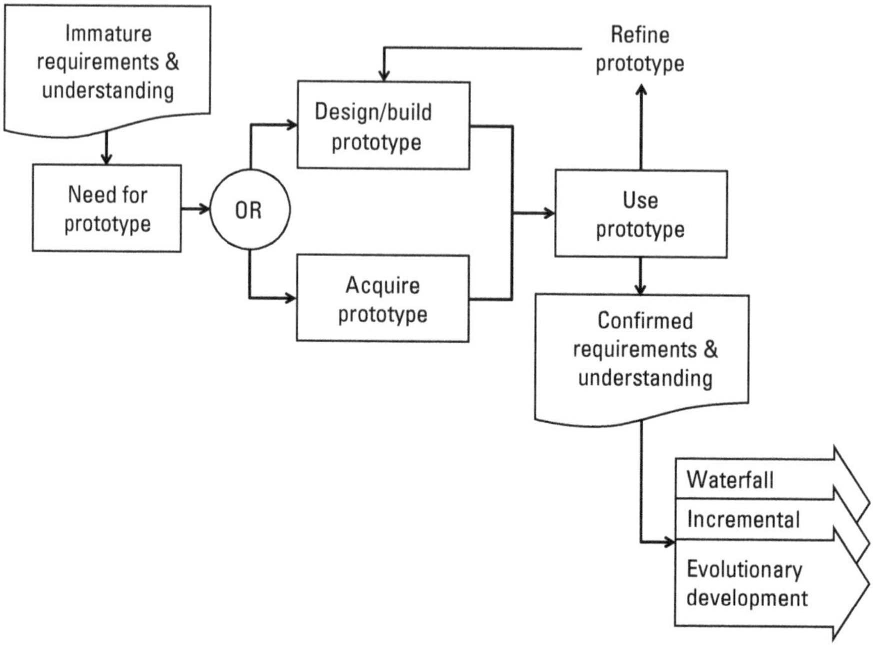

Figure 11.8 Prototyping preceding design and development.

11.6 Spiral Development Model

Boehm [11] pioneered the spiral model as a software development and enhancement framework. It is interpreted and presented here from a systems engineering perspective, rather than from a software perspective, and is illustrated in Figure 11.9.

The spiral model is not a development methodology per se but is a process by which the most appropriate development methodology can be selected for each system development effort. The spiral model does not presume any particular methodology at the start of the process, nor does it require the same methodology to be adopted for each cycle of the spiral. Spiral development recognizes that things change over the course of a development that may require strategy adjustment or change and supports adoption of the most appropriate process model such as waterfall, incremental, or evolutionary, depending on circumstances. For example, if aspects of the system development have many unresolved risks and poorly defined user requirements, evolutionary development augmented by prototyping might be employed. If the risks in other aspects of the system development are lower, and the requirements are well defined and stable, a more traditional waterfall approach may be adopted. The spiral development model is therefore best viewed as a meta-model [12] that generates other models.

Each cycle of the model follows the same sequence of activities:

- *Determine objectives, alternatives, and constraints* (top left of Figure 11.9). The key requirements of the system element under consideration are determined. The alternatives for achieving these requirements are identified. This may include options such as reuse of existing designs, use of off-the-shelf technology, or the need to proceed with some form of design and development. The constraints that are relevant to the system element are also identified and may include cost and time limitations, external systems and interfaces, environmental issues, and laws/legislation.
- *Evaluate alternatives; identify and resolve risks* (top right of Figure 11.9). The alternatives are considered in light of the requirements and the

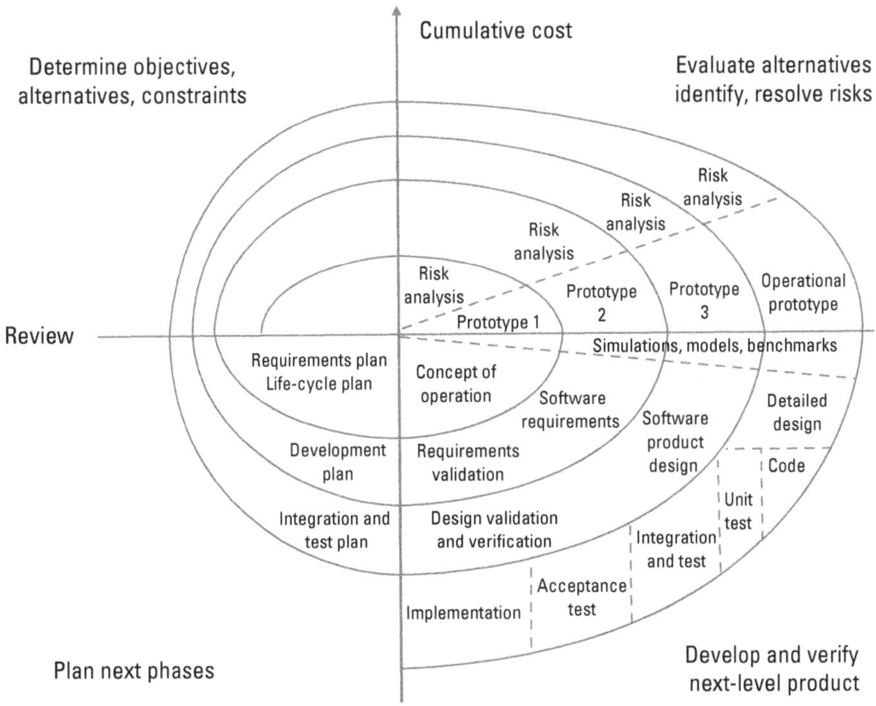

Figure 11.9 Spiral development model. (After: Boehm.)

constraints. Each alternative will typically come with a different set of risks. The spiral model incorporates prototyping, modeling, and simulation as examples of tools that can help to consider those risks and determine risk management strategies.

- *Develop and verify next-level product* (bottom right of Figure 11.9). The next level of product is then produced based on displacement from the center of the spiral. Close to the center of the spiral (the early stages of the development), the products might be sets of system-level requirements. Further out are the sets of subsystem requirements, and further still are the detailed design and integration products. Regardless of product, the spiral model includes relevant validation and verification activities before moving onto the next stage.
- *Plan next phases* (bottom left of Figure 11.9). In this phase, the verified and validated products from the previous phase are used to plan for the next iteration of the spiral. Early on, the plans will be project and system-level management plans, but, later, more detailed plans such as integration and test plans will be developed.
- *Review by stakeholders.* Each cycle of the spiral concludes with a period of review by relevant stakeholders (shown along the x-axis in Figure 11.9). This period of review allows the stakeholder to review the products and decisions made to this point and the plans for the subsequent phases. The reviews facilitate decision-making and renewed commitment to projects by focusing on risk and encouraging early and ongoing decisions about whether the system development should continue or be canceled.

Spiral development, with its flexibility to make ongoing decisions and make changes to the direction of system developments, has been found to be effective in situations where organizations are undertaking internal system developments. There are some challenges to be addressed when extending the model to commercial developments where a customer and contractor are bound by contractual terms and conditions relating to the development. In these cases, traditional contracting mechanisms such as fixed-price contracting may not sufficiently accommodate the changes to the development process encouraged under spiral development. Spiral developments may require some form of cost plus/risk sharing contracting mechanisms to be employed. This type of contracting mechanism is not universally popular with some customer organizations, which may rule out using spiral development models in those organizations.

11.7 Agile Development

Agile development refers to a range of approaches to software development approaches that aim to deliver value faster, with improved quality, and greater ability to respond to changes throughout the life cycle. Agile development therefore encompasses a number of development methodologies in which the approach to development is iterative or evolutionary, with adaptive planning, allowing requirements and solutions to evolve through closer collaboration of self-organizing, cross-functional teams and customers. Agile is a mindset defined in the Agile Manifesto [13] with four values, guided by 12 principles, and manifested through many different practices [14]. Although originally focused on software development, Agile approaches are now commonly considered to be applicable to all aspects of system development. It is worth noting that while the term "agile" became popularized after the manifesto, the approaches and techniques being used by project teams today existed before the Agile Manifesto, and the approaches existed by many years and, in some cases, decades.

Scrum is the most widely used Agile methodology that describes a process framework comprising short, time-boxed development cycles, called sprints, that have light documentation overhead with an emphasis on delivering early functionality and being agile to accommodate rapidly changing business environments; sprints are typically 1 to 4 weeks long. Scrum involves three main roles: ScrumMaster, Product Owner, and Developers, collectively the Scrum Team. Functionality (feature requests) are described in terms of a user story (a templated form of use case).

Shorter iterations, or sprints, aim to minimize the overall risk in development by ending each iteration with a working product that, through multiple iterations, will allow the product to adapt to changes and to reflect the stakeholders' increased understanding of the functions required of it. Development teams are normally small and colocated to facilitate efficient face-to-face communication and reduce the time required for each development cycle. The development team includes a customer representative, and each iteration ends with their review of progress and reevaluation of priorities.

Agile practices can be applied to each activity in the system life cycle and, in most cases, can be applied to one specific activity. For example, there is an Agile Extension to the Business Analysis Body of Knowledge [15], which provides BAs with seven key guidelines for the practice of business analysis within an Agile environment, as well as specific techniques that allow agile teams to deliver value. In project management, the PMI, in conjunction with the Agile Alliance, provides an agile practice guide that is an adjunct to the Project Management Body of Knowledge to provide tools, situational guidelines, and an understanding of the various agile practices to enable better project outcomes [16].

11.8 Contracting Approach Versus Development Approach

In some cases, such as in-house developments, the relationship between the acquirer and the supplier (the tasking activity and the performing activity) may be such that the development approach and the contracting approach align; that is, waterfall contracting approach aligns with a waterfall development approach; evolutionary with evolutionary, Agile with Agile; and so on. However, this does not necessarily have to be the case and often may not be in the larger government contracts.

Consider the case in which the customer and contractor's contract may be implemented using an evolutionary approach where, as shown in Figure 11.10, the requirements are collected and analyzed, and a subset is chosen to be formally contracted with contractor; the contract is therefore based on general conditions (covering skillsets, daily rates, deliverable formats, verification arrangements) and each build is then an amendment to that contact (in some form of deed of variation) that identifies the next subset of requirements, the associated SOW, and budget and schedule. The contractor develops that build, which is verified at an FQR, validated, and fielded. Feedback from that build is used to inform the pool of potential requirements (adding, deleting, and modifying requirements), from which the scope of the next build is developed, and an appropriate contract variation is negotiated for that next subset of requirements. However, as shown in Figure 11.11, in that build, the contractor may apply any system development approach, Agile, for example, because it suits the nature of the build. In the next build, the contractor may choose a linear sequential approach, and so on.

11.9 Summary

To facilitate the explanation of the systems engineering process throughout this book, we have used a waterfall approach to systems engineering without discussing in great detail the development and acquisition context within which it might be undertaken. As we have noted in this final chapter, however, there are a number of other development approaches, each of which has strengths

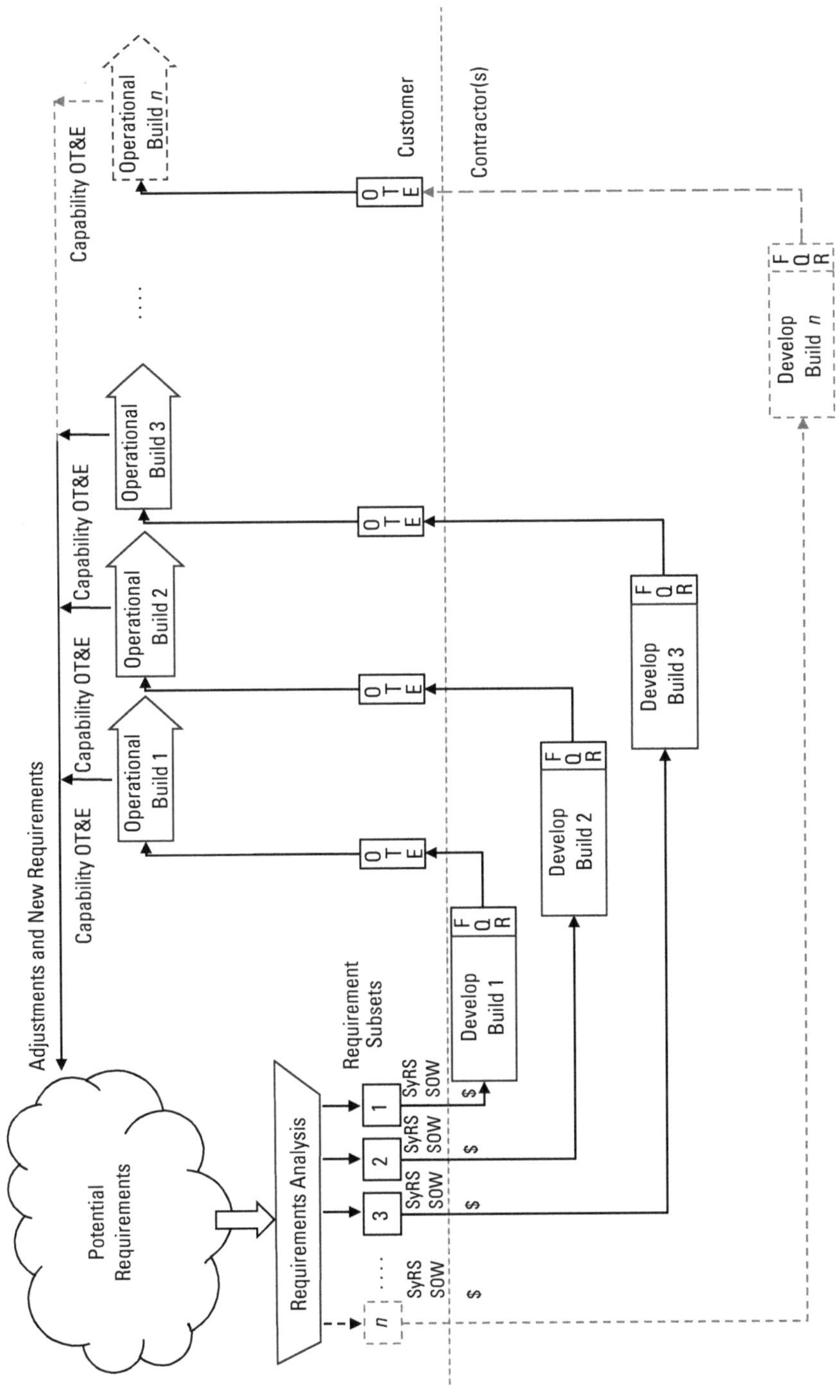

Figure 11.10 Contracting by build in an evolutionary system development.

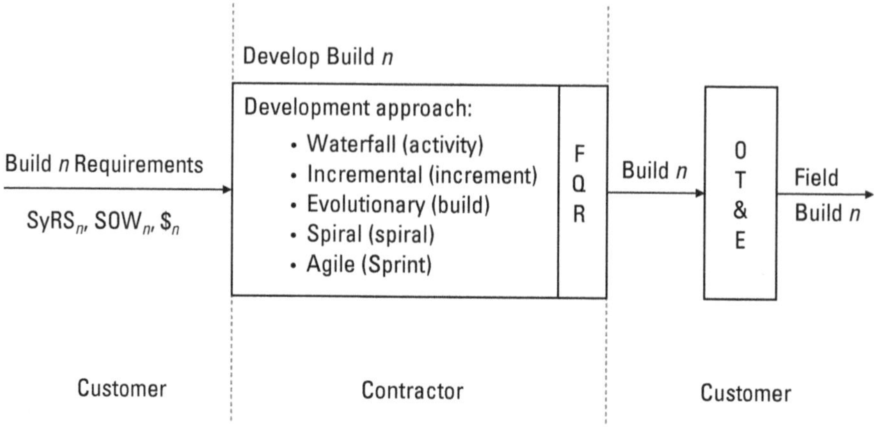

Figure 11.11 Development of a build does not necessarily need to employ the same development methodology as the contracting approach, nor does each build need to use the same development methodology.

and weaknesses depending on the nature of the system under development. The selection of a suitable development approach is critical activity early in a system life cycle.

While the linear-sequential waterfall approach is often used successfully in system development (particularly when used in conjunction with fixed-price contracting), the main reason it was used in this book is that it provides a logical, sequential flow of activities and deliverables that support the explanation of systems engineering. A solid understanding of waterfall development is also required since the waterfall approach can be considered as the basic building block underpinning the alternative approaches such as incremental, evolutionary, spiral, and Agile developments.

Review Questions

1. Briefly explain (with the aid of a diagram) the waterfall approach to system development and under what circumstances the use of the approach is appropriate or inappropriate.
2. Briefly explain (with the aid of a diagram) the incremental approach to system development and under what circumstances the use of the approach is appropriate or inappropriate.

3. Briefly explain (with the aid of a diagram) the evolutionary approach to system development and under what circumstances the use of the approach is appropriate or inappropriate.
4. Briefly explain (with the aid of a diagram) the spiral approach to system development and under what circumstances the use of the approach is appropriate or inappropriate.
5. Briefly explain (with the aid of a diagram) the Agile approach to system development and under what circumstances the use of the approach is appropriate or inappropriate.

Endnotes

[1] Forsberg, K., and H. Mooz, *Systems Engineering Overview*, Cupertino, CA: Center for Systems Management, 1996.

[2] Forsberg, K., and H. Mooz, "Application of the 'VEE' to Incremental and Evolutionary Development," *Proceedings of the NCOSE Conference*, St. Louis, MO, July 1995.

[3] Neill, C. J., and P. A. Lapante, "Requirements Engineering: The State of the Practice," *IEEE Software*, Vol. 20, No. 6, December 2003.

[4] Project Management Institute, *A Guide to the Project Management Body of Knowledge*, Upper Darby, PA: Project Management Institute, Version 6, 2017.

[5] Royce, W. W., *Managing the Development of Large Software Systems: Concepts and Techniques*, WESCON Technical Papers, Vol. 14, August 1970, pp. A/1-1–A/1-9.

[6] Forsberg, K., and H. Mooz, "The Relationship of Systems Engineering to the Project Cycle," *Proceedings of the NCOSE Conference*, Chattanooga, TN, October 1991.

[7] Davis, A. M., E. H. Bersoff, and E. R. Comer, "A Strategy for Comparing Alternative Software Development Life Cycle Models," *IEEE Transactions on Software Engineering*, Vol. 14, No. 10, October 1988, pp. 1453–1461.

[8] Dorfman, M., and R. H. Thayer, *Software Requirements Engineering*, New York: IEEE Computer Society Press, 2000, p. 9.

[9] Davis, A. M., E. H. Bersoff, and E. R. Comer, "A Strategy for Comparing Alternative Software Development Life Cycle Models," *IEEE Transactions on Software Engineering*, Vol. 14, No. 10, October 1988, pp. 1453–1461.

[10] Dorfman, M., and R. H. Thayer, (eds.), *System and Software Requirements Engineering*, Los Alamitos, CA: Computer Society Press, 1990, pp. 543–552.

[11] Boehm, B., "A Spiral Model of Software Development and Enhancement," *IEEE Computer*, Vol. 21, No. 5, May 1988, pp. 61–72.

[12] Martin, J. N., *Systems Engineering Guidebook: A Process for Developing Systems and Products*, Boca Raton, FL: CRC Press, 1997.

[13] *Manifesto for Agile Software Development*, 2001, http://agilemanifesto.org.

[14] Project Management Institute, *Agile Practice Guide*, Newtown Square, PA, 2017.

[15] International Institute of Business Analysis, *Agile Extension to the BABOK Guide*, Vol. 1, Toronto, ON.

[16] Project Management Institute, *Agile Practice Guide*, Newtown Square, PA, 2017.

Acronyms and Abbreviations

ABL	Allocated baseline
AI	Artificial intelligence
ANSI	American National Standards Institute
AT&E	Acceptance test and evaluation
ATR	Acceptance test report
BA	Business analyst
BC	Business case
BNR	Business needs and requirements
BRS	Business requirement specification
CAD	Computer-aided design
CAE	Computer-aided engineering
CAM	Computer-aided manufacturing
CASE	Computer-aided software environment
CCB	Configuration Control Board
CCP	Contract change proposal
CDR	Critical design review
CFD	Control flow diagrams
CI	Configuration item
CM	Configuration management
CMP	Configuration management plan
COD	Concept of operations document
COI	Critical operational issue
CONOPS	Concept of operations

COTS	Commercial-off-the-shelf
CSC	Computer software component
CSI	Critical strategic issue
CSCI	Computer software configuration item
CSU	Computer software unit
CWBS	Contract work breakdown structure
DDR	Detailed design review
DE	Digital engineering
DFD	Data flow diagram
DID	Data item description
DLOD	Defense lines of development
DoD	US Department of Defense
DOTMLPF	Doctrine, organization, training, materiel, leadership, personnel, facilities
DT&E	Developmental test and evaluation
ECP	Engineering change proposal
EIA	Electronics Industries Association
EMC	Electromagnetic compatibility
EMI	Electromagnetic interference
ERD	Entity relationship diagram
FBL	Functional baseline
FCA	Functional configuration audit
FEA	Finite element analysis
FFBD	Functional flow block diagram
FIC	Fundamental inputs to capability
FMECA	Failure mode effects and criticality analysis
FoS	Family of systems
FQR	Formal qualification review
FRACAS	Failure reporting, analysis, and corrective action system
GOTS	Government off-the-shelf
GPS	Global Positioning System
HDBK	Handbook

HMI	Human machine interface
HR	Human resources
HW	Hardware
HWCI	Hardware configuration item
ICD	Interface control document
ICL	Interface control list
ICWG	Interface (integration) Control Working Group
IEC	International Electrotechnical Commission
IEEE	Institute of Electrical and Electronics Engineers
IIBA	International Institute of Business Analysis
IIS	Introduction into service
ILS	Integrated logistics support
INCOSE	International Council on Systems Engineering
IS	Interim standard
ISO	International Organization for Standardization
IV&V	Independent verification and validation
LC	Life-cycle concept
LCC	Life-cycle cost
LSC	Logistic support concept
MBD	Model-based design
MBSE	Model-based systems engineering
MIL-HDBK	US Military Handbook
MIL-STD	US Military Standard
MOE	Measure of effectiveness
MOP	Measure of performance
MOS	Measure of suitability
MOTS	Military off the shelf
NCOSE	National Council on Systems Engineering
NDI	Nondevelopmental item
NRM	(INCOSE) *Needs and Requirements Manual*
OpsCon	Operational concept

OOA	Object-oriented analysis
OT&E	Operational test and evaluation
OTS	Off-the-shelf
PBL	Product baseline
PCA	Physical configuration audit
PDR	Preliminary design review
PERT	Program evaluation and review technique
PLC	Preliminary life-cycle concepts
PLM	Product life-cycle management
PM	Project management
PMI	Project Management Institute
PMP	Project management plan
PoS	Portfolio of systems
PRICIE	Personnel, research and development, infrastructure, concepts and doctrine, information technology, equipment
PSTN	Public Switched Telephone Network
PWBS	Program/project work breakdown structure
QA	Quality assurance
QFD	Quality function deployment
RAAF	Royal Australian Air Force
RAM	Reliability, availability, maintainability
RBS	Requirements breakdown structure
RF	Radio frequency
RFT	Request for tender
RMP	Risk management plan
RMT	Requirements management tool
SBP	Strategic business plan
SBS	System breakdown structure
SDD	System design document
SDG	Sustainable development goals
SDR	System design (definition) review
SE	Systems engineering

SEBoK	Systems Engineering Body of Knowledge
SEDS	Systems engineering detailed schedule
SEI	Software Engineering Institute
SEMP	Systems engineering management plan
SEMS	Systems engineering master schedule
SNR	Stakeholder needs and requirements
SOI	System of interest
SOP	Standard operating procedure
SoS	System of systems
SOW	Statement of work
SRD	Stakeholder requirement document
SRR	Systems requirements review
SS	System specification
SSADM	Structured system analysis and design
SSCC	Support system constituent capabilities
STD	Standard
StRS	Stakeholder requirements specification
SVR	System verification review
SW	Software
SWEBOK	Software engineering body of knowledge
SyRS	System requirements specification
SySML	System Modeling Language
T&E	Test and evaluation
TCO	Total cost of ownership
TEMP	Test and evaluation master plan
TPM	Technical performance measure
TRAP	Technical review and audit plan
TRA	Technical risk assessment
TRL	Technical readiness level
TRR	Test readiness review
UML	Unified Modeling Language
V&V	Verification and validation

VCRM	Verification cross-reference matrix
WBS	Work breakdown structure
WSOI	Wider system of interest

About the Authors

Ian Faulconbridge holds a bachelor's, a master's, and a doctorate in engineering, an MBA in project management, and a Master of Environmental Management (in conservation). Dr. Faulconbridge is a Fellow of Engineers Australia (FIEAust), an Engineering Executive (EngExec), a Chartered Professional Engineer (CPEng) in four colleges, and a registered professional engineer. He has over 35 years of applied professional engineering practice and has performed senior systems engineering roles within the defense and aerospace sectors, both in Australia and overseas. To augment and inform his professional practice, Dr. Faulconbridge has also spent periods as either a full-time or an adjunct engineering academic at the University of New South Wales, the Australian National University, the Queensland University of Technology, the University of Southern Queensland, and James Cook University. He has authored or coauthored seven books covering systems engineering, radar and electronic warfare, and avionics systems. Collectively, over 130,000 students have attended one of his courses and over 10,000 copies of his texts have been sold.

Michael Ryan holds bachelor's, master's, and PhD degrees in electrical engineering. Dr. Ryan is a Fellow of Engineers Australia (FIEAust), a Chartered Professional Engineer (CPEng) in systems engineering, electrical, and ITEE colleges, a Senior Member of the IEEE (SMIEEE), a Fellow of the International Council on Systems Engineering (FINCOSE), a Fellow of the Institute of Managers and Leaders (FIML), and a Fellow of the Royal Society of NSW (FRSN). He has over 45 years of experience in communications engineering, systems engineering, capability management, and management. From 1998 to 2020, he was a professor with the University of New South Wales, where he was most recently the director of the Capability Systems Centre. His research and teaching interests are in communications and information systems, requirements engineering, systems engineering, and project management, and he also regularly consults in those fields. He is a cochair of the Requirements Working Group (RWG) of INCOSE. He is also the author or coauthor of 14 books, 4 book chapters, and over 450 refereed journal and conference papers.

Index

Artech House Technology Management and Professional Development Library

Seth Sullivan, Series Editor

Actionable Strategies Through Integrated Performance, Process, Project, and Risk Management, Stephen S. Bonham

Advanced Systems Thinking, Engineering, and Management, Derek K. Hitchins

Applied Systems Engineering, Second Edition, Ian Faulconbridge and Michael Ryan

Applying Total Quality Management to Systems Engineering, Joe Kasser

Building Successful Virtual Teams, Francine Gignac

Critical Chain Project Management, Third Edition, Lawrence P. Leach

Decision Making for Technology Executives: Using Multiple Perspectives to Improve Performance, Harold A. Linstone

Designing the Networked Enterprise, Igor Hawryszkiewycz

Electrical Product Compliance and Safety Engineering, Volume 2, Steli Loznen and Constantin Bolintineanu

The Engineer's Guide to Authentic Leadership, Ronald J. Bennett and Elaine R. Millam

Engineer's and Manager's Guide to Winning Proposals, Donald Helgeson

Engineering and Technology Management Tools and Applications, B. S. Dhillon

Enterprise Release Management: Agile Delivery of a Strategic Change Portfolio, Louis Taborda

The Entrepreneurial Engineer: Starting Your Own High-Tech Company, R. Wayne Fields

Evaluation of R&D Processes: Effectiveness Through Measurements, Lynn W. Ellis

From Engineer to Manager: Mastering the Transition, Second Edition, B. Michael Aucoin

Global High-Tech Marketing: An Introduction for Technical Managers and Engineers, Jules Kadish

How to Become an IT Architect, Cristian Bojinca

Integrated IT Project Management, Kenneth R. Bainey

Introduction to Information-Based High-Tech Services, Eric Viardot

Introduction to Innovation and Technology Transfer, Ian Cooke and Paul Mayes

ISO 9001:2000 Quality Management System Design, Jay Schlickman

IT Project Portfolio Management, Stephen S. Bonham

Managing Complex Technical Projects: A Systems Engineering Approach, Second Edition, Ian Faulconbridge and Michael Ryan

Managing Successful High-Tech Product Introduction, Brian P. Senese

Managing Virtual Teams: Practical Techniques for High-Technology Project Managers, Martha Haywood

Mastering Technical Sales: The Sales Engineer's Handbook, Fourth Edition, John Care

The New High-Tech Manager: Six Rules for Success in Changing Times, Kenneth Durham and Bruce Kennedy

The Parameter Space Investigation Method Toolkit, Roman Statnikov and Alexander Statnikov

Planning and Design for High-Tech Web-Based Training, David E. Stone and Constance L. Koskinen

A Practical Guide to Managing Information Security, Steve Purser

Practical Model-Based Systems Engineering, Jose L. Fernandez and Carlos Hernandez

Practical Reliability Data Analysis for Non-Reliability Engineers, Darcy Brooker with Mark Gerrand

The Project Management Communications Toolkit, Second Edition, Carl Pritchard

Project Managment Process Improvement, Robert K. Wysocki

Reengineering Yourself and Your Company: From Engineer to Manager to Leader, Howard Eisner

The Requirements Engineering Handbook, Ralph R. Young

Requirements Practice in Conceptual Design, Second Edition: A Practical Guide, Michael Ryan

Running the Successful Hi-Tech Project Office, Eduardo Miranda

Successful Marketing Strategy for High-Tech Firms, Second Edition, Eric Viardot

Successful Proposal Strategies for Small Businesses: Using Knowledge Management to Win Government, Private Sector, and International Contracts, Sixth Edition, Robert S. Frey

Successful Proposal Strategies On-the-Go!, Robert S. Frey

Systems Approach to Engineering Design, Peter H. Sydenham

Systems Engineering Principles and Practice, H. Robert Westerman

Systems Reliability and Failure Prevention, Herbert Hecht

Team Development for High-Tech Project Managers, James Williams

Artech House
685 Canton Street
Norwood, MA 02062
Phone: 781-769-9750
Fax: 781-769-6334
e-mail: artech@artechhouse.com

Artech House
16 Sussex Street
London SW1V 4RW UK
Phone: +44 (0)20 7596-8750
Fax: +44 (0)20 7630-0166
e-mail: artech-uk@artechhouse.com

Find us on the World Wide Web at: www.artechhouse.com

www.ingramcontent.com/pod-product-compliance
Lightning Source LLC
LaVergne TN
LVHW051046210926
848486LV00001B/36